Soil Settlement and the Concept of Effective Stress and Shear Strength Interaction

Soil Settlement and the Concept of Effective Stress and Shear Strength Interaction

Mohd Jamaludin Md Noor
Faculty of Civil Engineering, Universiti Teknologi MARA, Shah Alam, Malaysia

CRC Press
Taylor & Francis Group
Boca Raton London New York

CRC Press is an imprint of the
Taylor & Francis Group, an **informa** business

First edition published 2021
by CRC Press
6000 Broken Sound Parkway NW, Suite 300, Boca Raton,
FL 33487-2742

and by CRC Press
2 Park Square, Milton Park, Abingdon, Oxon, OX14 4RN

CRC Press is an imprint of Taylor & Francis Group, LLC

Library of Congress Cataloging-in-Publication Data
Names: Noor, Mohd. Jamaludin Md, 1962- author.
Title: Soil settlement and the concept of effective stress and shear strength interaction / Mohd Jamaludin Md Noor.
Description: First edition. | Boca Raton : CRC Press, 2020. | Includes index.
Identifiers: LCCN 2020028972 (print) | LCCN 2020028971 (ebook)
Subjects: LCSH: Soil consolidation. | Shear strength of soils.
Classification: LCC TA749 .N66 2020 (ebook) | LCC TA749 (print) | DDC 624.1/51362—dc23
LC record available at https://lccn.loc.gov/2020028972

ISBN: 978-0-367-60811-8 (hbk)
ISBN: 978-0-367-63956-3 (pbk)
ISBN: 978-1-003-12150-3 (ebk)

DOI: 10.1201/9781003121503
https://doi.org/10.1201/9781003121503

Typeset in Times New Roman
by codeMantra

Contents

Preface

This book can be benefitted by both soil mechanics and rock mechanics communities. It presents the state-of-the-art soil and rock anisotropic deformation theories. It begins with the true understanding of the soil elastic–plastic behaviour assisted by the physical illustration of exaggerated soil particles rearrangement during settlement to ease the understanding especially to those undergraduates and postgraduates civil engineering students. This is the most fundamental concept of soil stress–strain behaviour that needs to be fully understood before embarking into more sophisticated theories of soil mechanics.

Then the conventional concept of effective stress introduced by the father of soil mechanics Terzaghi is applied in saturated soil mechanics and its limitations in explaining the various soil weird settlement behaviours are presented. This is followed by the introduction of the extended concept of effective stress known as the concept of effective stress and shear strength interaction, which is extended from saturated soil mechanics to unsaturated soil mechanics. This new concept utilises the development of the non-linear mobilised shear strength envelope as the settlement-resisting variable whenever the soil is subjected to anisotropic compression. This concept is called Rotational Multiple Yield Surface Framework (RMYSF) and it is the most important essence of this book. In this theory, the mobilised envelope is taken as the yield surface and whenever the stress Mohr circle representing the state of effective stress, which acts as the settlement-driving variable, extends higher than the yield surface, the soil will yield by undergoing settlement. This is indicated by the rotation of the mobilised envelope upwards to represent the increase in the mobilised strength, which resists the further settlement. The positions of the mobilised envelopes are representing a specific % axial strain and are the inherent property of the soil and valid for the whole effective stress range. This important intrinsic property of the soil is hidden and its direct relationship with the soil anisotropic compression is not realised. The subjected stress is represented by the Mohr circle and whenever it extends higher than the mobilised envelope through the increase in the vertical stress or reduction in suction, then the soil will respond by undergoing settlement. The settlement caused by the latter is referred to as wetting collapse. Then the concept of RMYSF is further refined by the introduction of the NSRMYSF, which can produce greater fidelity in the stress–strain prediction. This framework is a theoretical framework and is the first in soil mechanics. It is also the first soil volume change framework that incorporates the mobilised soil shear strength. This concept of effective stress and shear strength interaction is a soil settlement theory, which can model and quantify loading and inundation settlement plus the weird soil settlement behaviours and they are described. In addition, this NSRMYSF can make accurate soil stress–strain prediction and thus it is very good for soil settlement modelling. The applicability of NSRMYSF has been tested against various types of soil ranges from fine to coarse-grained soils. The keystone to this theory is the existence of the unique relationship between the magnitude of the rotation of the mobilized envelope (i.e., $\Delta\phi'_{\text{min}_{\text{mobilized}}}$) and the magnitude of the anisotropic compression (i.e., % axial strain), which is the fan shape mobilised shear strength envelopes itself.

The application of the theory is extended to rock mechanics and essentially, it proves the equivalent applicability as in soil mechanics. Except that in rock mechanics, there is the existence of mobilised cementation, which is proved to be increasing linearly with the rock anisotropic compression. In soil mechanics, there is no cohesion or cementation under effective stress analysis. The ability of the NSRMYSF to make an accurate prediction of the intact rock stress–strain response and the rock peak strength under different confining pressures are being substantiated.

About the author

Mohd Jamaludin Md Noor was born in Perlis, Malaysia in 1962. He received his primary education from Kangar English School (1969–1973) and secondary education from Derma English School (1974–1978), in Kangar, Perlis, Malaysia. In the following year, he did his "A level" at Blackburn College of Technology and Design, Blackburn, England (1979–1980) and subsequently obtained his first degree, B.Sc. Civil Engineering (Hons) from University College of Swansea, Wales, United Kingdom in 1984. He is graduated in M.Sc. Geotechnics by research from University MALAYA, Malaysia in 1996 with his thesis entitled "Modelling of Infiltration and Stability of Slope." He proceeded for his Ph.D. at the University of Sheffield, United Kingdom in 2001 and graduated in 2006 with his thesis entitled "Shear Strength and Volume Change Behaviour of Unsaturated Soils."

Among his significant contributions in soil mechanics are: (1) the true curved-surface envelope soil shear strength model that can replicate very well the non-linear shear strength behaviour relative to both net stress and suction, (2) the first soil settlement theory named as "Rotational Multiple Yield Surface Framework (RMYSF)" that characterises settlement from the standpoint of effective stress and the intrinsic shear strength developed within the soil mass when deformed and this can clarify all the complex soil volume change behaviour like settlement under effective stress decrease during wetting and massive settlement near saturation besides the normal settlement when effective stress increases upon loading, (3) a refined RMYSF called Normalised Strain RMYSF which can produce a closer prediction of soil stress–strain curves for saturated and unsaturated soils, (4) a state-of-the-art infiltration type slope stability method, which checks stability against infiltration by increasing incrementally the depth of wetting front until the threshold condition is achieved and applying the curved-surface envelope soil shear strength model and the resultant is the most conservative and reliable slope stability method that produces an excellent prediction of shallow rainfall-induced landslide and (5) state-of-the-art early warning slope monitoring system, which is the first to monitor rainfall-induced instability based on the advancement of wetting front rather than the elevation of groundwater table.

He worked with a contractor for one year upon getting his first degree and joined Universiti Teknologi MARA in 1985 as a lecturer until now. Whilst working as a lecturer, he went for industrial training with Geotechnical Design Division, Public Work Department of Malaysia at Bangi in 1988 where he gained the geotechnical design experiences, which include the designs of shallow and deep foundations, slope failure, building settlement investigations and reinstatement design of a failed slope. In 1989, he did his industrial training with Petaling Public Works Department, Selangor and gained the field experience on the supervision of buildings construction and slope reinstatement work using geotextile.

He has been actively involved in developments of new products related to geotechnical engineering. He introduced the V-armed flip hybrid anchor and later the J-hook hybrid anchor for effective slope strengthening using an active soil nail system. He insisted on the application of the active soil nail instead of the passive soil nail in Malaysia. Besides, he has developed a slope stability software known as "Slope-Rain" which deployed the infiltration type slope stability method he developed and utilises the non-linear soil shear strength enveloped he has developed. The software has been benefitted in many slope reinstatement projects. At the international level, he has won ONE Best Award from Malaysian Technology Exhibition 2010, ONE Best Design Award from Japan Intellectual Property Association 2010, THREE Gold medals and THREE Silver medals. At the national level, he has won TWO Gold medals, ONE Silver medal and THREE Bronze medals. He has also received the Deputy Vice-Chancellor (Research) Innovation Award, Universiti Teknologi MARA in 2010.

He has been a member of the International Advisory Committee for Asia Pacific Unsaturated Soils Conference in Newcastle, Australia, 2009 and Pattaya, Thailand, 2011. Besides, he is also a member of the Technical Committee for the drafting of Malaysian Annex to Eurocode 7 for Geotechnical Works under SIRIM Berhad.

Chapter 1

The true soil stress–strain response and shear strength behaviour

1.1 Introduction to soil stress–strain response and shear strength behaviour

A stress–strain curve is one of the very important characteristics of soil. It reflects how a soil would respond to the change in the subjected stress. It is the basic property of a soil. It governs how a soil would respond when it is subjected to a stress increase or decrease. In contrast to the stress–strain curve of a metal where under a low-stress range, the response is linear-elastic, in the soil, even though the curve is linear at low-stress levels, the response is not purely elastic.

The soil shear strength failure envelope is defined from the stress–strain curves where the maximum deviator stress is taken as the failure condition. The failure envelope can be either linear or non-linear. The maximum deviator stresses will be the diameter of the Mohr circles at various effective confining pressures. The output shear strength parameters will be the unique intrinsic properties of the soil whether it is for a linear or a non-linear failure envelope. Different soils would have a different set of stress–strain curves and obviously, they produced shear strength parameters, which are specific for the soil only.

The effective confining pressure or the subjected effective stress is the prime factor that influences the way the response of soil to stress increases or decreases. In other words, the depth below the ground, which governs the magnitude of the confining effective stress, would have a strong influence on how the soil responds to stress. The deeper the depth, the higher the confining effective stress. Thence, it is very important for any soil volume change framework to be developed or formulated based on the soil stress–strain curves at various effective stresses.

Figures 1.1–1.3 show stress–strain curves for various soils, which are (1) saturated Ham River sand at normal and elevated net confining pressures (Bishop, 1966), (2) greywacke rockfill in large-scale triaxial testing by Indraratna et al. (1993) and (3) gneiss rock residual soil at horizon C by Futai and Almeida (2005) at a suction of 100 kPa, respectively. The works by Bishop (1966) and by Indraratna et al. (1993) dealt with granular soil and obviously, in the interpretation of the shear strength failure envelope, there will be no cohesion intercept. However, the same goes to the work of Futai and Almeida (2005) where under the saturated condition, there will be no cohesion base on effective stress analysis. Nonetheless, the failure envelope would intercept the shear strength axis at a certain apparent shear strength, because the soil

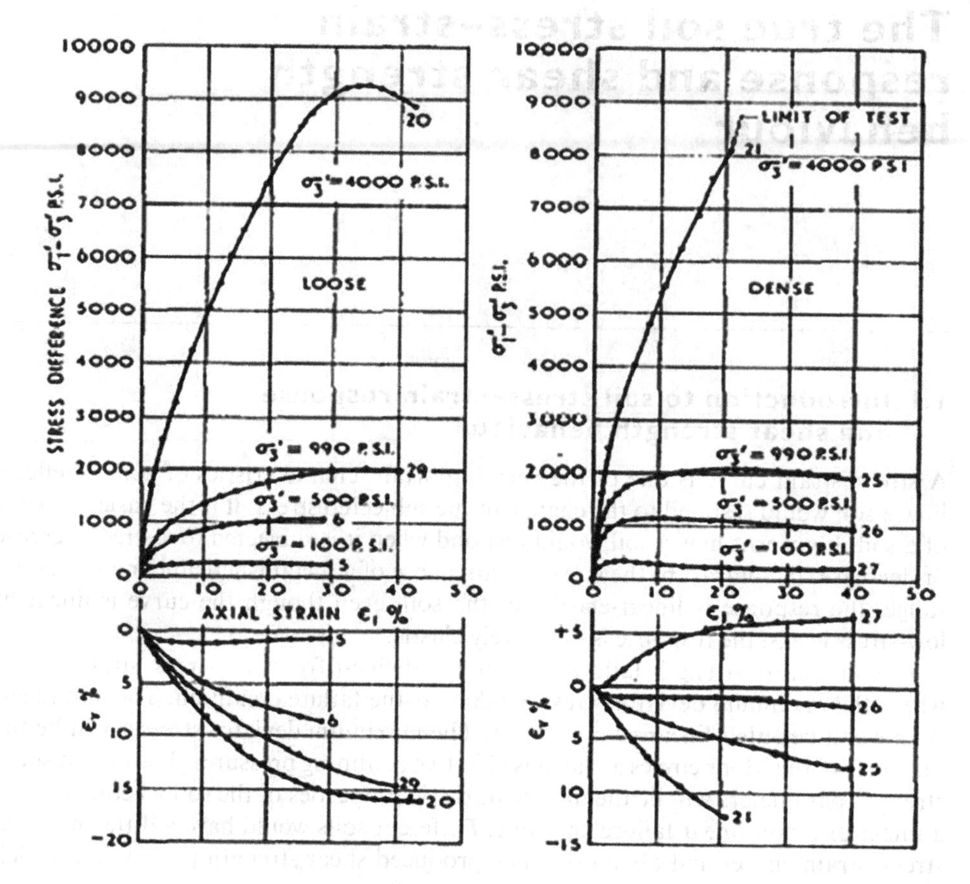

Figure 1.1 Stress–strain and volume change relationships for drained triaxial tests on saturated Ham River sand at normal and elevated net confining pressures (Bishop, 1966). (Note: 1 lb/in.^{2}. = 6.896 kN/m^2.)

is partially saturated. This apparent shear strength may be mistakenly thought of as the cohesion c' when viewed in two dimension, i.e., shear strength versus net stress.

Mostly, it can be noted that these stress–strain curves exhibit in common the normal soil characteristics, which are:

1. Stress–strain curves become steeper as the confining pressure increases.
2. Axial strain at failure increases with the increase of the effective confining pressure. In other words, the axial strain at failure is not the same for each stress–strain curve. The maximum deviator stress occurs at different axial strains for each stress–strain curve.
3. Maximum deviator stresses increase with the increase in the effective confining pressures.
4. The responding strain is higher under the same increase in stress at a lower effective confining pressure. This will be further illustrated in the following paragraph.

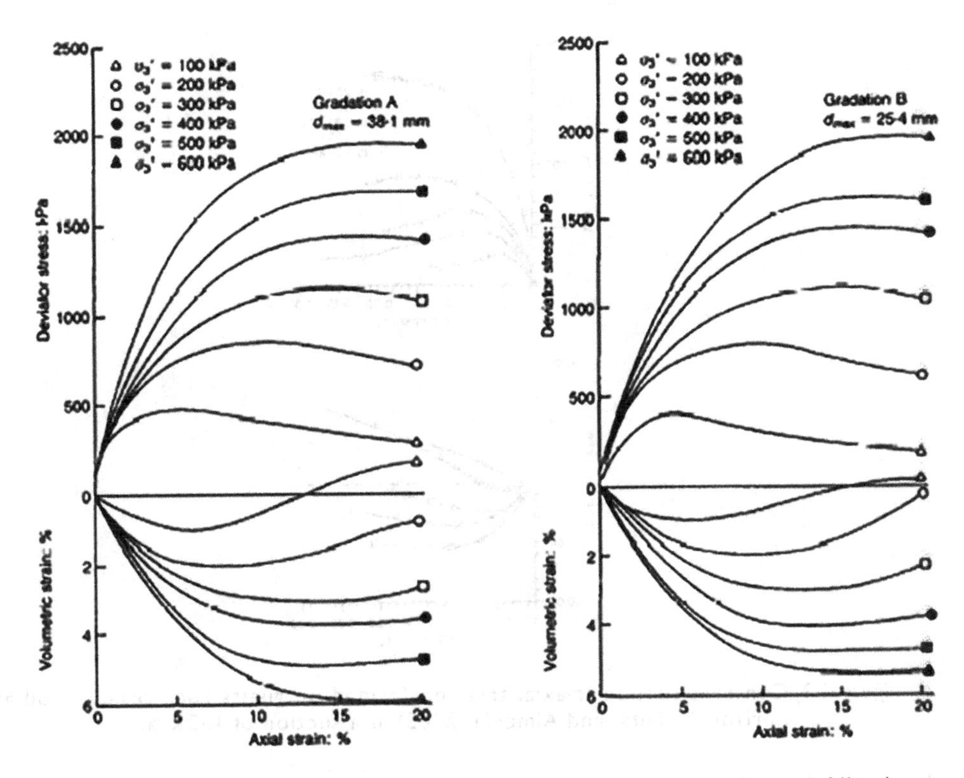

Figure 1.2 Stress–strain and volume change behaviour of greywacke rock fill in large-scale triaxial testing (Indraratna et al., 1993).

The slope of the stress–strain curve is termed as stiffness but for soil, it cannot be regarded as the modulus of elasticity, because the behaviour is not purely elastic even though the curve is linear at the low-stress range. The fact that it is not purely elastic can be seen when unloading is carried out. Essentially, the unloading curve does not retrace the initial loading curve. Thence, soil behaviour is a unique behaviour and it is not comparable to the stress–strain curve of a metal.

Figure 1.4 shows the typical stress–strain curves for silty CLAY at effective confining pressures of 100, 200 and 300 kPa. Notice that the axial strain at failure marked with the points A, B and C increases with the increase of the effective confining pressure. Besides, the maximum deviator stresses increase with the increase in the effective confining pressures. In addition, another very important attribute of the stress–strain curve is that it becomes steeper as the confining pressure increases.

It is important to note that for the same magnitude of stress increase from 200 to 300 kPa, the soil would respond differently under different effective confining pressures. For the effective stresses of 100, 200 and 300 kPa (refer Figure 1.4), the strain increase would be as labelled as a, b and c, respectively. Essentially, the increase in the strain is higher for the lower effective confining pressure, even though the stress increase is the same, i.e., increase from 200 to 300 kPa. In other words, the response is bigger when the confining pressure is less, whenever there is a stress change. This is in

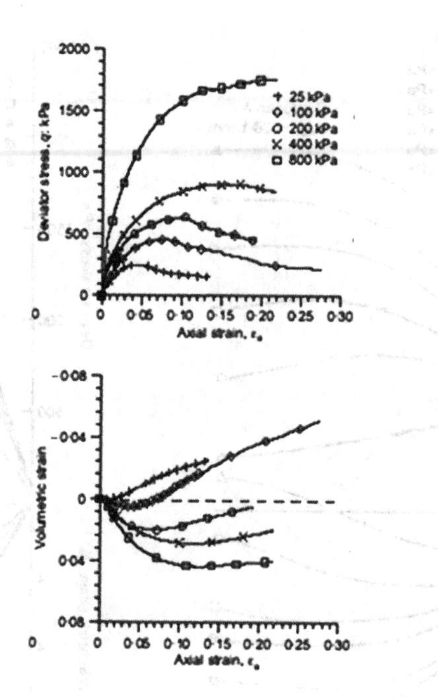

Figure 1.3 Constant suction triaxial tests performed on gneiss rock residual soil at horizon C (Futai and Almeida, 2005) at a suction of 100 kPa.

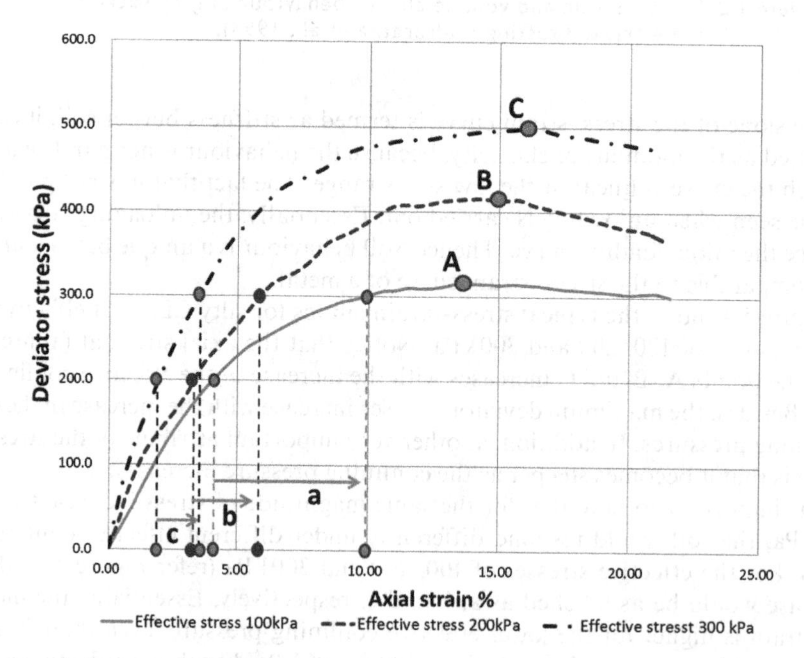

Figure 1.4 Typical soil stress–strain curves under different confining pressures for silty CLAY.

fact very distinctive soil stress–strain behaviour and this unique characteristic needs to be embedded in a good soil volume change framework. This attribute would be very much related to the weird soil volume change behaviour whereby there is a bigger inundation settlement under low effective confining pressure compared to higher effective confining pressure. In other words, a single-storey house would settle more compared to the adjacent double-storey house where both are supported by shallow foundation when the groundwater table rises. This is also known as inundation settlement or wetting collapse. This weird settlement behaviour is very difficult to be incorporated in the soil settlement framework.

1.2 Soil elastic–plastic and fully elastic strain responses to stress

In the shearing stage of a triaxial test where the deviator stress is being applied, the axial strain increases. This resulted axial strain comprises elastic and plastic components. This section will differentiate the elastic and the plastic strains and proves that the soil shows an elastic–plastic response whenever the applied stress is less than the past maximum stress, i.e., a normally consolidated case. Figure 1.5 shows the typical stress–strain curves for sand under effective stresses of 50, 100, 200 and 300 kPa. Each curve is subjected to unloading and reloading. When being unloaded, it does not track back the initial loading curve. In other words, the soil is not elastic. The unloading curves do not drop vertically downward and this indicates that the soil is not a fully plastic material. The unloading curve drops slightly inclined towards the origin. Notice that the unloading curves are linear and parallel.

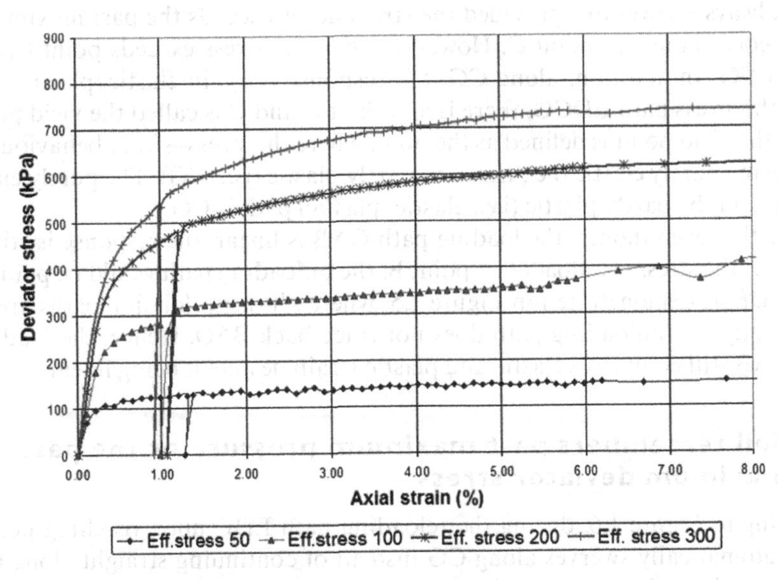

Figure 1.5 Stress–strain curves showing soil response when being unloaded under different effective stresses.

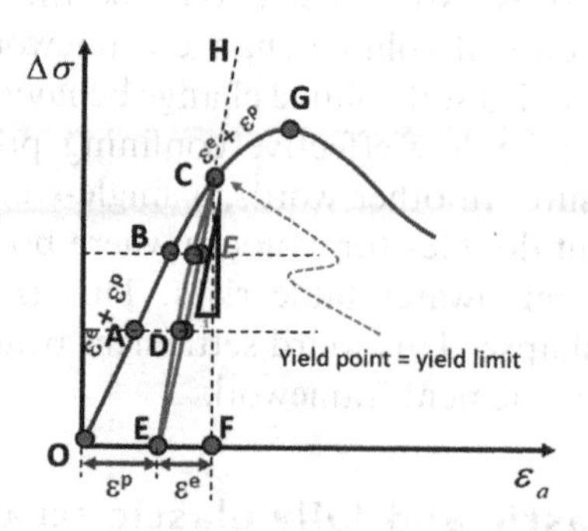

Figure 1.6 Stress–strain curve demonstrating the different conditions where the soil behaves elastic–plastic and fully elastic when subjected to stress.

The true soil response to stress is illustrated in this section. Figure 1.6 shows the typical response of soil to stress, where at times, it behaves elastic–plastic and at another condition, it behaves fully elastic but never responds fully plastic. Consider when the deviator stress is an increase from 0 at the origin to point A, B and C. The soil response is elastic–plastic. This elastic and plastic response can be seen when the stress is unloaded at C. The total strain when the stress reached a point C is OEF. In addition, OEF comprises elastic and plastic strain. The actual unloading path from point C is CDE. If the soil is a fully elastic material, the unloading path would be CBAO. In addition, if the soil is a fully plastic material, the unloading path would be vertically downward, CF. Since the unloading path is CDE the strain, OE cannot be recovered and it is the plastic strain. The strain EF is recoverable and this is the elastic strain. If the stress is repeatedly unloaded and reloaded along EDC, this elastic strain EF is always recoverable provided the stress never exceeds the past maximum stress, which corresponds to point C. However, when the stress exceeds point C, the stress path is CG. In addition, along CG, the response is again elastic–plastic. Note that along the stress path EDCG, there is a kink at C and C is called the yield point. That is why the yield point is defined as the point where the stress–strain behaviour changes from being purely elastic (i.e., EDC) to partly plastic (i.e., CG). The purely elastic path is EDC and the partly plastic (i.e., elastic–plastic) path is CG.

Note that even though the loading path OAB is linear, the response is still elastic–plastic. If the stress is unloaded at point B, the unloading path would be parallel to the path CDE as demonstrated in Figure 1.5, where all the unloading paths are parallel. Apparently, the unloading path does not trace back BAO. Hence, the loading stress path OAB still comprises elastic and plastic strain despite it being linear.

1.3 Soil remembers past maximum pressure or the past maximum deviator stress

Referring to Figure 1.6, during the reloading path EDC upon reaching point C, the path automatically swerves along CG instead of continuing straight along CH. This is because the soil remembers the past maximum pressure that it has been subjected

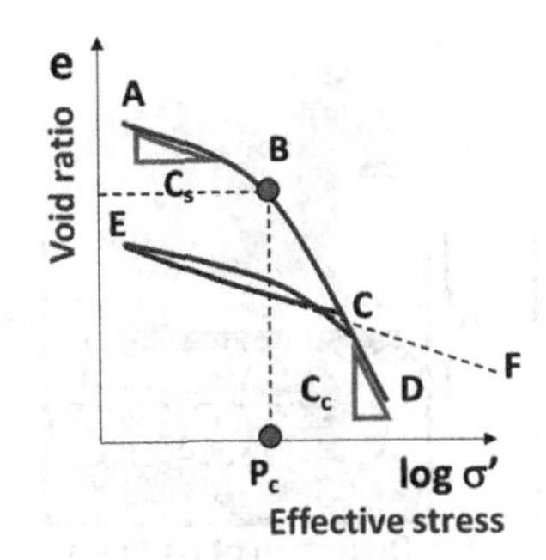

Figure 1.7 Consolidation curve and the pre-consolidation pressure.

to (i.e., corresponding to the deviator stress at point C in Figure 1.6) and immediately when the stress exceeded that value, it will automatically respond elastic–plastic instead of responding fully elastic before that. This is a unique soil behaviour.

This similar phenomenon is also encountered in soil consolidation curve of Terzaghi (1943) as shown in Figure 1.7. The consolidation curve is a plot of void ratio, e, against log effective stress, log σ'. This consolidation curve is considered to be obtained from conducting a consolidation test on an undisturbed soil specimen sampled at a certain depth in the ground. The gradient AB, C_s is slightly gentle compared to gradient BCD, C_c. AB is the over-consolidated curve while BCD is a normally consolidated curve. During loading from point A, the curve automatically swerves down at B. Point B corresponds to the pre-consolidation pressure, P_c, which is the past maximum pressure that the soil has experienced. Thence from the consolidation curve, the past maximum pressure that the soil experienced in history can be detected. This is because the soil always remembers the past maximum pressure that it has been subjected to. This can be further substantiated by repeating the unloading process upon reaching point C. The unloading curve is CE, which is parallel to BA. When reloaded, the curve follows the path ECD where upon reaching C, which is the current past maximum pressure, the curve automatically swerves down to D instead of proceeding to F. This proves that the soil always remembers the past maximum pressure it has experienced.

1.4 Physical meaning of the soil elastic and plastic strains

This section will explain how the soil elastic and plastic strain comes about. Seeing its physical meaning is important for the lasting understanding. Consider a horizontal footing platform resting on a soil structure with exaggerated rounded soil particles arranged as shown in Figure 1.8a. Before being loaded, the soil is considered to be rounded in shape. However, when a load is placed on the platform and the soil structure is compressed as shown in Figure 1.8b, notice that the soil structure has been rearranged and the rounded shape has been compressed into an oval shape. The total settlement is ρ and this total settlement comprises elastic, ρ_e, and plastic, ρ_p, components. These elastic and plastic settlement components can be seen when the load is being removed and the soil bounces back partly as shown in Figure 1.8c.

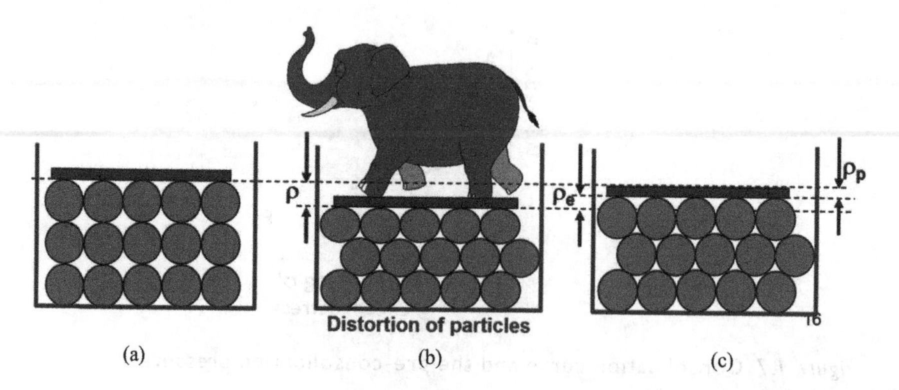

Figure 1.8 Soil particles distortion and rearrangement associated with the elastic and plastic settlement, respectively. (a) Initial particles arrangement. (b) Particles distorted and rearranged due to the applied loading. (c) Particles rearrangement maintained when unloaded but the distorted shape reverts to the initial shape.

The rearrangement of the particles cannot be reverted to the initial arrangement as in Figure 1.8a and this produces the permanent settlement, which is the plastic settlement, ρ_p. However, the rebound settlement, ρ_e, is the recoverable settlement, which is gained because the compressed shape of the particles has been reverted to the rounded shape when the pressure is released. Note that the plastic settlement, ρ_p labelled in Figure 1.8c, has resulted from the particle rearrangement, which can be seen by comparing the soil structure in Figure 1.8a and 1.8c, where the shape of the particles is rounded in both cases. Only the particle arrangement differs. Thence, the plastic settlement is due to the rearrangement of the soil particles.

Therefore, it can be summarised that:

1. Plastic deformation ρ_p refers to the greater part of the deformation, which is due to the slippage between the soil particles as the soil skeleton rearranges itself to accommodate higher loads. This component of deformation is irrecoverable or plastic.
2. Elastic deformation ρ_e takes place along the unloading and reloading line where the change in stress can be accommodated without the need for the rearrangement of soil particles. Deformation is primarily due to the distortion of the soil particles and can be recovered on unloading.
3. A material yields when its stress–strain behaviour changes from being purely elastic to partly plastic OR when the deformation stops being recoverable upon unloading. This is often marked by an abrupt change in the slope (i.e., stiffness) of the stress–strain curve. Yielding does not necessarily mean failure.
4. Failure, in Mohr–Coulomb theory failure, is the onset of mobilising the maximum shear stress where the Mohr stress circle (i.e., representing normal and shear stresses on a slip plane) touches the failure envelope.

1.5 Triaxial test and the conventional soil shear strength model

Shear strength is an important property of soil. It governs the soil mechanical behaviour like slope failure and settlement. Not understanding the true shear strength behaviour causes the problem to understand those soil mechanical behaviour. Therefore, it is important to make the right interpretation from the soil stress–strain curves discovered from conducting field or laboratory tests. The common laboratory tests conducted for the determination of soil shear strength are the consolidated drained (CD) triaxial test and the consolidated undrained (CIU) triaxial test (Bishop, 1966, 1971; Head, 1981). Nevertheless, the CIU test with pore pressure (PWP) measurement is often practised because it is quicker compared to a CD test. The typical set up of these triaxial tests is shown in Figure 1.9. In the CIU test, the valves A and B are always closed during the shearing stage to prevent specimen water from escaping the specimen. In the CD test, valve A is opened and valve B is closed during shearing. The specimen water that has been squeezed out from the specimen through valve A is monitored at the Imperial College Volume Change Unit. Three CIU triaxial tests have been conducted labelled as Test A, B and C and the consolidation and shearing data are presented in Figures 1.10 and 1.11, respectively. Table 1.1 shows the data at failure obtained from the shearing stage. The data will allow the determination of the minor and major principle

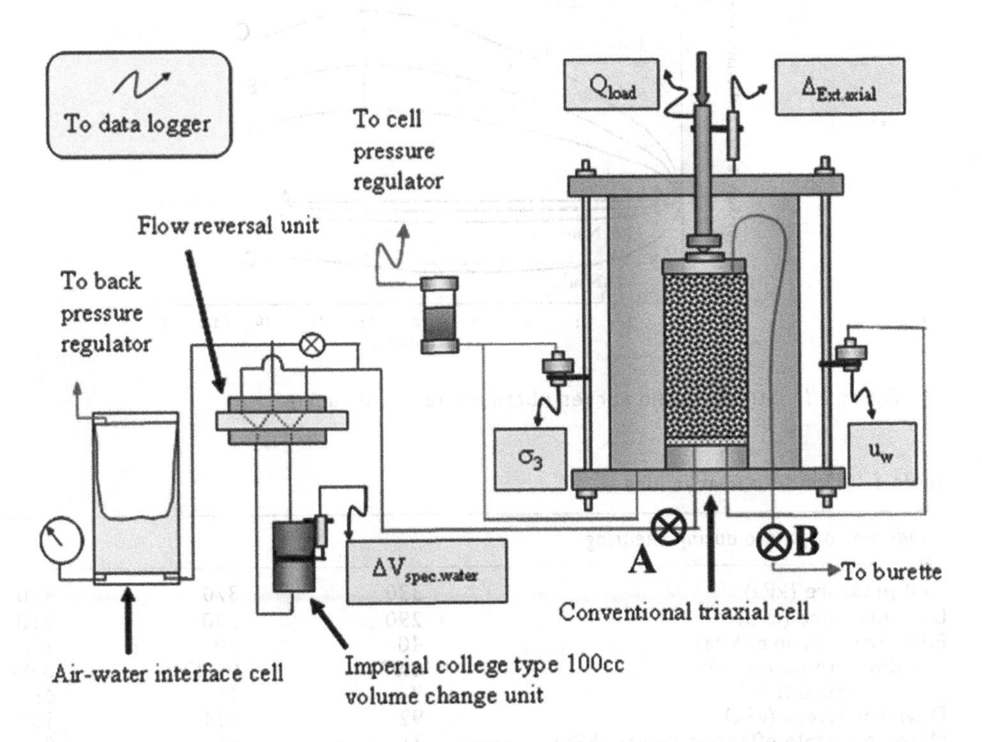

Figure 1.9 Typical set up of the CD and CIU triaxial tests.

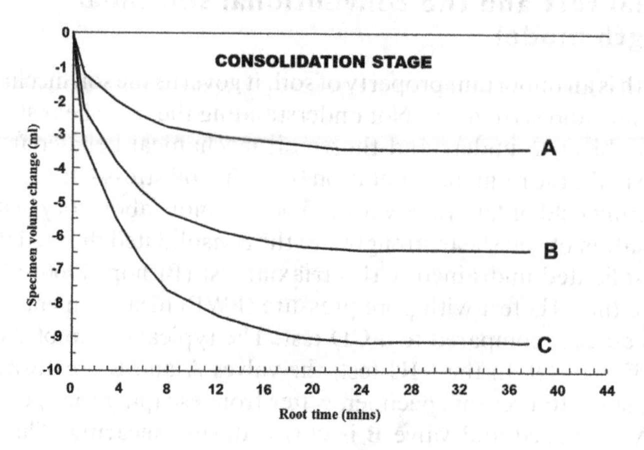

Figure 1.10 Consolidation curves under effective stresses of 40, 80 and 160 kPa.

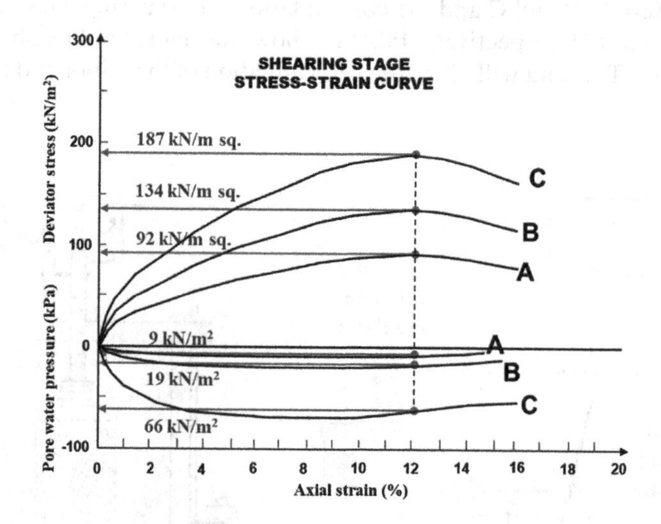

Figure 1.11 Stress–strain curves obtained during shearing.

Table 1.1 Conditions at failure

Conditions at failure during shearing			
Cell pressure (kPa)	330	370	450
Back pressure (kPa)	290	290	290
Effective pressure (kPa)	40	80	160
Loading rate (mm/min)	0.032	0.032	0.032
Pore pressure (kPa)	9	19	66
Deviator stress (kPa)	92	134	187
Minor principle effective stress (kPa)	31	61	94
Major principle effective stress (kPa)	123	195	281

effective stresses so that the respective Mohr circles can be drawn. There are three stages involved in each test, namely:

1. Saturation stage
   Saturation stage is the process of forcing water into the specimen to achieve a certain degree of saturation base on Skempton's B value (Skempton, 1961). This B value is defined as the ratio of the change in the pore water pressure, ΔU, and the change in cell pressure, $\Delta\sigma_3$.

   i.e., $B = \dfrac{\Delta U}{\Delta\sigma_3}$

   The water is forced into the specimen from the Imperial College Volume Change unit by applying the PWP by 5kPa less than the applied cell pressure. By this way, the specimen is not consolidated. The PWP and the cell pressure are elevated until the Skempton's B value is greater than 0.95. As the PWP is increased during the process of saturation, the water is forced harder into the specimen. Alternately, the Skempton's B value check is carried out until the required minimum value is achieved. The Skempton's concept is that when the specimen is fully saturated, then the increase in the cell pressure, $\Delta\sigma_3$, will produce an equal increase in the PWP, ΔU, i.e., B equal to 1.0. During the B value check, the cell pressure is elevated while valve A and B are closed and the PWP increase is monitored until it becomes constant. This is followed by the increase in the PWP to a value less by 5kPa from the cell pressure and subsequently, valve A is opened to force more water into the specimen. However, it is important to be cautious during the process that the specimen is not consolidated, i.e., when valve A has opened the difference between the cell pressure and the PWP is always at 5kPa.
2. Consolidation stage
   Consolidation stage is a process of consolidating the specimen under a certain consolidation pressure or a certain effective stress. In consolidating the specimen, the cell pressure is an increase to a targeted value higher than the PWP by the targeted magnitude of effective stress while the valves A and B are closed. When the consolidation is ready, then valve A is opened to begin consolidation. The consolidation is considered done when the graph becomes horizontal. Figure 1.10 shows the typical consolidation curves.
3. Shearing stage
   Once the specimen is fully consolidated, the deviator stress is applied to begin the shearing stage. The typical rate of applying the deviator stress is like 0.01–0.5 mm/min. In the CD test, the rate of applying the deviator stress must be slow enough so that the PWP is maintained constant. The typical output stress–strain curves are as shown in Figure 1.11. The maximum value of the deviator stress is considered as the failure condition for the interpretation of the shear strength envelope at failure. Subsequently, the corresponding value of the PWP, back pressure, BP, and the cell pressure is noted.

Based on effective stress analysis, the corresponding Mohr circles are drawn as in Figure 1.12. From here, the linear failure envelope is deduced by drawing a straight

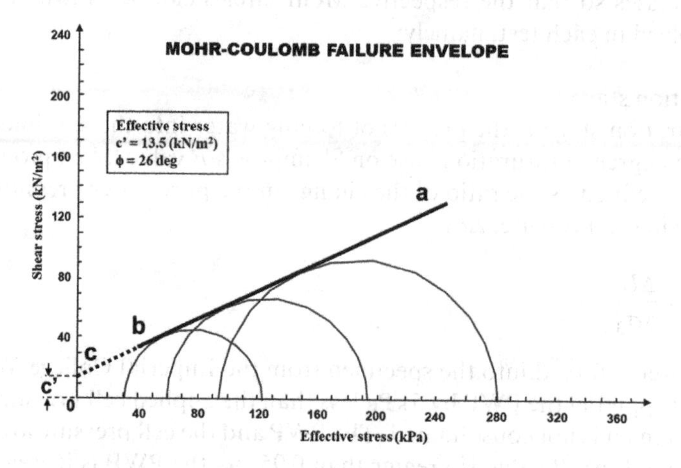

Figure 1.12 Interpretation of the linear Mohr–Coulomb failure envelope from CIU triaxial tests.

line ab for the higher range of effective stress. Then the linear failure envelope is extrapolated to c indicated by the dotted line, bc. In other words, the cohesion c' is achieved based on the extrapolation and not based on the factual data. Thence, to determine the actual value of the cohesion c', further test is needed at lower effective stresses of magnitude like 10 or 20 kPa.

1.6 The interpretation of the true non-linear soil shear strength behaviour from stress–strain curves

There have been many reports that soil does not have cohesion c' even for clays when interpreted according to effective stress analysis. The reports include those of Henkel (1958), Bishop (1966, 1971), Atkinson and Farrar (1985) and Day and Axten (1989). Figure 1.13 shows the curvilinear shear strength envelope for Mission Valley

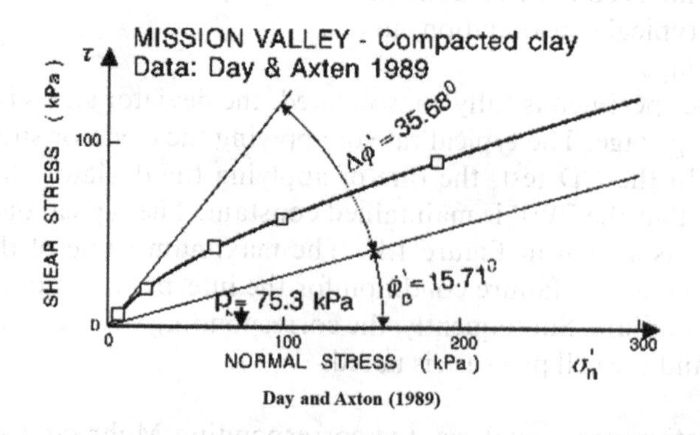

Figure 1.13 Curvilinear shear strength envelope with zero cohesion intercept for Mission Valley compacted clay (Day and Axten, 1989).

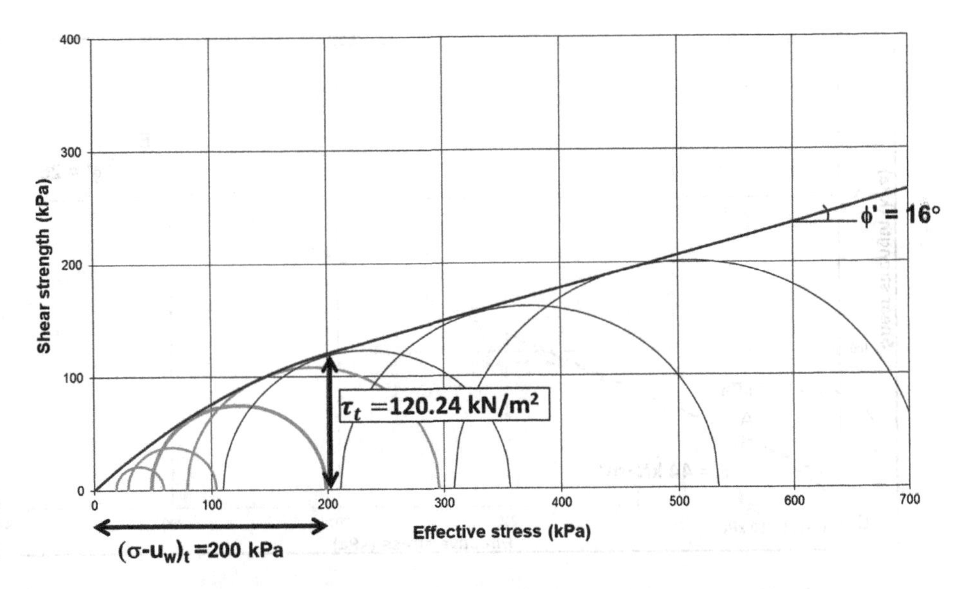

Figure 1.14 Curvilinear shear strength envelope for granitic residual soil grade IV from Kuala Kubu Bharu, Malaysia (Md Noor and Derahman, 2012).

compacted clay reported by Day and Axten (1989). Md Noor and Derahman (2012) have reported that granitic residual soil grade IV from Kuala Kubu Bharu, Malaysia has curvilinear shear strength envelope as shown in Figure 1.14. Their works focussed on low-stress levels and substantiated that the cohesion intercept is at the origin.

In addition, there are a number of equations for non-linear failure envelope proposed by several authors like De Mello (1977), Hoek and Brown (1980), Maksimovic (1996), Md Noor and Anderson (2006) and Lade (2010). The equation of Md Noor and Anderson (2006) will be applied for the definition of the non-linear shear strength failure envelope.

$$\tau_f = (K + \alpha\sigma)^{\beta} \qquad \text{(1.1) (De Mello, 1977)}$$

$$\tau = a\sigma^{n}. \qquad \text{(1.2) (Charles and Watts, 1980)}$$

$$\tau = A(\sigma_n - \sigma_m)^{B}. \qquad \text{(1.3) (Hoek and Brown, 1980)}$$

$$\tau = P_a A\left(\frac{\sigma}{P_a} + T\right)^{n} \qquad \text{(1.4) (Hoek and Brown, 1980)}$$

$$\tau_f = \sigma_n' \tan\left(\phi' + \frac{\Delta\phi'}{1 + \sigma_n'/P_N}\right) \qquad \text{(1.5) (Maksimovic, 1996)}$$

$$\left(\frac{s}{P_a}\right) = a\left(\frac{\sigma'}{P_a}\right)^{b} \qquad \text{(1.6) (Lade, 2010)}$$

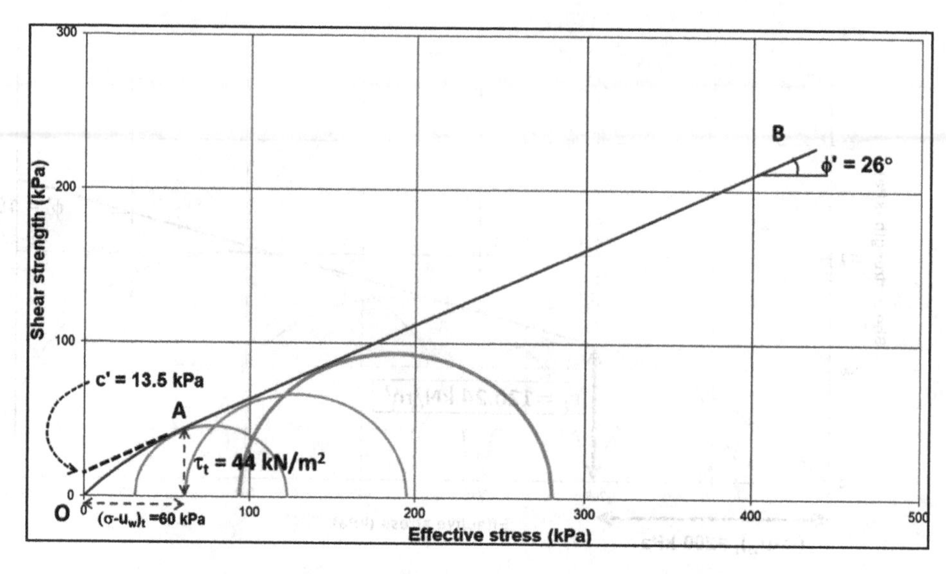

Figure 1.15 Non-linear failure envelope with zero cohesion intercept.

The result of the CIU triaxial test described in Section 1.5 can be interpreted according to the non-linear failure envelope of Md Noor and Anderson (2006) as shown in Figure 1.15. The non-linear shear strength equation of Md Noor and Anderson (2006) under the fully saturated condition with respect to the effective stress axis along the curve OA and the linear section AB is represented by Equations 1.7 and 1.8, respectively. Equation 1.9 gives the value of N in Equation 1.7.

$$\tau_{\text{satf}} = \frac{(\sigma - u_{\text{w}})}{(\sigma - u_{\text{w}})_{\text{t}}}\left[1 + \frac{(\sigma - u_{\text{w}})_t - (\sigma - u_{\text{w}})}{N(\sigma - u_{\text{w}})_{\text{t}}}\right]\tau_{\text{t}} \tag{1.7}$$

$$\tau_{\text{satf}} = (\sigma - u_{\text{w}})\tan\phi'_{\text{minf}} + \left[\tau_{\text{t}} - (\sigma - u_{\text{w}})_{\text{t}}\tan\phi'_{\text{minf}}\right] \tag{1.8}$$

$$N = \frac{1}{1 - \left[(\sigma - u_{\text{w}})_{\text{t}}\dfrac{\tan\phi'_{\text{minf}}}{\tau_{\text{t}}}\right]}, \tag{1.9}$$

where $(\sigma - u_{\text{w}})_{\text{t}}$ is the transition effective stress beyond which the shear strength is assumed to behave linearly, τ_{t} is the transition shear strength that corresponds to transition effective stress, $\tan\phi'_{\text{minf}}$ is the gradient of the straight section of the graph and N is a constant. The shear strength parameters according to this non-linear failure envelope are given in Table 1.2.

Table 1.2 Shear strength parameters for the non-linear failure envelope at saturation according to the equation of Md Noor and Anderson (2006)

No.	*Shear strength parameters according to the non-linear failure envelope*	*Symbol*	*Value*
1	Transition effective stress (kPa)	$(\sigma - u_w)_t$	60
2	Transition shear strength (kN/m^2)	τ_t	44
3	Minimum friction angle (°)	ϕ'_{minf}	26°
4	Constant, *N*	N	2.9858

1.7 Summary

Essentially stress–strain curves of a soil, being the most fundamental property of soil, define how a soil would react when subjected to a stress and it reacts differently under different effective confining stresses. It is also very important to note that most of the time, the response of soil to stress is elastic–plastic and when the stress condition is less than the past previous maximum stress, the reaction is fully elastic. Moreover, there is no time that the soil behaves fully plastic when interpreted from the stress–strain curves. Therefore, the common practice that assumes the soil to behave elastic–perfectly plastic as shown in Figure 1.16 is not in accordance with its actual stress–strain behaviour. At a small strain OA, the soil is considered to behave purely elastic and at a higher strain AB, the behaviour is considered to be fully plastic. This is nowhere near the actual stress–strain behaviour of soil. This elastic–perfectly plastic behaviour is considered probably to simplify the stress–strain property since the actual behaviour is non-linear. It is very important to consider the actual elastic and elastic–plastic behaviour of soil accordingly to avoid any misinterpretation of the soil mechanical behaviour, especially when modelling soil movement.

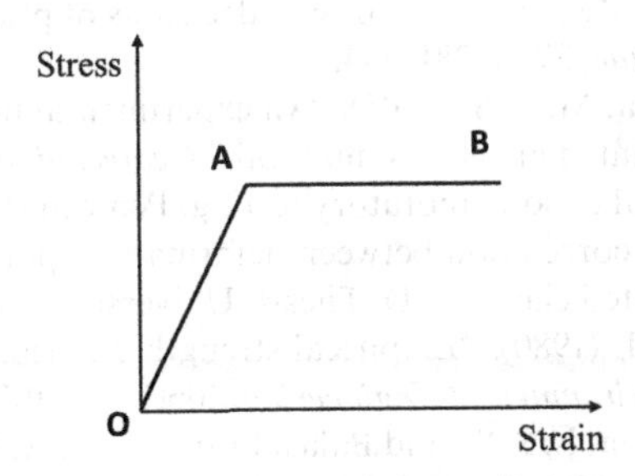

Figure 1.16 Elastic–perfectly plastic stress–strain behaviour commonly applied in soil modelling.

Application of the true elastic–plastic behaviour is thus very important to characterise soil mechanical behaviour. It has already been applied for the definition of the soil shear strength failure envelope. The same stress–strain curves allowed the definition of soil shear strength according to both linear and non-linear failure envelopes. The application of the non-linear failure envelope refined the shear strength behaviour at low-stress levels. Moreover, this would reveal the understanding of the shallow rainfall-induced slope failure and inundation settlement, which occurred under a low-stress range. The application of the linear failure envelope does not fit the shear strength behaviour at low stress but it is a mere extrapolation from the behaviour at the high-stress range and this over-estimates the magnitude at low stress. This could be the reason why the shallow rainfall-induced landslide is very difficult to be back analysed by geotechnical engineers. The application of the true stress–strain curves and the true elastic–plastic characteristic in soil volume change framework would reveal the actual mechanics in the occurrence of various soil complex volume change behaviours.

The true soil shear strength behaviour at low stress is non-linear with zero cohesion intercept when interpreted based on effective stress analysis. Section 1.6 shows examples that a compacted clay soil and undisturbed granitic residual soil have zero cohesion intercept.

References

Atkinson, J. H. and D. M. Farrar (1985) Stress path tests to measure soil strength parameters for shallow slips. *Proc. of the 11th Int. Conf. SMFE, San Francisco*, Vol. 2, 983–986.

Bishop, A. W. (1959). "The principle of effective stress." *Teknisk Ukeblad*, 106(39), 859–863.

Bishop, A. W. (1966). "The strength of soils as engineering materials." *Geotechnique*, 16(2), 91–130.

Bishop, A. W. (1971). "Shear strength parameters for undisturbed and remolded soil specimens." *Stress Strain Behavior of Soils: Proceedings of the Roscoe Memorial Symposium, Cambridge University*, Henley on Thames, UK, 3–58

Charles, J. A., and Watts, K. S. (1980). "The influence of confining pressure on the shear strength of compacted rockfill." *Geotechnique*, 30(4), 353–367.

Day, R. W., and Axten, G. W. (1989). "Surficial stability of compacted clay slopes." *Journal ASCE*, 115(4), 577–580.

De Mello, V. F. B. (1977). "Reflections on design decisions of practical significance to embankment dams." *Geotechnique*, 27(3), 281–354.

Futai, M. M., and Almeida, M. S. S. (2005). "An experimental investigation of the mechanical behaviour of an unsaturated gneiss residual soil." *Geotechnique*, 55(3), 201–213.

Head, K. H. (1981). Manual of soil laboratory testing, Pentech Press, London.

Henkel, D. J. (1958). "The correlation between deformation, pore-water pressure and strength characteristics of saturated clay." Ph.D. Thesis, University of London.

Hoek, E., and Brown, E. T. (1980). "Empirical strength criterion for rock masses." *Journal of Geotechnical and Geoenvironmental Engineering*, 106(GT9), 1013–1035.

Indraratna, B., Wijewardena, L. S. S., and Balasubramaniam, A. S. (1993). "Large-scale triaxial testing of greywacke rockfill." *Geotechnique*, 43(1), 37–51.

Lade, P. V. (2010). "The mechanics of surficial failure in soil slopes." *Engineering Geology*, 114(1/2), 57–64.

Maksimovic, M. (1996). "A family of nonlinear failure envelopes for non-cemented soils and rock discontinuities." *Electronic Journal Geotechnical Engineering*, 1, ppr9607.

Md Noor, M. J., and Anderson, W. F. (2006). "A comprehensive shear strength model for saturated and unsaturated soils." *Proceedings of the 4th International Conference Unsaturated Soils*, ASCE Geotechnical Special Publication No. 147, Carefree, Arizona, Vol. 2, 1992–2003.

Md Noor, M. J., and Derahman, A. (2012). "Curvi-linear shear strength envelope for granitic residual soil grade VI." *5th Asia-Pacific Conference on Unsaturated Soils*, Pattaya, Thailand, 227–232.

Skempton, A. W. (1961). "Effective Stress in Soils, Concrete and Rocks." Proceedings Conference Pore Pressure *and Suction in Soils*, Butterworths, London, England, 4–16.

Terzaghi, K. (1943). "Theoretical soil mechanics". New York, Wiley Publications.

Chapter 2

Concept of effective stress and shear strength interaction in governing soil settlement

2.1 Conventional concept of effective stress and its limitations

The effective stress concept has been the most fundamental concept in soil mechanics, and it was first introduced by Terzaghi (1936). Since then, soil mechanical behaviour is being characterised based on this concept. Soil shear strength models and settlement models are being developed based on the effective stress concept. Terzaghi (1936) wrote that "all the measurable effects of a change in stress, such as compression, distortion, and a change in shearing resistance, are exclusively due to changes in effective stress." In brief, the concept dictates that shear strength increases with the increase in the effective stress and ground settlement gets bigger when the effective stress is higher.

Effective stress σ' is defined as the soil grain-to-grain stress and its relationship to the total stress σ and the pore water pressure u_w is given by Equation 2.1 of Terzaghi (1936) and illustrated in Figure 2.1. The total stress is the applied stress or load to the soil structure like from a footing resting on the ground. This total stress will be supported partly by the soil skeleton in the ground and partly by the pore water pressure if the ground is flooded. The stress supported by the soil skeleton is termed as the effective stress. However, when the pore water pressure is being fully dissipated and becomes zero, then the total stress will be fully supported by the pressure transferred to the soil skeleton. In other words, the effective stress is the stress between the soil grains.

$$\sigma' = \sigma - u_w \tag{2.1}$$

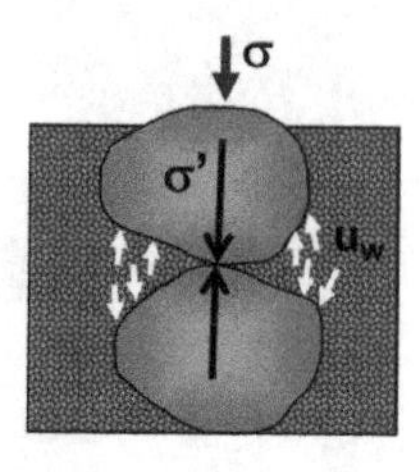

Figure 2.1 Effective stress is the soil grain-to-grain stress.

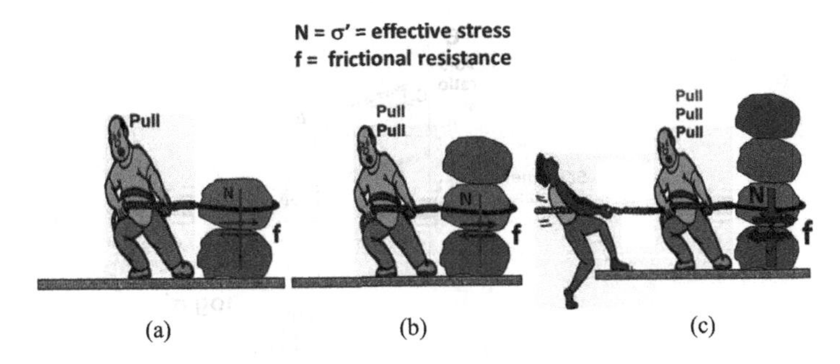

Figure 2.2 Concept of effective stress in soil shear strength. (a) The normal force, N exerted by the top boulder produce friction, f acting in the direction resisting the pulling force. (b) Topping another boulder increase the normal force, N and produce subsequently increase in the friction force, f. (c) Topping the third boulder will further increase the normal force, N and produce higher frictional force, f making the pulling effort more difficult.

Figure 2.2 illustrates the concept of effective stress in relation to soil shear strength behaviour. The minute soil particles are represented in an exaggerated boulder form. In Figure 2.2a, a boulder on a platform is stacked with another boulder. The man is trying to pull down the top boulder. The top boulder will exert a normal force N due to its weight on the bottom boulder. This will subsequently produce a frictional force f acting over the contact surface between the boulders in the direction perpendicular to the normal force and resisting the pulling effort by the man. This normal force N resembles the effective stress σ' and the frictional force resembles the shear stress or the soil shear strength τ. In Figure 2.2b and c, more boulders are stacked and they produce higher normal force N and thus give rise to a higher frictional force f. This will make the man's effort to pull down the second boulder more difficult. This explains the concept of effective stress in soil shear strength, where the higher the effective stress the higher the shear stress to resist the movement of the second boulder. This is demonstrated in the linear soil shear strength model of Terzaghi (1936) as shown in Figure 2.3 where the shear strength equation is as in Equation 2.2. The model demonstrates the linear increase in shear strength with the increase in the effective stress. However, the soil shear strength failure envelope was first introduced as a non-linear envelope by Coulomb (1776).

$$\tau = c' + (\sigma - u_w)\tan\phi' \qquad (2.2)$$

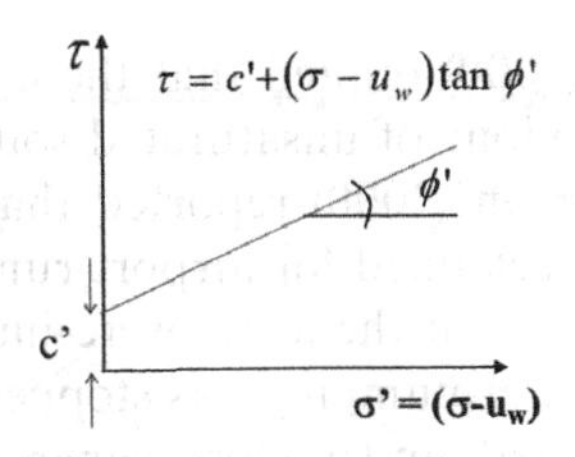

Figure 2.3 Shear strength model of Terzaghi (1936).

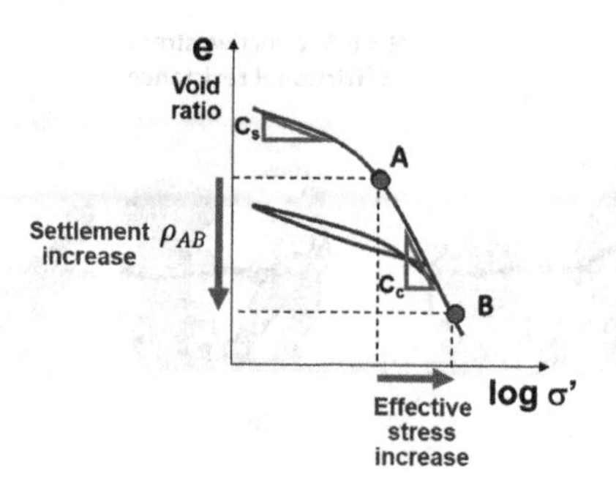

Figure 2.4 Consolidation curve of Terzaghi (1943).

The consolidation curve of Terzaghi (1943) shown in Figure 2.4 demonstrates the effective stress concept in soil settlement behaviour. As the effective stress is increased from point A to B on the graph, there will be a subsequent increase in the settlement ρ_{AB}, indicated by the vertical downward arrow showing the reduction in the void ratio e. In other words, whenever there is an increase in the effective stress, then there will be a subsequent settlement. For the stress increase within the normally consolidated graph denoted by the gradient c_c, the magnitude of settlement is given by Equation 2.3 (Terzaghi, 1943), where Δp is the increase in the effective stress. Note that the settlement is calculated based on the increase in the effective stress. Many similar empirical equations have been introduced and they quantified settlement based on effective stress or load increase considering the soil is either clay or sand. Equations 2.3–2.5 are developed to characterize the soil settlement in clay soil. Equations 2.6 and 2.7 are developed for sands. Note that all these equations quantify settlement based on pressure increase, Δp, or applied pressure, q, or stress increase, $\Delta\sigma$. In other words, when there is no load increase, then automatically there will be no settlement. However, in the case of inundation settlement or wetting collapse, which is triggered by the rise of the groundwater table, the effective stress decrease and by the concept of Terzaghi (1936), there should not be any settlement. This complexity regarding the influence of effective stress on soil settlement behaviour is illustrated in Figure 2.5. Soil settlement can be triggered by either an increase or a decrease of effective stress. Therefore, the effective stress concept is not comprehensive enough to resolve this complexity. In other words, it is not the effective stress by itself that govern the soil settlement behaviour. There has to be another stress state variable that governs the settlement and this variable must be the resisting variable, because effective stress is the driving variable.

To attest this, Fredlund (2000) quoted that there is still no closed-form solution for soil volume change behaviour of unsaturated soils up to the year 2000. In addition, Blanchfield and Anderson (2000) reported that ex-coalfields in Sheffield and Rotherham that have been reclaimed for airport runway and new housing developments underwent settlement when the areas were inundated when the groundwater table rises after the mines water pumping has stopped. This is despite the backfilled have been properly compacted under close supervision according to the British standard.

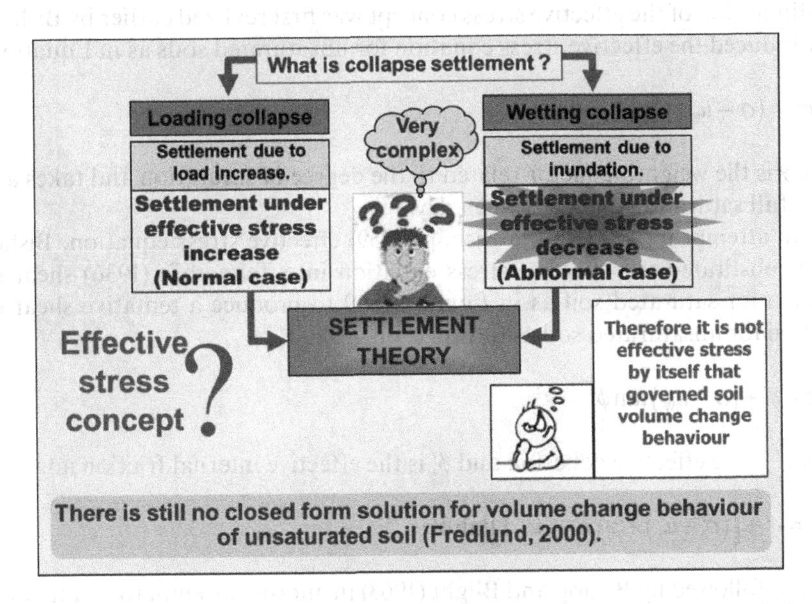

Figure 2.5 The complexity involved in characterizing the soil settlement behaviour based on the effective stress concept.

$$S = \frac{C_c H}{1+e_0}\log\left(\frac{p_0+\Delta p}{p_0}\right) \quad (2.3)\ \text{(Terzaghi, 1943)}$$

$$S_i \text{ or } \rho = \frac{qB}{E}\left(1-v^2\right)I_p \quad (2.4)\ \text{(Steinbrenner, 1934)}$$

$$S_i \text{ or } \rho = \frac{\mu_o \mu_1 q B}{E_u}\left(1-v^2\right) \quad (2.5)\ \text{(Janbu et al., 1956)}$$

$$s = \frac{H}{C}\ln\frac{\sigma'_o+\Delta\sigma}{\sigma'_o} \quad (2.6)\ \text{(De Beer and Martens, 1951)}$$

where, Equation 2.6 (De Beer and Martens, 1951)

$$C = 1.5\frac{q_c}{\sigma'_o}$$

$$s_i = C_1 C_2 q_n \sum \frac{I_z}{E}\Delta z \quad (2.7)\ \text{(Schmertmann et al., 1978)}$$

Thence, there is a limitation for this concept of effective stress in characterizing soil settlement behaviour. This concept cannot explain the occurrence of wetting collapse, because when the groundwater is inundated, there is a drop in the effective stress due to the buoyancy effect, i.e., the soil grain-to-grain stress reduces. By the effective stress concept, there should not be any settlement since the effective stress has decreased.

This limitation of the effective stress concept was first realized earlier by Bishop (1959); he introduced the effective stress equation for unsaturated soils as in Equation 2.8.

$$\sigma' = (\sigma - u_a) + \chi(u_a - u_w) \tag{2.8}$$

where χ is the weighting factor related to the degree of saturation and takes a value of 1.0 for full saturation and zero when dry.

In an attempt to validate the Bishop (1959) effective stress equation, Bishop et al. (1960) substituted the effective stress equation into Terzaghi's (1936) shear strength equation for saturated soil as in Equation 2.9 to produce a tentative shear strength equation for unsaturated soil as in Equation 2.10;

$$\tau = c' + (\sigma - u_w)\tan\phi' \tag{2.9}$$

where c' is the effective cohesion and ϕ' is the effective internal friction angle.

$$\tau = c' + \left[(\sigma - u_a) + \chi(u_a - u_w)\right]\tan\phi' \tag{2.10}$$

This was followed by Bishop and Blight (1963) in another attempt to restate the Bishop (1959) effective stress equation where a graphical illustration for volume change behaviour of unsaturated soil was expressed in the form of the two independent stress state variables as shown in Figure 2.6. This further reinstates the use of the two stress state variables. Burland (1964, 1965) further questioned the validity of the Bishop (1959) effective stress equation but insisted that the use of those stress state variables in formulating the mechanical behaviour of unsaturated soil is inevitable. This was then followed by Matyas and Radhakrishna (1968) who described the volume change

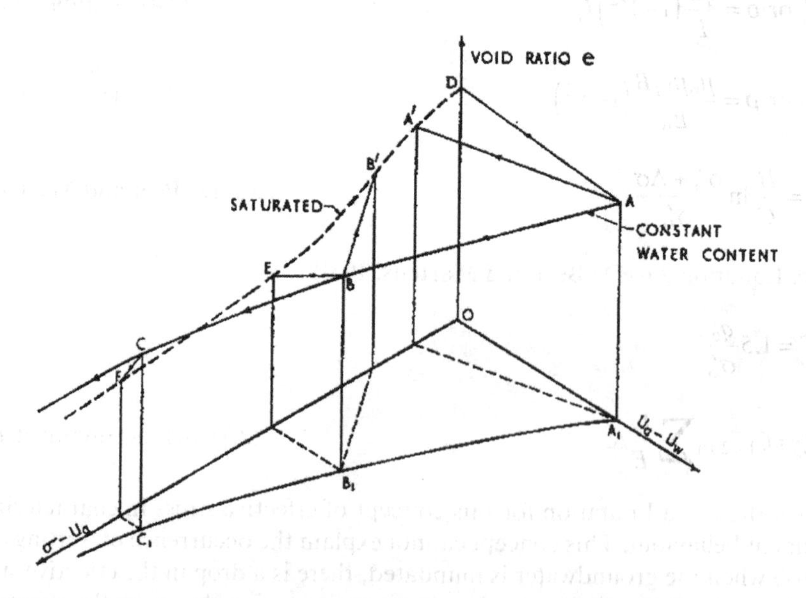

Figure 2.6 Volume change in terms of the void ratio against net stress, σ', and suction, $(u_a - u_w)$ (Bishop and Blight, 1963).

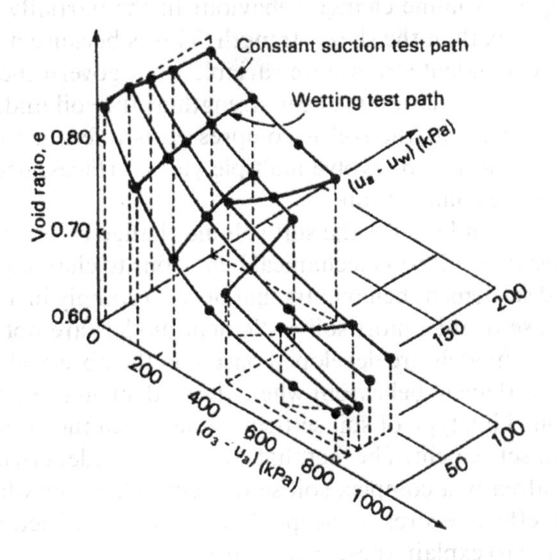

Figure 2.7 Warped surface representing the volume change behaviour of a mixture of flint (80%) and kaolin (20%) under isotropic stress conditions (Matyas and Radhakrishna, 1968).

behaviour of unsaturated soil in the form of a three-dimensional surface with respect to the two independent stress state variables as shown in Figure 2.7. Barden et al. (1969) came up with the same idea as Burland (1964, 1965) in the use of the two independent stress state variables in accessing volume change behaviour. Brackely (1971) studied the application of the effective stress principle to the volume change behaviour of unsaturated soil and concluded that the use of a single-valued effective stress equation, i.e., Equation 2.8, is limited. Since then, the application of the two independent stress variables, i.e., net stress and suction, in accessing the mechanical behaviour of unsaturated soil has intensified and they were used to formulate constitutive equations for shear strength and volume change behaviour of unsaturated soils. Subsequently, Fredlund and Morgenstern (1976) and Fredlund et al. (1978) have come up with a semi-empirical volume change, and shear strength equation, for unsaturated soil using the two independent stress state variables.

2.2 Soil complex settlement behaviour

Wetting collapse is the settlement that occurs when the soil is wetted. The process takes place under effective stress decrease. This behaviour contradicts the effective stress concept. This behaviour is categorised as the complex soil volume change behaviour. Nevertheless, this is not the only soil complex volume change behaviour. There are several more soil complex volume change behaviours, which will be briefly discussed here. The combination of these complex soil volume change behaviours has made it difficult to formulate a framework that can comprehensively characterise the soil volume change behaviour and complies with each of those behaviours.

Characterising soil volume change behaviour in the partially saturated condition is much more complex than the shear strength. This is because it is not just the variation of the two independent stress state variables that govern the soil volume change behaviour but also the variation in shear strength as the soil undergoes compression. The increase in strength as the soil is compressed is referred to as the soil hardening. This means that it has to involve multiple yield surfaces, because the yielding is continuous during the compression.

According to Fredlund (2000), the soil volume change behaviour or settlement behaviour is the most difficult soil mechanical behaviour to characterise. This is because of the many weird settlement behaviours that need to comply in a good soil settlement framework. Because of this, most soil settlement models are not comprehensive and empirical where the models are developed based on the observed laboratory and field one-dimensional settlement behaviour where the settlement coefficient is incorporated in the formulation. This type of soil settlement model cannot synthesise between the different modes of settlement. The fact that soil settled under effective stress increases and decreases is already a complex soil settlement behaviour when viewed from the standpoint of the effective stress concept. The concept of effective stress of Terzaghi (1936) is inadequate to explain these behaviours.

Among the weird soil settlement behaviours are as follows:

2.2.1 Settlement under effective stress decrease

Inundation settlement or wetting collapse or collapse compression is the phenomenal soil mechanical behaviour under low-stress levels, which is very difficult to reason. Besides, the incidence is taking place under a constant applied vertical load; in other words, there is no load increase.

The major cause of collapse compression in unsaturated soils is the reduction in suction upon inundation, whether it is a fine or coarse-grained soil (Matyas and Radhakrishna 1968; Escario and Saez 1973; Cox, 1978; Lloret and Alonso, 1980; Maswoswe, 1985; Blanchfield and Anderson 2000). The unique massive volume change near saturation is a phenomenon in both fine and coarse-grained soils. The former is as reported by Alonso et al. (1990) for clay soils and the latter is reported by Goodwin (1991), Tadepalli et al. (1992) and Blanchfield and Anderson (2000).

Goodwin (1991) carried out one-dimensional inundated compression tests in the same Rowe cell used in this research on coarse-grained soil of well-distributed particle size between 37.5 mm and 63 μm. The specimen was 254 mm in diameter and had a thickness of 110 mm. Suctions were measured at the bottom and top of the specimens. On achieving equalization, the specimens were subjected to monotonic incremental vertical net stress up to a maximum of 120 kPa. Between each loading stage, the incremental stress was maintained until equilibrium in terms of suction, and axial strains were established before proceeding to the next incremental stress. At the end of the final loading stage, the specimen was inundated. Figure 2.8 shows typical graphs of total axial strain and the vertical net stress, and the massive compression collapse upon inundation was apparent indicated by the last two points on the graphs.

Tadepalli et al. (1992) conducted collapse tests using Indian Head silty sand in an oedometer with three small-tip tensiometers installed along the side of the oedometer to measure suction during the test. Vertical pressure was applied under constant water

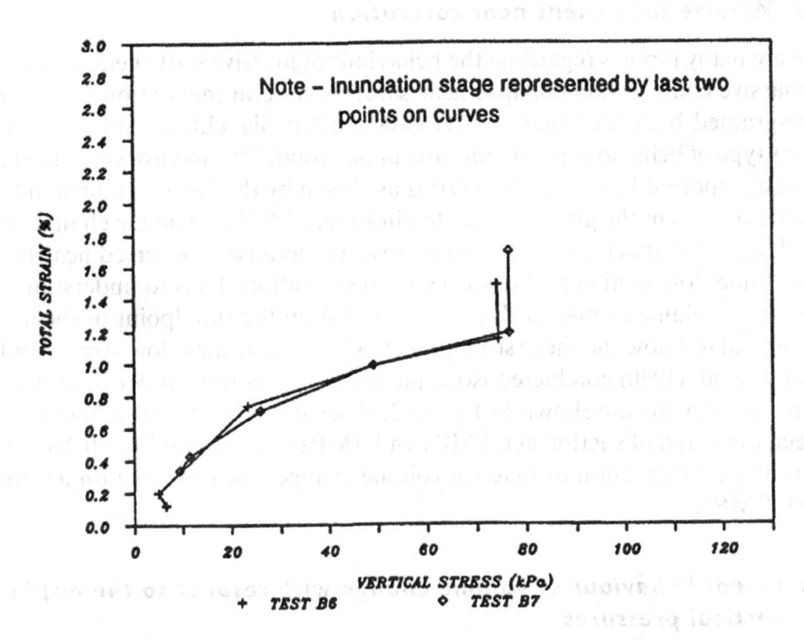

Figure 2.8 Significant one-dimensional compression of coarse-grained soils due to inundation indicated by the last two data points on the graphs (Goodwin 1991).

content conditions. On achieving equilibrium, the specimen was inundated. The result was as shown in Figure 2.9. Once inundated, all the tensiometers showed a progressive drop in the suction and subsequently, the total volume of the specimen is progressively decreased. The settlement stopped when the suction dropped to zero throughout the entire specimen. In other words, this is an inundation settlement and the settlement stopped when the specimen is fully saturated.

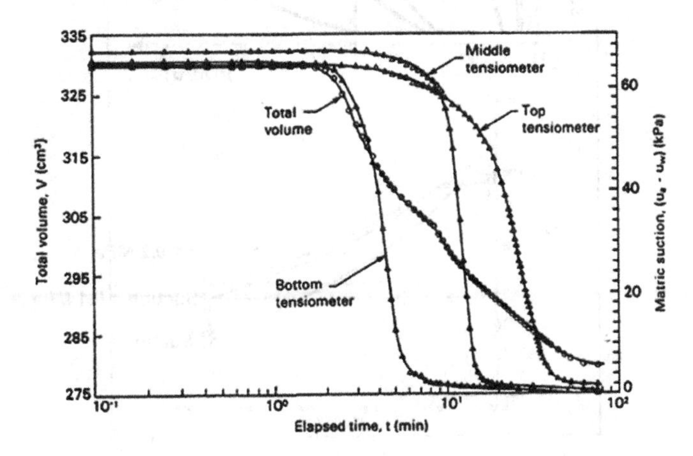

Figure 2.9 Volume changes versus time during inundation of a compacted specimen of Indian Head silty sand (Tadepalli et al., 1992).

2.2.2 *Massive settlement near saturation*

There are many reports regarding the behaviour of massive settlement near saturation. The massive compression collapse near saturation upon inundation usually involved coarse-grained backfilled material (Goodwin 1991; Blanchfield and Anderson 2000) and this type of behaviour is still not fully understood. The massive settlement near saturation as reported by Goodwin (1991) is as shown by the last two points indicated by two vertical lines in the graphs presented in Figure 2.8. This volume change behaviour was taking place under a low confining pressure, because it occurred near the ground surface under low suction and it was close to saturation. Thus to understand the phenomenon of volume change in this type of soil from the standpoint of shear strength, it is essential to know the shear strength behaviour under these low-stress conditions.

Alonso et al. (1990) conducted isotropic compression tests under different suctions and the stress paths are shown in Figure 2.10. Essentially, there is a bigger difference between the graph of suction at 0.1 MPa and 0 MPa and the graphs of 0.2 and 0.1 MPa. This is also an indication of massive volume change when the suction approaches 0, i.e., 0.1–0 MPa.

2.2.3 *Linear behaviour of volume change with respect to the applied vertical pressures*

Blanchfield and Anderson (2000) conducted one-dimensional consolidation tests using specimens of diameter 254 mm with an initial suction of 200 kPa in the study of collapse behaviour of opencast coalmine backfill when inundated. The specimens

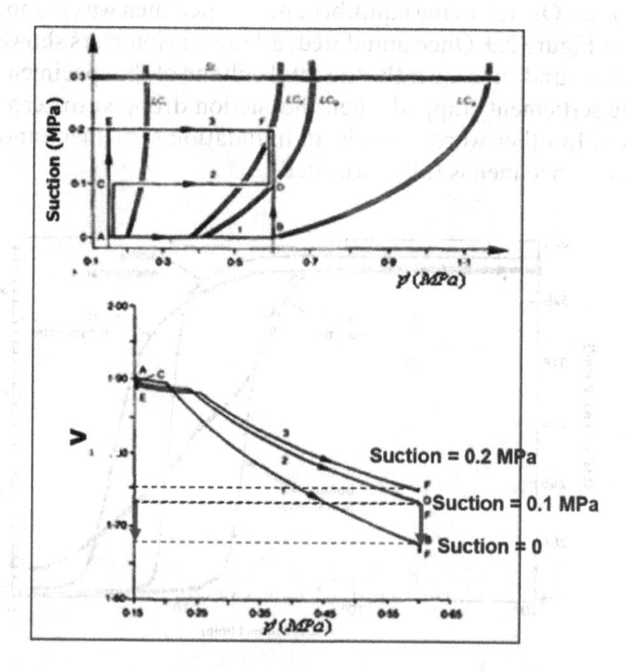

Figure 2.10 Various responses of specific volume with mean net stress, p', under wetting and drying stress paths (Alonso et al., 1990).

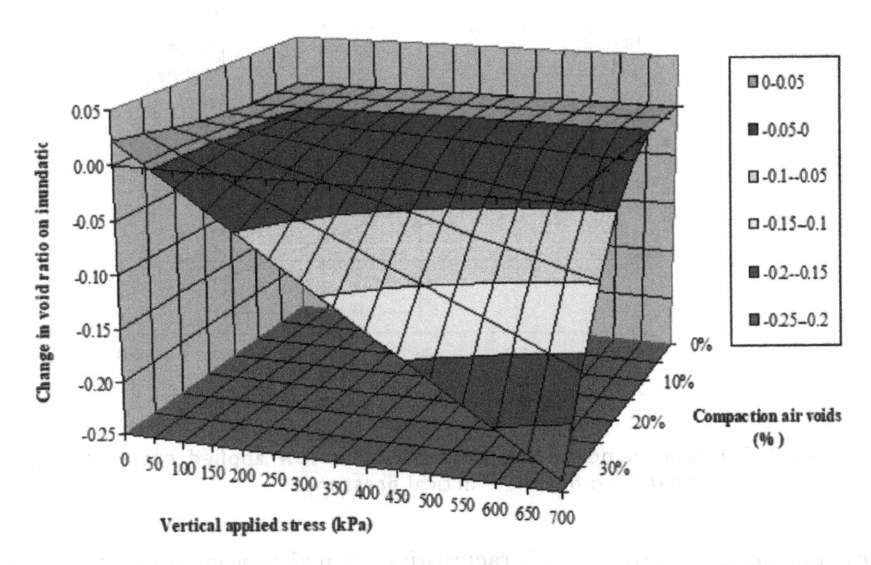

Figure 2.11 Three-dimensional representation of void ratio change with compaction air voids and vertical stress at inundation (Blanchfield and Anderson, 2000).

were inundated at vertical pressures of 10, 160 and 700 kPa in different series of tests. The next stage of the test was to apply increments of vertical pressures of 10, 20, 40, 80, 160, 200, 400 and 700 kPa every 24 hours to simulate the build-up of the overburden pressure on the backfill and the volume changes were recorded after each increment. The results are presented in Figure 2.11. The significant feature of the results is the linear behaviour of volume change with respect to the applied vertical pressures. This agrees with the findings of Cox (1978), Feda (1988), Brandon et al. (1990), Basma and Tuncer (1992) and Vilar (1995).

2.2.4 *Wetting collapse is greater under low applied net vertical pressure compared to high net vertical pressure*

Mine tailing sand from Puchong, Selangor, Malaysia was tested in a large Rowe cell of 250 mm diameter by Md Noor et al. (2008). The specimen of thickness 100 mm was equalised at a suction of 200 kPa. Incremental vertical pressures of 50, 100, 200, 300, 400 and 500 kPa were applied. The settlement was allowed to be stabilised under each pressure stage before the next incremental pressure was applied. In the first test, the specimen was inundated at 200 kPa net pressure, and in the second test, it was inundated at 400 kPa and the result is shown in Figure 2.12. Essentially, the tests proved that there is a bigger inundation settlement for 200 kPa net pressure compared to 400 kPa. This is another complex soil settlement behaviour. Common sense would tell that settlement should be bigger when the net vertical pressure is higher and the result is vice-versa.

All these complex soil settlement behaviours need to comply in a good and comprehensive soil volume change framework. The framework must be developed from the soil stress–strain behaviour and the true non-linear shear strength behaviour with respect to net stress and suction needs to be incorporated. The combination of the

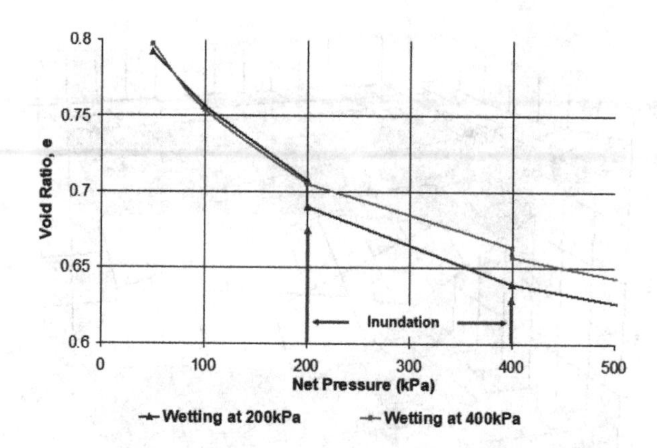

Figure 2.12 Bigger inundation settlement under low applied net vertical pressure compared to high net vertical pressure.

above four complex settlement characteristics will make the circumstances worse and thereby increase the level of difficulty in the framework formulation. This has posed a great difficulty to geotechnical researchers since 1950s. A framework that can comply with all these characteristics must be very idyllic and would be able to make excellent settlement prediction. Characterising the soil volume change behaviour based on net stress and suction without the incorporation of shear strength would be inadequate to arrive at a comprehensive volume change theory. The role of shear strength has to be incorporated in the formulation when it is apparent that a denser soil has a greater resistance to settlement compared to a loose soil subjected to the same load increase. This is because essentially the denser soil has a greater shear strength compared to the loose soil and greater shear strength provides a greater resistance to settlement.

All these weird behaviours need to be made coherent and synthesized in a single framework. In other words, a good and comprehensive volume change model must be able to model these behaviours in a single framework. The second, third and sixth characteristics are the most difficult to comply. This complexity has been noted by Fredlund (2000) as the most difficult soil behaviour to characterize and there is still no closed-form solution at the time that paper was published.

This puzzled behaviour is a unique soil volume change behaviour and very difficult to perceive under the concept of effective stress. Nevertheless, these behaviours must be linked to the basic soil stress–strain response; otherwise, it can never be understood. Thence a good framework must be formulated from the standpoint of the soil stress–strain behaviour, and automatically, these weird behaviours will be justified and matched. This chapter will demonstrate qualitatively the ability of the RMYSF to model those weird behaviours.

2.3 Initial concept of Mohr circle and mobilised shear strength

During the shearing stage of a triaxial test where the deviator stress is applied, the soil undergoes anisotropic compression where the particles are compressed and rearrange to a denser state. This is also commonly known as strain hardening or work hardening.

This is an elastic–plastic straining, which results in an increase in shear stress between the grains, which in turn increases the internal shear strength. This shear strength is recognised as mobilised shear strength.

When shear strength at failure is reached, an incline failure plane is developed and shearing occurs along this plane. At this failure condition, there is no more particle re-arrangement and thus there is no more strain hardening. At this condition, the shortening of the specimen is purely due to the shearing along the incline failure plane. At this point, the stress–strain curve is curving downwards. Therefore, strain hardening is only happening when the soil is undergoing particle rearrangement to a denser state and at this point, the mobilised friction angle is increasing. The subsequent increase in the mobilised shear strength is to compensate for the application of the higher deviator stress.

Initially, as the deviator stress Δq is increased from zero to the deviator stress corresponding to point A, i.e. Δq_A, as shown in Figure 2.13a, the Mohr stress circle starts growing up to Mohr circle 1, shown in Figure 2.14. At this point, the effective minor principal stress is equal to $\sigma'_{3A} = \sigma_{3A} - u$ and the effective major principal stress is equal to $\sigma'_{1A} = \sigma_{3A} + q_A - u$. The maximum stress ratio at this point is τ_m / σ' and the mobilized shear stress τ_m in the specimen is given by Equation 2.11, which is the function of the effective major principal stress σ' and the mobilized friction angle ϕ'_{mob}. Note that soil does not have the cohesion c' when interpreted based on effective stress analysis even for clay soil (Henkel, 1958; Bishop, 1966). The Mohr stress circle and the mobilized friction angle corresponding to point A in Figure 2.13a are represented by Mohr stress circle 1 in Figure 2.14 and the mobilized friction angle ϕ'_{mob} can be determined according to Equations 2.12–2.15.

$$\tau_m = \sigma' \tan \phi'_{mob} \tag{2.11}$$

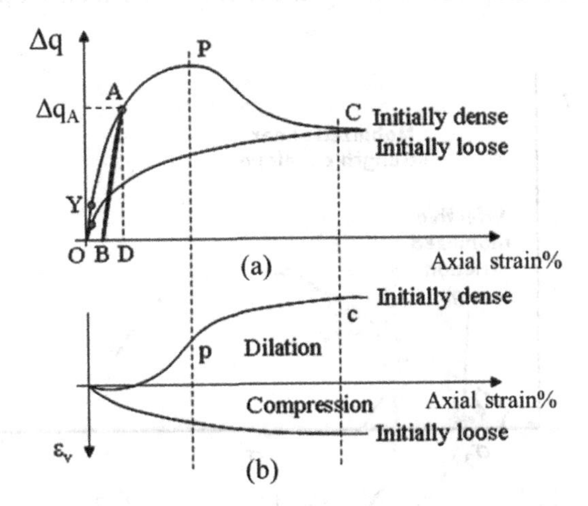

Figure 2.13 Typical stress–strain curves for loose and dense sand. (a) Dense sand has well defined peak strength and loose sand shows a gradual increase in strength towards critical state strength. (b) Dense sand undergo dilation i.e. volume increase during shearing and loose sand continues to be compressed when sheared.

$$\sin\phi'_{mob} = \frac{(\sigma'_{1A} - \sigma'_{3A})}{(\sigma'_{1A} + \sigma'_{3A})} \tag{2.12}$$

$$\sin\phi'_{mob} = \frac{(\sigma_{3A} + q_A - u - \sigma_{3A} + u)}{(\sigma_{3A} + q_A - u + \sigma_{3A} - u)} \tag{2.13}$$

$$\sin\phi'_{mob} = \frac{q_A}{(2\sigma_{3A} + q_A - 2u)} \tag{2.14}$$

$$\phi'_{mob} = \sin^{-1}\frac{q_A}{(2\sigma_{3A} + q_A - 2u)} \tag{2.15}$$

In Equation 2.15, the value of ϕ'_{mobf} can be determined by substituting q_A with the maximum deviator stress, i.e., deviator stress at failure. Note that the specimen would only undergo elastic–plastic yielding whenever the applied stress state is higher than the mobilized shear strength represented by the Mohr stress circle 1 (refer to Figure 2.14) that represents the stress state in the specimen corresponding to yield point A (refer to Figure 2.13a). When the reloading path travels along BA in Figure 2.13a, the behaviour is fully elastic until the stress exceeds the yield stress at A and the response automatically reverts to elastic–plastic. The Mohr circle 2 in Figure 2.14 represents the higher mobilized shear strength under a higher effective stress that would produce the same irrecoverable volume change on the specimen as the Mohr circle 1; therefore, they touch the same yield envelope or the mobilized shear strength envelope.

It is to be noted that the above derivation of the mobilised shear strength envelope is based on a linear envelope. Essentially, the shear strength envelope and the mobilised

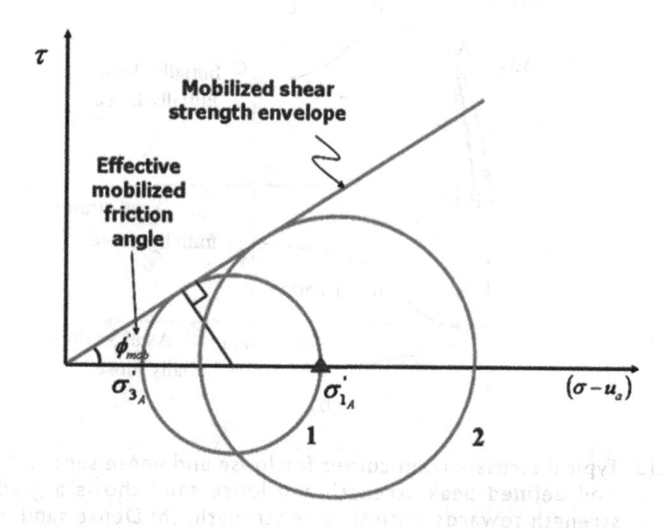

Figure 2.14 Mohr circles representing different mobilized shear strengths under the same mobilized shear strength envelope or the yield envelope.

shear strength envelope are non-linear. Besides, the mobilised shear strength envelope is referred to as the yield envelope, because whenever the deviator stress exceeds point A, the response change from purely elastic to elastic–plastic, i.e., point A, is the yield point and the corresponding mobilised shear strength envelope at A is the yield locus. Later in the development of the Rotational Multiple Yield Surface Framework, which is a new theoretical soil volume change framework, the true non-linear or the curved surface mobilised shear strength envelope with respect to net stress and suction is applied. Section 2.7 presents the derivation of the non-linear or curvilinear mobilised shear strength envelope with respect to effective stress.

2.4 Non-linear shear strength variation relative to effective stress at saturation

The shear strength under saturated conditions is the soil minimum shear strength due to the absence of the water meniscus that provides suction forces that give the extra apparent shear strength to the soil. The typical soil shear strength behaviours with respect to effective stress for coarse and fine-grained soils are presented in Figures 2.15 and 2.16, respectively. Essentially, mostly the shear strength behaviour of soils with respect to effective stress under saturated conditions can be generalised as curvilinear and zero cohesion intercept as shown in Figure 2.17. It is non-linear at low-stress levels when the effective stress approaches zero and become linear at higher stress levels. Special attention is given to the behaviour at low-stress levels to try to understand the modes of wetting collapse. The reviewed shear strength behaviour is summarized below. Along the effective stress axis, the shape of the curve OAB sketched in Figure 2.17 would have the following characteristics (Md Noor, 2006):

1. At zero net or effective stress, the shear strength, τ, should be zero.
2. The shear strength under saturated conditions should increase non-linearly from point O up to the transition effective stress, $(\sigma - u_w)_t$, represented by point A in Figure 2.17. Beyond point A, it increases linearly as represented by the line AB.
3. The shear strength under saturated conditions that corresponds to the transition effective stress, $(\sigma - u_w)_t$, is named the transition shear strength, τ_t.
4. The gradient of the curve (i.e., $\tan\phi'$) at the transition effective stress, $(\sigma - u_w)_t$, which is at point A, must be equal to the gradient of the linear section AB (i.e., $\tan\phi'_{minf}$) to achieve a smooth transition.

The saturated shear strength behaviour along the effective stress axis according to the above criteria is represented by Equations 2.16 and 2.17 for the curved line, OA, and the linear segment, AB, in Figure 2.17, respectively. Equation 2.18 gives the value of N in Equation 2.16 (Md Noor and Anderson, 2006).

$$\tau_{satf} = \frac{(\sigma - u_w)}{(\sigma - u_w)_t}\left[1 + \frac{(\sigma - u_w)_t - (\sigma - u_w)}{N(\sigma - u_w)_t}\right]\tau_t \tag{2.16}$$

$$\tau_{satf} = (\sigma - u_w)\tan\phi'_{minf} + \left[\tau_t - (\sigma - u_w)_t \tan\phi'_{minf}\right] \tag{2.17}$$

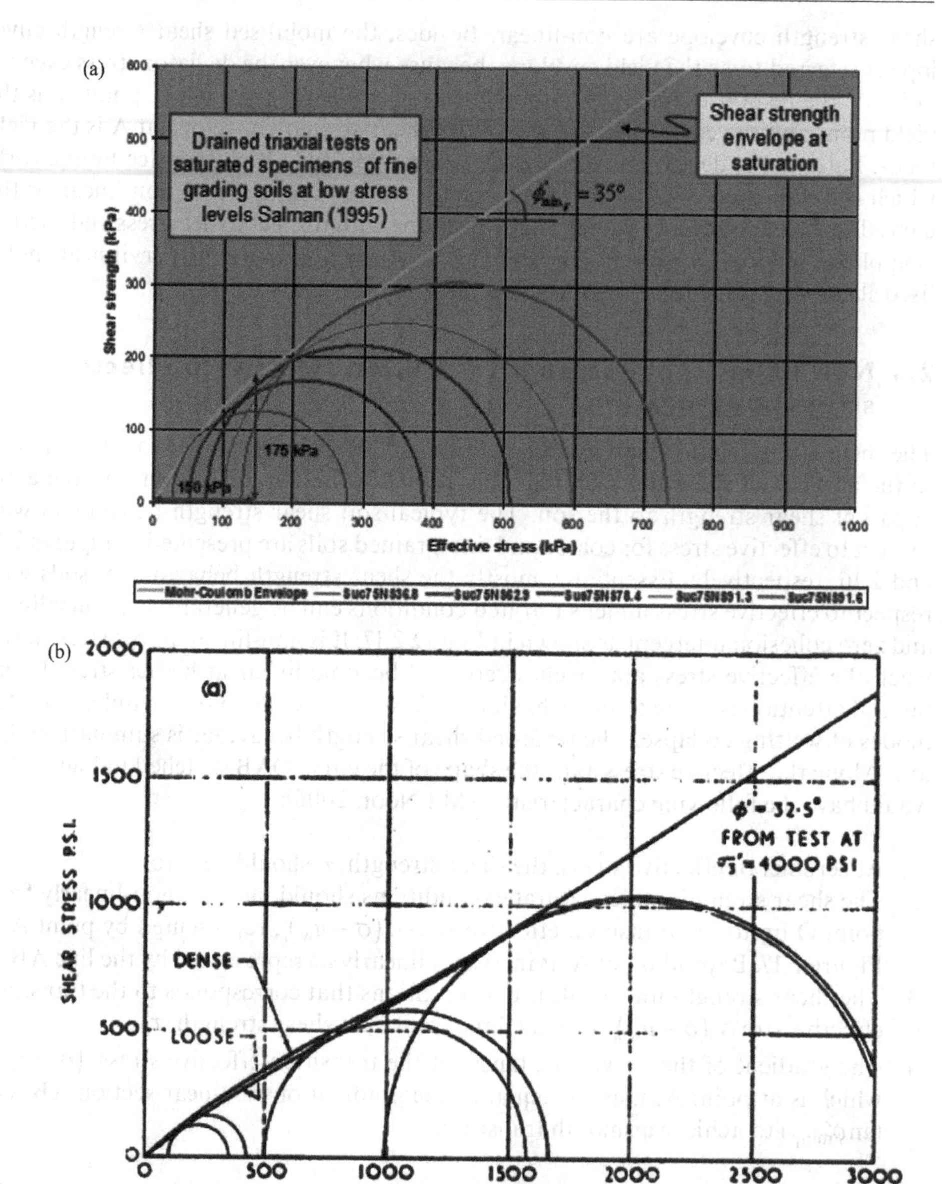

Figure 2.15 (CONTINUED) Reported curvilinear type of Mohr–Coulomb envelope for coarse-grained soils. (a) The curvilinear type of Mohr–Coulomb envelope reported by Salman (1995) for fine grading limestone gravel. (b) Mohr–Coulomb envelopes for loose and dense sand (Bishop, 1966). (c) Mohr–Coulomb envelope for granular soils of minimum particle size of 38 mm (Charles and Watts, 1980).

(Continued)

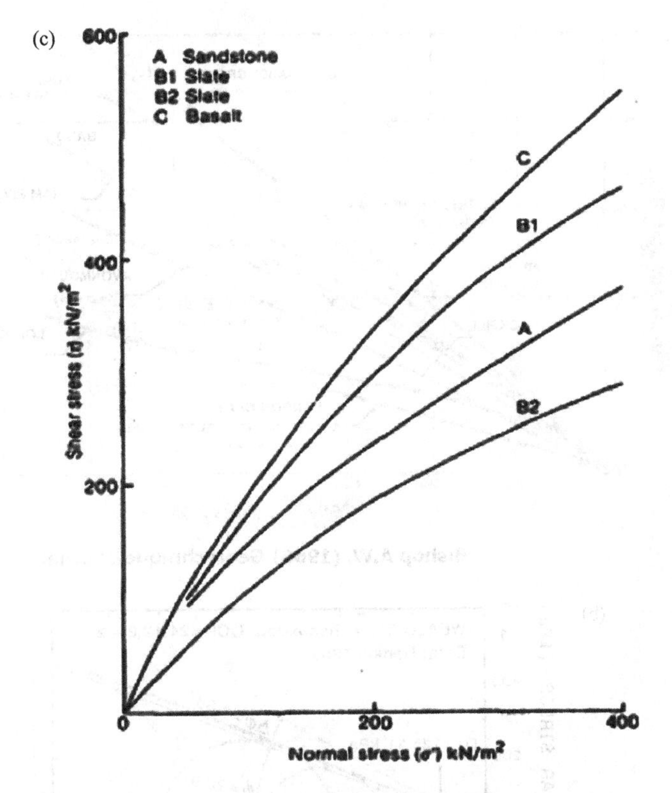

Figure 2.15 (CONTINUED) Reported curvilinear type of Mohr–Coulomb envelope for coarse-grained soils. (a) The curvilinear type of Mohr–Coulomb envelope reported by Salman (1995) for fine grading limestone gravel. (b) Mohr–Coulomb envelopes for loose and dense sand (Bishop, 1966). (c) Mohr–Coulomb envelope for granular soils of minimum particle size of 38 mm (Charles and Watts, 1980).

$$N = \frac{1}{1 - \left[(\sigma - u_w)_t \dfrac{\tan\phi'_{minf}}{\tau_t} \right]} \tag{2.18}$$

where $(\sigma - u_w)_t$ is the transition effective stress beyond which the shear strength is assumed to behave linearly, τ_t is the transition shear strength that corresponds to transition effective stress, $\tan\phi'_{minf}$ is the gradient of the straight section of the graph and N is a constant.

The following mathematics is to substantiate the fulfilments of the above criterion 1, 2, 3 and 4 (Md Noor, 2006). When the effective stress equals zero, the shear strength given by Equation 2.16 becomes zero. This, therefore, satisfies criterion no. 1. When the effective stress equals the transition effective stress, the shear strength, τ_{satf}, in Equation 2.16 and 2.17, is equal to the transition shear strength, τ_t. This satisfies criterion no. 3. Thus, the linear and the non-linear segments of the graph will meet at point

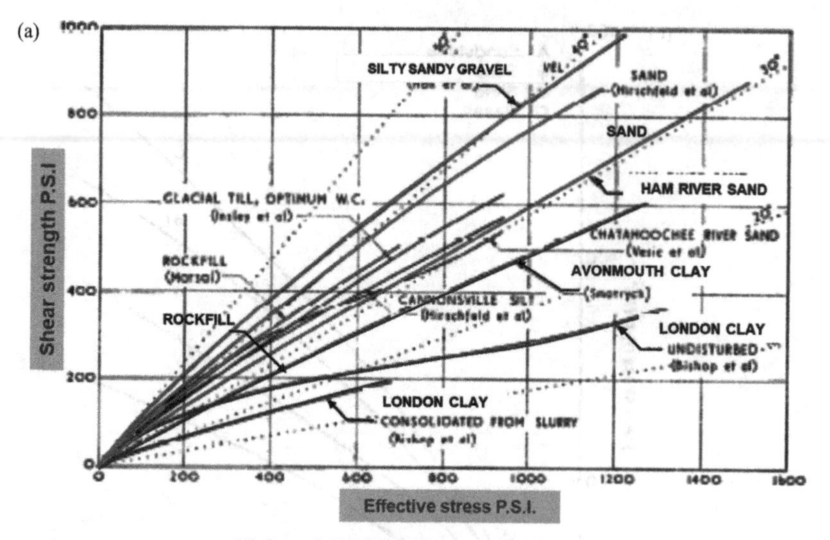

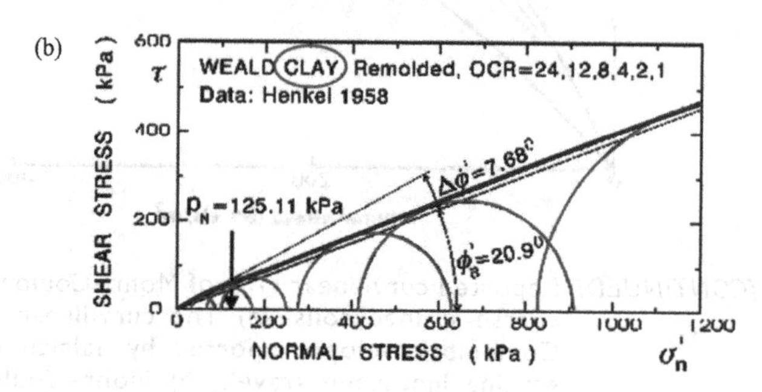

Figure 2.16 Reported curvilinear type of Mohr–Coulomb envelope for fine-grained soils. (a) Curvilinear failure envelope for London Clay and Avonmouth Clay (Bishop, 1966). (b) Curvilinear failure envelope for Weald clay (Henkel, 1958).

A with coordinates $\left[(\sigma - u_w)_t, \tau_t\right]$ as shown in Figure 2.16. Therefore, the combination of both Equations 2.16 and 2.17 is fulfilling criterion no. 2.

However, for Equation 2.16 to be valid at the transition effective stress, $(\sigma - u_w)_t$, its gradient must be equal to the gradient of the linear segment (i.e., $\tan\phi'_{min}$). Expanding Equation 2.16 produces Equation 2.19;

$$\tau_{satf} = \frac{(\sigma - u_w)}{(\sigma - u_w)_t}\tau_t + \frac{(\sigma - u_w)}{N(\sigma - u_w)_t}\tau_t - \frac{(\sigma - u_w)^2}{N(\sigma - u_w)_t^2}\tau_t \tag{2.19}$$

Differentiating Equation 2.19 with respect to effective stress produces

$$\frac{d\tau_{satf}}{d(\sigma - u_w)} = \frac{\tau_t}{(\sigma - u_w)_t} + \frac{\tau_t}{N(\sigma - u_w)_t} - \frac{2(\sigma - u_w)\tau_t}{N(\sigma - u_w)_t^2} \tag{2.20}$$

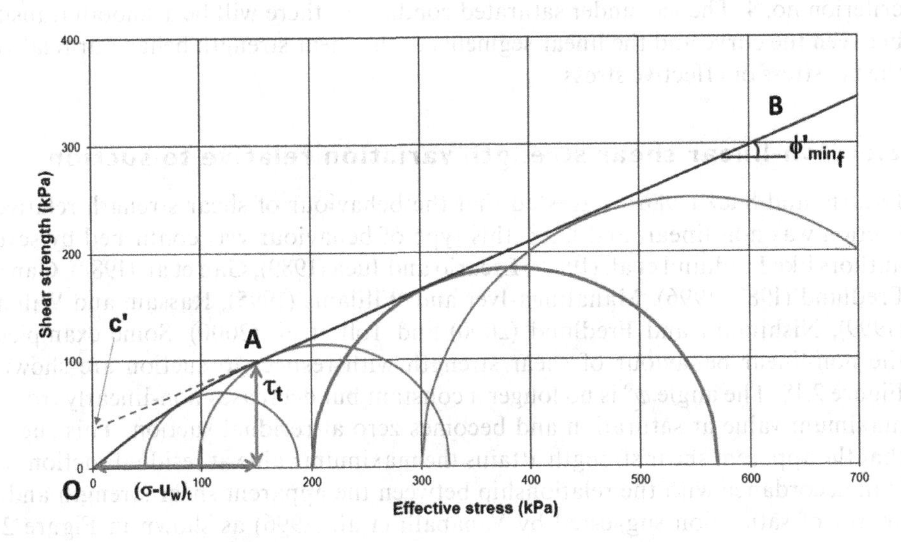

Figure 2.17 The generalised curvilinear soil shear strength behaviour under saturated condition.

Note that $\dfrac{d\tau_{satf}}{d(\sigma - u_w)} = \tan\phi'$ in the case of the curved segment, OA, in Figure 2.17. The value of ϕ' is a maximum at the origin when the net stress is zero.

When $(\sigma - u_w) = (\sigma - u_w)_t$, $\dfrac{d\tau_{satf}}{d(\sigma - u_w)}$ becomes

$$\left[\frac{d\tau_{satf}}{d(\sigma - u_w)}\right]_{(\sigma - u_w)_t} = \frac{\tau_t}{(\sigma - u_w)_t} + \frac{\tau_t}{N(\sigma - u_w)_t} - \frac{2\tau_t}{N(\sigma - u_w)_t}$$

$$\Rightarrow \left[\frac{d\tau_{satf}}{d(\sigma - u_w)}\right]_{(\sigma - u_w)_t} = \frac{\tau_t}{(\sigma - u_w)_t} - \frac{\tau_t}{N(\sigma - u_w)_t} \tag{2.21}$$

Substituting the expression for N gives

$$\left[\frac{d\tau_{satf}}{d(\sigma - u_w)}\right]_{(\sigma - u_w)_t} = \frac{\tau_t}{(\sigma - u_w)_t} - \frac{\tau_t\left[1 - (\sigma - u_w)_t \dfrac{\tan\phi'_{minf}}{\tau_t}\right]}{(\sigma - u_w)_t}$$

$$\Rightarrow \left[\frac{d\tau_{satf}}{d(\sigma - u_w)}\right]_{(\sigma - u_w)_t} = \frac{\tau_t}{(\sigma - u_w)_t} - \left[\frac{\tau_t}{(\sigma - u_w)_t} - \tan\phi'_{minf}\right] = \tan\phi'_{minf} \tag{2.22}$$

Therefore, the gradient of the curve, OA, formed by Equation 2.16 equals $\tan\phi'_{minf}$ when the effective stress equals the transition effective stress, $(\sigma - u_w)_t$. This satisfies criterion no. 4. Thence, under saturated conditions, there will be a smooth transition between the curve and the linear segments of the shear strength behaviour relative to the net stress or effective stress.

2.5 Non-linear shear strength variation relative to suction

Escario and Saez (1986) suggested that the behaviour of shear strength relative to suction was non-linear, and later, this type of behaviour was confirmed by several authors like Fredlund et al. (1987), Escario and Juca (1989), Gan et al. (1988), Gan and Fredlund (1988, 1996), Mahalinga-Iyer and Williams (1995), Rassam and Williams (1999), Nishimura and Fredlund (2000) and Toll et al. (2000). Some examples of the non-linear behaviour of shear strength with respect to suction are shown in Figure 2.18. The angle ϕ^b is no longer a constant but decreases non-linearly from the maximum value at saturation and becomes zero at residual suction. This suggests that the apparent shear strength attains the maximum value at residual suction. This is in accordance with the relationship between the apparent shear strength and the degree of saturation suggested by Vanapalli et al. (1996) as shown in Figure 2.19, where maximum shear strength is attained at residual saturation, which corresponds to residual suction.

It was also reported that shear strength decreases slightly beyond residual suction under lower confining pressures (Escario and Juca, 1989; Mahalinga-Iyer and Williams, 1995; Gan and Fredlund, 1996; Vanapalli et al., 1998). At higher confining pressures, shear strength was reported to remain constant at suctions exceeding residual suction (Escario and Juca, 1989; Gan and Fredlund, 1996; Rassam and Williams, 1999; Nishimura and Fredlund, 2000; Toll et al., 2000). Nevertheless, there are also reports that under high confining pressure, there is a slight drop in shear strength at high suction (Escario and Juca, 1989; Gan and Fredlund, 1996). These reports would imply that shear strength does decrease beyond residual suction and there are reports that claim that it remains constant probably due to the maximum suction applied to be close to residual suction. At this locality, the reduction in shear strength is not apparent and the curve is almost flat.

Thus, mostly, shear strength increases non-linearly up to residual suction and reduces at higher values of suction (Vanapalli et al. 1998). Perhaps the decrease of shear strength at this stage is getting more gradual as confining pressure increases. Ultimately, the effect of suction on shear strength will cease at one point when the soil becomes dry. Perhaps, the reduction in the apparent shear strength from the maximum value to zero is a gradual process. This is true even though the suction gets very high, because the water contact area with the soil particles relative to the overall particle size would become infinitely small. This could be the reason why the same magnitude of suction is not giving the same apparent shear strength between soils with different grain sizes.

Furthermore, Nishimura and Fredlund (2000) and Gan and Fredlund (1996) reported that the angle ϕ^b increases sharply towards ϕ' at saturation. However, in most reported cases, the angle ϕ^b is less than ϕ' at zero suction as listed in Table 2.1. Therefore, if there was to be a case that ϕ^b equals ϕ' at zero suction, it can be considered as a special case whereby, in general, the angle ϕ^b is considered to be less than ϕ' at zero suction.

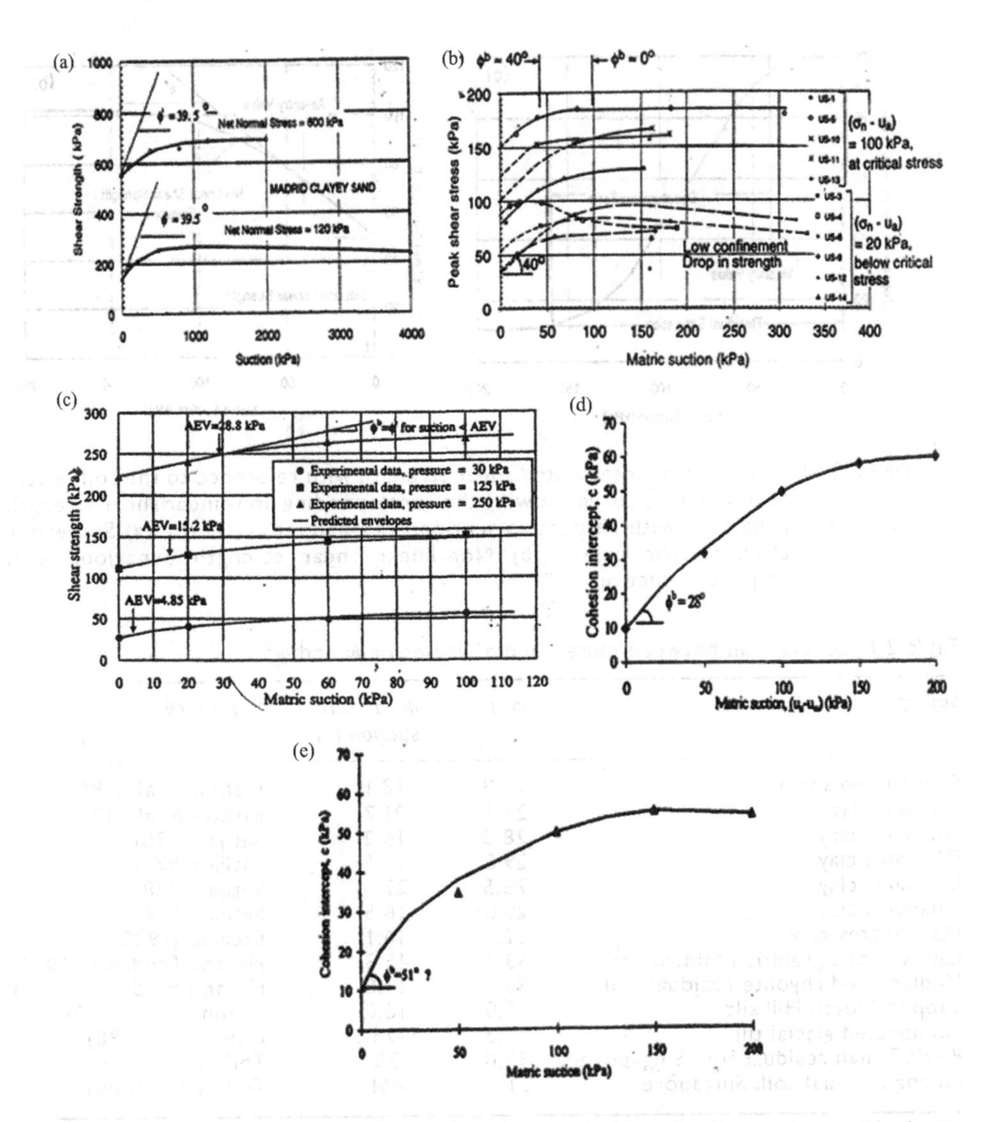

Figure 2.18 Non-linear behaviour of shear strength at suction lower than residual suction and remaining constant beyond that. (a) Madrid clayey sand (Escario and Juca, 1989). (b) Fine ash tuff (Gan and Fredlund, 1996). (c) Sandy silt (Rassam and Williams, 1999). (d) Bukit Timah residual soil, Singapore (Toll et al., 2000). (e) Jurong residual soil, Singapore (Toll et al., 2000).

Additional shear strength that arises due to the presence of suction is termed as apparent shear strength or apparent cohesion. It is termed as apparent since this additional strength will vanish when the conditions become saturated or dry. The shear strength at a fully saturated or dry condition is the basic minimum shear strength of the soil. The many reports on this shear strength behaviour with respect to suction have shown that along the suction axis, apparent shear strength increases non-linearly

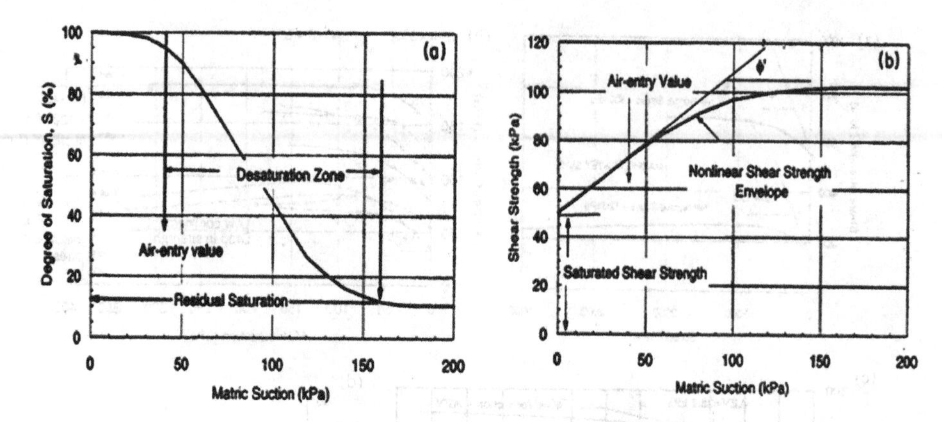

Figure 2.19 Behaviour of unsaturated shear strength with reference to the soil–water characteristic curve showing the corresponding non-linear shear strength behaviour with respect to suction (Vanapalli et al., 1996). (a) Soil-water characteristic curve. (b) Non-linear shear strength behaviour with respect to suction.

Table 2.1 Comparison between experimental values of ϕ' and ϕ^b

Soil type	ϕ' (°)	ϕ^b *at zero suction* (°)	*Reference*
Compacted shale	24.8	18.1	Bishop et al. (1960)
Boulder clay	27.3	21.7	Bishop et al. (1960)
Dhanauri clay	28.5	16.2	Satija (1978)
Dhanauri clay	29.0	12.6	Satija (1978)
Dhanauri clay	28.5	22.6	Satija (1978)
Dhanauri clay	29.0	16.5	Satija (1978)
Madrid grey clay	22.5	16.1	Escario (1980)
Undisturbed granitic residual soil	33.4	15.3	Ho and Fredlund (1982)
Undisturbed rhyolite residual soil	35.3	13.8	Ho and Fredlund (1982)
Tappen-Notch Hill silt	35.0	16.0	Krahn et al. (1989)
Compacted glacial till	25.3	7–25.5	Gan et al. (1988)
Bukit Timah residual soil, Singapore	32.0	28	Toll et al. (2000)
Jurong residual soil, Singapore	51	<51	Toll et al. (2000)

up to the residual suction and then undergoes a gradual decrease non-linearly to zero at the corresponding ultimate suction under the respective net stress.

Based on the shear strength behaviour with respect to suction outlined, the shape of the curve in this direction under a constant effective stress should have the following characteristics (Md Noor, 2006):

1. At zero suction, the apparent cohesion, c_s, should be zero.
2. At residual suction, $(u_a - u_w)_r$, the apparent cohesion or the apparent shear strength should be at its maximum value, c_s^{max}.
3. The differentiation of the shear strength equation with respect to suction must give the tangent value of angle, ϕ^b, which represents the gradient of the curve.

4. The angle ϕ^b should be at a maximum value when the suction equals zero. It need not necessarily be equal to ϕ' at this point.
5. The angle ϕ^b should be zero when the suction equals the residual suction to support characteristic no. 2.
6. The shear strength should drop steeply near saturation; therefore, the angle ϕ^b must increase sharply as suction approaches zero.
7. When the residual suction is exceeded, the apparent cohesion should decrease non-linearly from c_s^{max} at residual suction until it becomes zero at the corresponding ultimate suction, $(u_a - u_w)_u$, in a much more gradual manner than when it was increasing.
8. The gradient of the curve (i.e., $\tan\phi^b$) that represents the shear strength beyond residual suction must be zero at residual suction $(u_a - u_w)_r$. This is to achieve a smooth transition of the curves before and after residual suction.
9. The dimension of the right-hand side of the proposed shear strength equation must be those of stress, e.g., kN/m^2.

For values of suction ranging from zero to residual suction, as represented by the curve OF in Figure 2.20, the proposed shear strength equation is as Equation 2.23. This equation must satisfy the listed shape characteristics numbered 1–9 above (Md Noor, 2006).

$$c_s = \frac{(u_a - u_w)}{(u_a - u_w)_r}\left[1 + \frac{(u_a - u_w)_r - (u_a - u_w)}{(u_a - u_w)_r}\right] c_s^{max} \tag{2.23}$$

where $(u_a - u_w)_r$ is the residual suction, which is the suction that corresponds to the maximum apparent shear strength, i.e., a zero value of angle ϕ^b and c_s^{max} is the maximum apparent cohesion.

When the suction is zero, Equation 2.23 produces zero apparent cohesion, c_s, which satisfies characteristic no. 1, and when the suction equals the residual suction, Equation 2.23 indicates that the apparent cohesion c_s becomes c_s^{max}, which satisfies the characteristic no. 2.

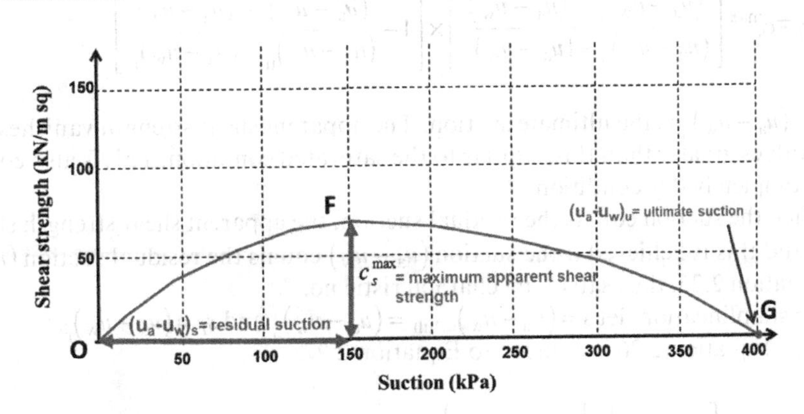

Figure 2.20 Variation of apparent shear strength with respect to suction (Md Noor, 2006).

Then, applying the characteristic no. 3 by differentiating Equation 2.23 with respect to suction should give the tangent value of angle ϕ^b, i.e., $\tan\phi^b$, which is the expression for the gradient of the curved surface, as follows;

$$\tan\phi^b = -\frac{2c_s^{max}}{(u_a - u_w)_r^2}(u_a - u_w) + \frac{2c_s^{max}}{(u_a - u_w)_r} \tag{2.24}$$

When the suction is zero, Equation 2.24 becomes

$$\tan\phi^b = \frac{2c_s^{max}}{(u_a - u_w)_r} \tag{2.25}$$

Equation 2.25 gives the value of the maximum gradient that occurs at zero suction and this is true, because $\tan\phi^b$ cannot be any bigger than the value on the right-hand side of the Equation 2.25, as c_s^{max} is the maximum possible value of the apparent cohesion. Therefore, this satisfies the required shape characteristics no. 3 and 4. When the suction equals the residual suction, Equation 2.24 gives a zero value of $\tan\phi^b$, which means that the tangent line has become horizontal indicating that the apparent shear strength is at its maximum value. This fulfils requirement no. 5. Furthermore, the equation for the variation of the tangent to the curved surface, as given by Equation 2.24, is giving a non-linear increase as suction approaches zero and reaches the maximum value at suction equals zero as given by Equation 2.25. This satisfies the shape requirement no. 6.

Above the residual suction up to the ultimate suction, the shear strength criteria with respect to suction are characteristics no. 7 and 8. This is represented by the curve FG in Figure 2.20 and the proposed equation for the apparent shear strength c_s is Equation 2.26,

$$c_s = c_s^{max}\left[\frac{(u_a - u_w)_u - (u_a - u_w)}{(u_a - u_w)_u - (u_a - u_w)_r}\right] \times \left[1 - \frac{(u_a - u_w)_r - (u_a - u_w)}{(u_a - u_w)_u - (u_a - u_w)_r}\right] \tag{2.26}$$

where $(u_a - u_w)_u$ is the ultimate suction. The apparent shear strength vanishes for suction values greater than this, similar to the saturated condition. This is also equivalent to a completely dry condition.

When the suction equals the residual suction, the apparent shear strength should be c_s^{max} and this is achieved if the suction $(u_a - u_w)$ equals the residual suction $(u_a - u_w)_r$ in Equation 2.26, thus satisfying characteristic no. 7.

For simplification, let $s = (u_a - u_w)$, $s_{ult} = (u_a - u_w)_u$ and $s_r = (u_a - u_w)_r$. Then Equation 2.26 is reduced to Equation 2.27.

$$c_s = c_s^{max}\left[\frac{s_{ult} - s}{s_{ult} - s_r}\right] \times \left[1 - \frac{s_r - s}{s_{ult} - s_r}\right] \tag{2.27}$$

Breaking up the square brackets, Equation 2.27 becomes Equation 2.28,

$$c_s = \frac{c_s^{max}}{s_{ult} - s_r} s_{ult} - \frac{c_s^{max}}{s_{ult} - s_r} s - \frac{c_s^{max}}{(s_{ult} - s_r)^2} \left[s_{ult} \times s_r - s_{ult} \times s - s \times s_r - s^2 \right] \tag{2.28}$$

Differentiating Equation 2.28 with respect to suction s

$$\frac{dc_s}{ds} = -\frac{c_s^{max}}{s_{ult} - s_r} + \frac{c_s^{max}}{(s_{ult} - s_r)^2} \times s_{ult} + \frac{c_s^{max}}{(s_{ult} - s_r)^2} \times s_r - \frac{c_s^{max}}{(s_{ult} - s_r)^2} \cdot 2s \tag{2.29}$$

Another requirement for the proposed Equation 2.26 to be valid is that

$$\frac{dc_s}{ds} = 0 \text{ when } s = s_r \text{. or } \tan\phi^b = \left(\frac{dc_s}{ds} \right)_{s=s_r} = 0$$

This is checked by substituting $s = s_r$ into Equation 2.29, which gives $\tan\phi^b$ or $\frac{dc_s}{ds}$ a zero value as follows:

$$\left(\frac{dc_s}{ds} \right)_{s=s_r} = -\frac{c_s^{max}}{s_{ult} - s_r} + \frac{c_s^{max}}{(s_{ult} - s_r)^2} \times s_{ult} + \frac{c_s^{max}}{(s_{ult} - s_r)^2} \times s_r - 2\frac{c_s^{max}}{(s_{ult} - s_r)^2} \times s_r$$

$$\Rightarrow \left(\frac{dc_s}{ds} \right)_{s=s_r} = -\frac{c_s^{max}}{s_{ult} - s_r} + \frac{c_s^{max}}{(s_{ult} - s_r)^2} \times s_{ult} + \frac{c_s^{max}}{(s_{ult} - s_r)^2} \times s_r$$

$$\Rightarrow \left(\frac{dc_s}{ds} \right)_{s=s_r} = -\frac{c_s^{max}}{s_{ult} - s_r} + \frac{c_s^{max}}{(s_{ult} - s_r)^2} \times (s_{ult} - s_r)$$

$$\Rightarrow \left(\frac{dc_s}{ds} \right)_{s=s_r} = -\frac{c_s^{max}}{s_{ult} - s_r} + \frac{c_s^{max}}{s_{ult} - s_r}$$

$$\Rightarrow \left(\frac{dc_s}{ds} \right)_{s=s_r} = 0 \tag{2.30}$$

Therefore, this satisfies characteristic no. 8. Finally, the dimension of the right-hand side of Equation 2.23 takes the dimension of c_s^{max}, which is kN/m^2, and this is identical to the dimension for apparent shear strength, c_s, on the left-hand side of the equation. Thus, Equation 2.26 can be accepted as the equation for shear strength with respect to suction beyond the residual suction. The combination of Equations 2.23 and 2.26 comprehensively defines the shear strength behaviour along the suction axis.

It should be noted that Equation 2.26 assumes that there is a gradual decrease in the apparent cohesion as the suction increases from residual suction as the soil dries up. This is a logical behaviour, because the apparent cohesion must diminish towards zero, as is the case when the soil becomes dry. The experimental data from the literature (e.g., Escario and Juca, 1989; Gan and Fredlund, 1996) indicate this type of behaviour at the lower confining pressures. However, at higher confining pressures, this effect is less obvious but the fading effect is still observable (e.g., Escario and Juca, 1989; Gan and Fredlund, 1996; Toll et al., 2000). This implies that the ultimate suction must have been increasing with the confining pressure. Therefore, there is a need to define the variation of ultimate suction with respect to net confining stress. A linear variation is assumed, and the proposed equation is Equation 2.31.

$$(u_a - u_w)_u = \zeta(\sigma - u_a) + (u_a - u_w)_u^{\sigma'=0} \tag{2.31}$$

where

ζ is a constant, which is the rate of change of ultimate suction with respect to net stress, and $(u_a - u_w)_u^{\sigma'=0}$ is the ultimate suction when net stress is zero.

2.6 Three-dimensional curved surface envelope soil shear strength model

The shear strength behaviour with respect to suction, defined by Equations 2.23 and 2.26, and with respect to effective stress in fully saturated conditions, defined by Equations 2.16 and 2.17, is represented by the lines indicated on the shear surface envelope as shown in Figure 2.21. Note that the shear strength at saturation is the minimum shear strength. The presence of suction owing to being unsaturated provides extra shear strength. Because this shear strength diminishes as suction decreases, it is then referred to as apparent shear strength c_s. Therefore, the total shear strength at any point on the surface envelope under specific net stress and suction is obtained by adding the shear strength under saturated conditions τ_{satf} and the apparent shear strengths c_s.

The shear strength curved surface envelope shown in Figure 2.21 consists of four distinct zones. The zones are divided as suction equals residual suction and net stress equals transition net stress. These are the points where the shear strength changes in behaviour with respect to the respective stress state variables. The shear strength equation to represent each zone is formed from the sum of the equations representing saturated and unsaturated conditions that represent the respective area. However, the term effective stress, $(\sigma - u_w)$, in Equations 2.16 and 2.17 has to be changed to net normal stress, $(\sigma - u_a)$, to form a general representation of the model. However, the term $(\sigma - u_w)_t$ has to be maintained, because it is a constant. In this manner, the shear strength equations representing Zone 1, 2, 3 and 4 are derived as Equations 2.32–2.35, respectively (Md Noor and Anderson, 2006).

$$\tau_f = \frac{(\sigma - u_a)}{(\sigma - u_w)_t}\left[1 + \frac{(\sigma - u_w)_t - (\sigma - u_a)}{N(\sigma - u_w)_t}\right]\tau_t + \frac{(u_a - u_w)}{(u_a - u_w)_r}\left[1 + \frac{(u_a - u_w)_r - (u_a - u_w)}{(u_a - u_w)_r}\right]c_s^{max} \tag{2.32}$$

Valid for zone 1, where suction $0 \geq (u_a - u_w) \leq (u_a - u_w)_r$ and net stress $0 \geq (\sigma - u_a) \leq (\sigma - u_w)_t$.

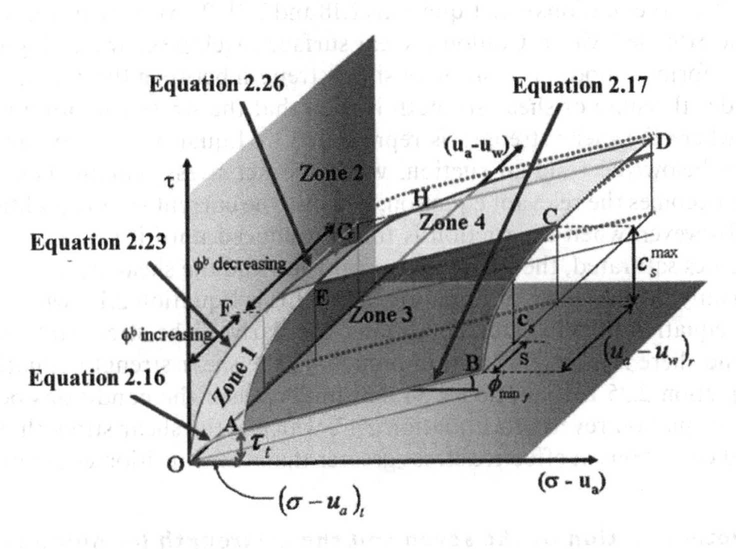

Figure 2.21 Three-dimensional curved surface envelope soil shear strength model (Md Noor and Anderson, 2006).

$$\tau_f = \frac{(\sigma-u_a)}{(\sigma-u_w)_t}\left[1+\frac{(\sigma-u_w)_t-(\sigma-u_a)}{N(\sigma-u_w)_t}\right]\tau_t + c_s^{max}\left[\frac{(u_a-u_w)_u-(u_a-u_w)}{(u_a-u_w)_u-(u_a-u_w)_r}\right] \times\left[1-\frac{(u_a-u_w)_r-(u_a-u_w)}{(u_a-u_w)_u-(u_a-u_w)_r}\right] \tag{2.33}$$

Valid for zone 2, where suction $(u_a-u_w)_r \geq (u_a-u_w) \leq (u_a-u_w)_u$ and net stress $0 \geq (\sigma-u_a) \leq (\sigma-u_w)_t$.

$$\tau_f = (\sigma-u_a)\tan\phi'_{min} + \left[\tau_t-(\sigma-u_w)_t\tan\phi'_{min}\right] + \frac{(u_a-u_w)}{(u_a-u_w)_r}\left[1+\frac{(u_a-u_w)_r-(u_a-u_w)}{(u_a-u_w)_r}\right]c_s^{max} \tag{2.34}$$

Valid for zone 3, where suction $0 \geq (u_a-u_w) \leq (u_a-u_w)_r$ and net stress $(\sigma-u_a) \geq (\sigma-u_w)_t$.

$$\tau_f = (\sigma-u_a)\tan\phi'_{min} + \left[\tau_t-(\sigma-u_w)_t\tan\phi'_{min}\right] + c_s^{max}\left[\frac{(u_a-u_w)_u-(u_a-u_w)}{(u_a-u_w)_u-(u_a-u_w)_r}\right]\times\left[1-\frac{(u_a-u_w)_r-(u_a-u_w)}{(u_a-u_w)_u-(u_a-u_w)_r}\right] \tag{2.35}$$

Valid for zone 4, where suction $(u_a-u_w)_r \geq (u_a-u_w) \leq (u_a-u_w)_u$ and net stress $(\sigma-u_a) \geq (\sigma-u_w)_t$.

Therefore, six equations (i.e., Equations 2.18 and 2.31–2.35) are required to completely define the extended Mohr–Coulomb shear surface envelope shown in Figure 2.21. The equations form a smooth transition of shear strength between the zones.

Consider the state of shear strength is such that the stress condition is initially in zone 2, where the shear strength is represented by Equation 2.33. When the suction decreases below the residual suction, while the net stress remains constant, Equation 2.32 becomes the relevant equation, because the current stress condition is within zone 1. However, when the suction is further reduced until it becomes zero and the soil becomes saturated, the suction term vanishes and the shear strength equation reduces to only the first term of Equation 2.32. This is Equation 2.16, which is the shear strength equation for saturated conditions. Similarly, if the stress state is initially in zone 4 and there is a similar suction reduction, the shear strength equation changes from Equation 2.35 to Equation 2.34 and finally when the conditions become saturated, the equation reverts to Equation 2.17, which is the shear strength equation for saturated conditions at effective stress greater than the transition effective stress.

2.6.1 Determination of the seven soil shear strength parameters

There are seven soil shear strength parameters involved in defining the proposed shear strength model (Md Noor and Anderson, 2006). Three parameters define the saturated shear strength behaviour with respect to effective stress and four parameters define the shear strength behaviour with respect to suction.

The transition shear strength, τ_t, transition effective stress, $(\sigma - u_w)_t$, and the effective minimum friction angle at failure, ϕ'_{minf}, are the three parameters that can be determined by conducting triaxial compression tests at different effective stresses on saturated specimens. The more the tests at different effective stresses, the easier it is to define the curvilinear line of the shear strength envelope at saturation. Along the curvilinear line, the effective stress that corresponds to the point that the curve changes to a straight line is taken as the transition effective stress $(\sigma - u_w)_t$ and the corresponding shear stress is taken as the transition shear strength τ_t. The inclination of the linear section of the envelope to the horizontal is the effective minimum friction angle at failure ϕ'_{minf}.

The maximum apparent cohesion, c_s^{max}, residual suction, $(u_a - u_w)_r$, ultimate suction at zero effective stress, $(u_a - u_w)_u^{\sigma'=0}$, and the rate of change of ultimate suction with respect to net stress, ζ, are the four parameters that can be determined by conducting triaxial compression tests at different suctions on unsaturated specimens, and for each value of suction, tests are carried out at different net stresses. The tests at different net stresses with identical suction give the shear strength envelope corresponding to the suction under consideration. The intersection of the curvilinear line envelope with the plane $\tau:(u_a - u_w)$ determines the apparent shear strength, c_s, for the considered suction. By having a series of curvilinear line envelopes at different suctions, the curve for the variation of the apparent shear strength with respect to suction at different net stresses can be mapped onto the $\tau:(u_a - u_w)$ plane. The maximum apparent shear strength c_s^{max} is taken at the apex of the curve formed on the $\tau:(u_a - u_w)$ plane and the corresponding suction is taken as the residual suction. In other words, residual suction is the value of suction that produces the maximum apparent shear strength.

The ultimate suction at zero net stress and the rate of change of ultimate suction with respect to net stress are determined by drawing the variation of the apparent shear strength (as the ordinate) with respect to suction (as the abscissa) at different net stresses. The contribution of the shear strength at the saturated condition is excluded in this procedure. The graphs should be overlapping for suctions less than residual suction, and at suctions higher than residual suction, the graphs cut the suction axis at different values of ultimate suction. The value of the ultimate suction corresponding to each net stress is noted. Then the points are plotted on to the graph of ultimate suction as the ordinate and net stress as the abscissa and the best straight line is drawn through the points. The gradient of the straight line is taken as the rate of change of ultimate suction with respect to net stress, ζ, and the point at which the line intersects the ordinate is taken as the ultimate suction at zero effective stress, $(u_a - u_w)_u^{\sigma'=0}$. This procedure will be applied for the interpretation of the new warped-surface shear strength envelope.

Alternatively, residual suction can be obtained by conducting pressure plate tests. The procedure for determining the residual suction from the soil–moisture characteristic curve is described by Fredlund and Xing (1994).

2.6.2 *Procedure for drawing the shear strength contours of the shear strength surface envelope in $\tau:(u_a - u_w)$ space*

One way to present the three-dimensional shear strength surface envelope in two-dimensional space is to draw the shear strength contour lines at different net stress in $\tau:(u_a - u_w)$ space (Md Noor, 2006). When the space between the contours is large, it indicates a steep drop in shear strength with respect to net stress and vice versa. The primary requirement for this exercise is to obtain the seven soil shear strength parameters according to the proposed soil shear strength model as described in Section 2.6. The six shear strength equations required to plot the contours are Equations 2.18 and 2.31–2.35. The following is the systematic procedure for drawing the contour lines.

1. Select the values of net stress that correspond to the contour lines to be drawn, e.g., 0, 50, 100, 150 and 200 kPa. These values of net stress will become the values of effective stress when the condition becomes saturated.
2. Determine the shear strength at saturation that corresponds to the selected values of net stresses in step 1 by using Equations 2.16–2.18.
3. Select the values of suction to be plotted on each contour line, e.g., 0, 20, 40 and 60 kPa. Use Equations 2.23, 2.26 and 2.31 to determine the apparent shear strength at the selected suction values. Use Equation 2.31 to determine the ultimate suction in Equation 2.26. Note that the value of ultimate suction varies with net stress.
4. Calculate the total shear strength at the respective suctions and net confining stresses by summing their corresponding values of shear strength at failure under saturated conditions, τ_{satf}, and the apparent shear strength, c_s, determined in steps 2 and 3, respectively.
5. In the $\tau:(u_a - u_w)$ space, for each respective net stress considered, plot the total shear strength that corresponds to the selected suction. Repeat this step for the rest of the selected net stresses. In this way, the whole contour lines for the selected net stresses can be drawn.

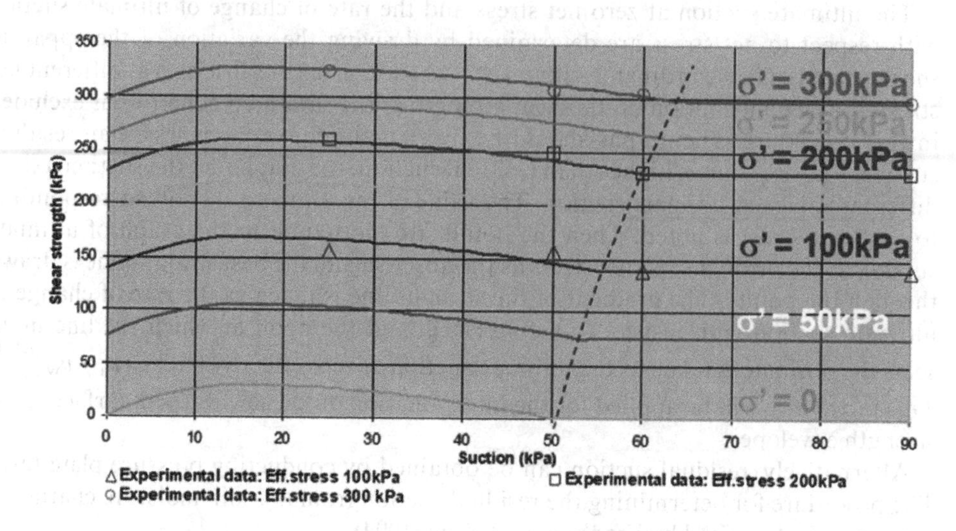

Figure 2.22 Graphs for variation of shear strength with respect to suction at various net stresses (Md Noor, 2006).

This procedure is applied in the drawing of the shear strength contours as shown in Figure 2.22.

2.7 Development of mobilised shear strength and anisotropic compression

The Rotational Multiple Yield Surface Framework utilises the development of the mobilised shear strength as the soil undergoes anisotropic compression. The mobilised shear strength envelope in the form of curvilinear envelope rotates towards the failure shear strength envelope as the soil undergoes anisotropic compression through particle rearrangement. This is the concept that relates the interaction between the mobilised shear strength and the applied load in the form of effective stress. The mobilised shear strength is represented by the curvilinear mobilised shear strength envelope and the applied effective stress is represented by the effective stress Mohr circle. This section will present the rotation of the curvilinear or the non-linear mobilised shear strength envelope as the soil is compressed anisotropically in a consolidated drained triaxial test. This non-linear mobilised shear strength envelope rotates upwards to indicate the increase in the mobilised shear strength. This increase in the mobilised shear strength is required to compensate with the increase in the deviator stress. Moreover, this strength is increased through the rearrangement of the particles to a denser state. Note that at this early stage of the shearing, the incline failure plane is not yet being developed, so it must be the rearrangement of the particles that provide the strength to compensate with the increasing deviator stress. This non-linear mobilised shear strength envelope is the soil intrinsic property where its position is related to the amount of compression. Its position is related to the amount of axial strain that the specimen has been compressed. Thence, the relationship between the position of the non-linear mobilised shear strength envelope and the corresponding axial strain that it represents is a very important characteristic in this concept of effective stress and shear strength interaction. The position of the non-linear mobilised shear strength envelope is denoted by the minimum mobilised friction angle $\phi'_{\min_{mob}}$ of the envelope.

Three saturated consolidated drained triaxial tests were conducted at effective stresses of 100, 200 and 300 kPa using undisturbed soil specimens taken by dipping the split spoon samplers into the ground and excavating around the samplers to retrieve the samplers. The test soil was taken from Kuala Kubu Baru, Selangor, Malaysia. The specimen consists of 50.70% clay, 5.67% silt, 41.73% sand and 1.90% gravel and the soil is classified as sandy CLAY. It is considered a grade V weathering as the soil matrix still contained the whitish colour partly weathered feldspar. The size of the specimen is 50 mm diameter and its height is 100 mm. The stress–strain curves obtained are shown in Figure 2.23.

The overall failure axial strain for the three stress–strain curves is taken at 12%. The mobilised shear strength envelopes will be drawn for axial strains of 1%–11%. The values of deviator stresses for each axial strain at effective stresses of 100, 200 and 300 kPa are presented in Table 2.2.

The Mohr circles, the corresponding mobilised shear strength envelopes and the envelope at failure are shown in Figure 2.24a–c.

The magnitudes of the three shear strength parameters used to define the curvilinear mobilised shear strength envelopes and the envelope at failure as in Figure 2.25 are shown in Table 2.3. Notice that some of the Mohr circles did not touch the specified curvilinear envelopes, i.e., the Mohr circles are located below the envelopes and some are higher than the envelopes. In other words, the envelopes are defined to produce the best fitting to the Mohr circles. The mobilised and the failure envelopes are considered as the inherent property of the soil, which relates the state of soil shear strength developed in the soil mass with respect to the vertical deformation, i.e., the axial strain of the soil under anisotropic stress condition. By this approach, it can be seen that the soil volume change behaviour is very much related to its state of shear strength.

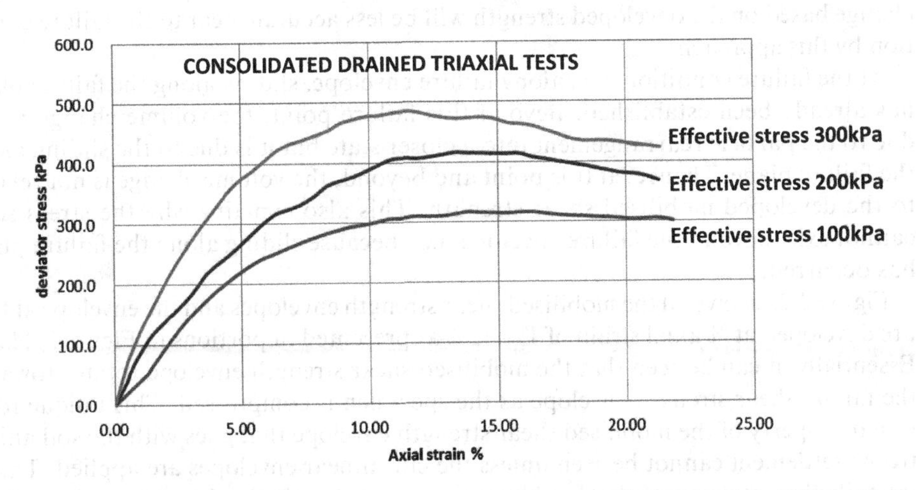

Figure 2.23 Stress–strain curves for granitic residual soil grade V from Kuala Kubu Baru, Malaysia at effective stresses of 100, 200 and 300 kPa obtained through consolidated drained triaxial tests.

Table 2.2 Magnitudes of deviator stresses derived from stress–strain curves

Axial strains (%)	*Deviator stress (kN/m²)*		
	Eff. stress 100 kPa	*Eff. stress 200 kPa*	*Eff. stress 300 kPa*
1	40	85	122
2	97	142.2	203
3	150	183	273
4	188	233	330
5	220.7	266	374
6	244	304.6	407.5
7	262.2	325	433
8	277.9	349.4	447
9	290.2	372	467
10	298.6	388.8	477.4
11	306.6	405.9	482.9
12	310.2	407.3	482

Note that the axial strain increment considered here is 1%. The layout of the envelopes shows that their separation becomes closer as it approaches the failure envelope even though under the same increase in the axial strain. This behaviour indicates that as the compactness of the soil increases, i.e., as the soil becomes hard, there is only a slight increase in the shear strength under the same increase in the axial strain. This is due to the approaching of the development of the shear plane nearing the failure envelope. Part of the strength is released at the sliding plane. At this stage, the increase in the strength due to the particle rearrangement is small when some of the strength is lost because of the sliding along the failure plane. In other words, the prediction of volume change based on the developed strength will be less accurate near to the failure condition by this approach.

At the failure condition, i.e., along failure envelope, sliding along the failure plane has already been established. Beyond this failure point, the volume change is not due to the particle rearrangement into a closer state but it is due to the sliding along the failure plane. Thence, at this point and beyond, the volume change is not related to the developed mobilised shear strength. This also explains why the stress state cannot exist beyond the failure envelope, i.e., because sliding along the failure plane has occurred.

Figure 2.25 shows all the mobilised shear strength envelopes and the envelope at failure developed at % axial strain of 1%–12% as presented in portions in Figure 2.24a–c. Essentially, it can be seen that the mobilised shear strength envelope rotates towards the failure shear strength envelope as the specimen is compressed. This unique rotational property of the mobilised shear strength envelope that goes with the soil anisotropic settlement cannot be seen unless the curvilinear envelopes are applied. This is the soil inherent property that has been ignored, while only the shear strength envelope at failure is being focussed. Most importantly, the degree of settlement is indicated by the position of the mobilised shear strength envelope.

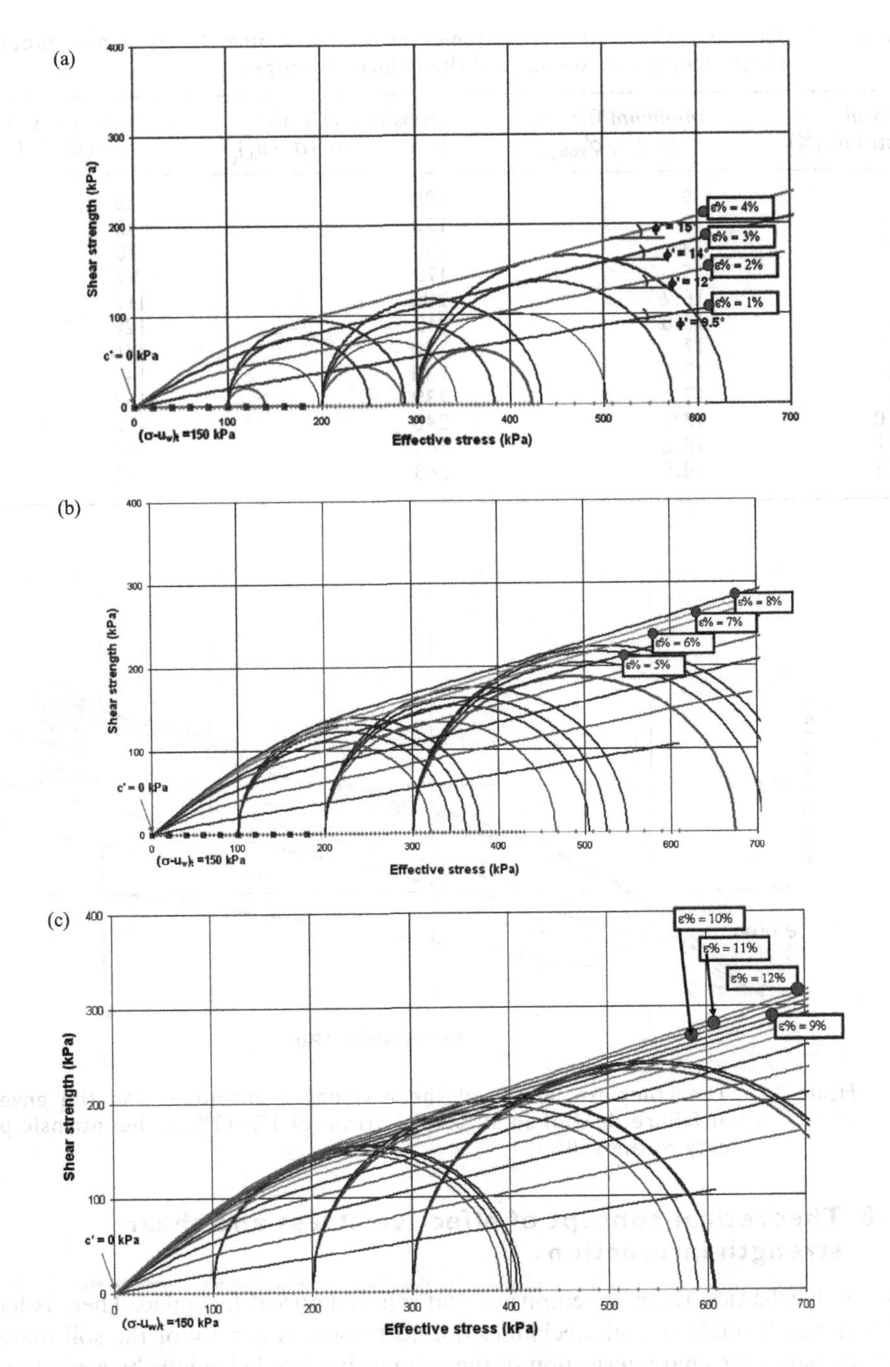

Figure 2.24 Mobilised shear strength envelopes and the envelope at failure. (a) Mobilised shear strength envelopes for axial strains of 1%, 2%, 3% and 4%. (b) Mobilised shear strength envelopes for axial strains of 5%, 6%, 7% and 8%. (c) Mobilised shear strength envelopes for axial strains of 9%, 10% and 11% and failure envelope at 12% axial strain.

Table 2.3 The three shear strength parameters used to define the curvilinear mobilised shear strength envelopes and the failure envelopes

Axial strains (%)	*Minimum friction angle (°), $\phi'_{mob_{min}}$*	*Transition effective stress (kPa), $(\sigma - u_a)_t$*	*Transition shear strength (kN/m²), τ_t*
1	9.5	100	20
2	12	120	45
3	14	150	70
4	15	170	92
5	16.2	200	112
6	16.8	215	128
7	17	230	140
8	17.2	238	148
9	17.4	239	153
10	17.9	240	156
11	18.2	242	160
12	18.5	243	163

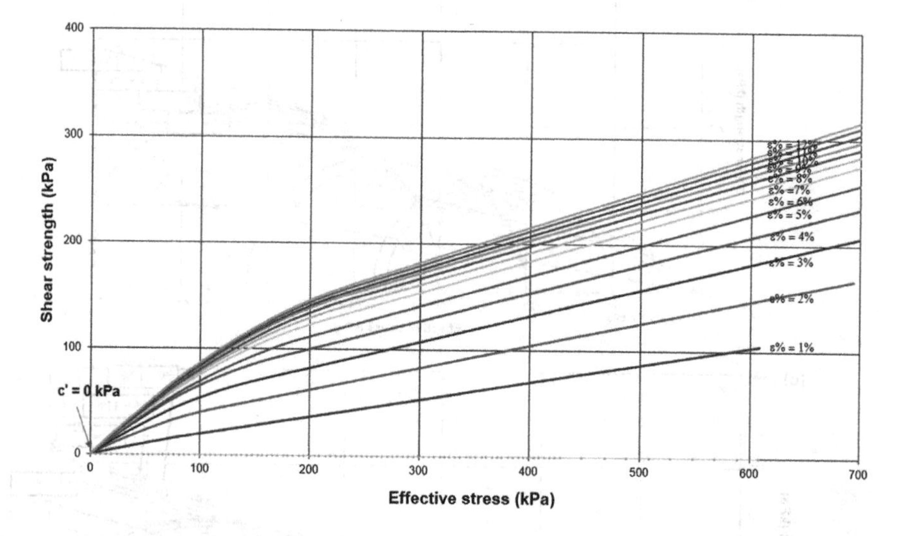

Figure 2.25 The complete mobilised shear strength envelopes and the envelope at failure developed at % axial strain of 1%–12% as the intrinsic property of the soil.

2.8 Theoretical concept of effective stress and shear strength interaction

In soil mechanics, there are empirical and semi-empirical formulae. There is hardly theoretical formula in soil mechanics due to the heterogeneity of the soil material, which makes the characterisation of the soil mechanical behaviour by a single equation very difficult. This is because even for a homogeneous soil, the properties change with depth or the effective stress. The empirical formula is derived based purely on the experimental data or through observation and does not involve any theoretical or hypothetical concept. Whereas the semi-empirical formula is the combination of mathematical relationship and experimental evidence. This is like the consolidation equation by Terzaghi (1943) as in Equation 2.3, which involves the mathematical relationship for the calculation of soil settlement, which incorporates the consolidation constants c_c and c_r determined from laboratory tests. However, this equation cannot be applied for the whole soil depth as the consolidation constants c_c and c_r change with depths. On the other hand, the scientific theoretical formula is a plausible or scientifically acceptable mathematical relationship or a general conceptual principle that can quantify a certain behaviour, which incorporates the material constant. The material constant must not change with the material responses to load or stress.

The most famous scientific theoretical formula in engineering is the equation for spring extension as shown in Equation 2.36 and is illustrated in Figure 2.26 where k is the spring stiffness, which is a constant for the spring irrespective of the extension x. Therefore, when any load F is hook on to the spring, the extension x can be determined directly. This is considered as a theoretical formulation, because the stiffness k remains constant and it is the inherent property of the spring.

$$F = kx \tag{2.36}$$

However, for soil, it is very difficult to get a theoretical relationship for settlement. The best that researchers can produce is semi-empirical formula like the Terzaghi (1943) equation for settlement as shown in Equation 2.3. This difficulty is because soil is a very heterogeneous material. The soil constant in Equation 2.3 is the coefficient of consolidation, c_c. In contrary to the spring stiffness k, this soil constant c_c varies with depth even for the same soil. Thus, it is very difficult to make a single formulation to be applicable for the whole soil depth. So the most the researchers can get is the semi-empirical formulation for soil settlement where the formula is based on the soil property c_c determined from conducting laboratory test combined with a certain soil mechanics formulation and verified with the field settlement observation. At every single discretised soil layers, the coefficient of consolidation c_c needs

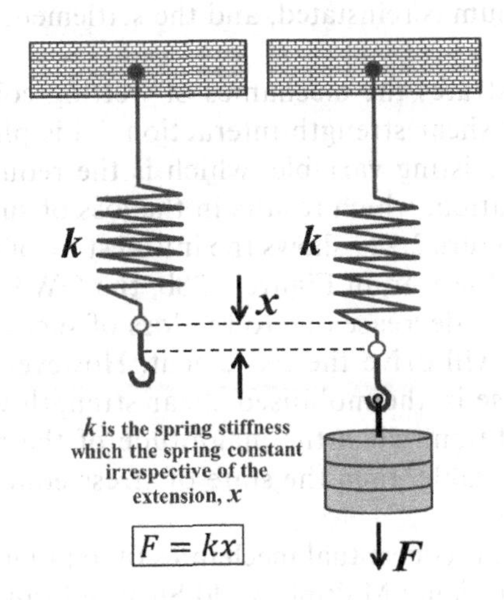

Figure 2.26 The theoretical formula for the spring extension.

to be determined. The settlement is calculated by summing the settlements in each soil layer considered. Thence, there is no single settlement formulation for the whole soil formation.

This concept of effective stress and shear strength interaction, which makes use of the development of the mobilised shear strength in characterising the soil settlement behaviour, is the first theoretical formulation for soil settlement. The stiffness k in the spring resembles the mobilised shear strength envelopes of the soil, which is the intrinsic property of the soil and is unique for any value of effective stress. In other words, the set of the mobilised shear strength envelopes is a unique soil property irrespective of the depth in the ground in the case of soft clay formation that exists along the coastal deposits.

In the spring system, F is the external driving variable and the stiffness k is the intrinsic spring property. In this context of soil, the applied effective stress is the external driving variable and the set of the mobilised shear strength envelopes is the intrinsic property, which is unique for any depth in the ground as illustrated in Figure 2.27a. Figure 2.27a–c demonstrates the mechanics of loading collapse in the concept of effective stress and shear strength interaction. Note that the set of the mobilised shear strength envelopes is unique for the specific soil and its position in the set represents the degree of the anisotropic settlement. In the spring system, the resultant output is the extension x and similarly, in this context of soil, the resultant output is settlement ρ. Figure 2.27a represents the initial state of stress equilibrium between the driving variable and the resisting variable. The loading gives rise to an effective stress state and thus it is the external driving factor and the mobilised shear strength is developed within the soil mass and it is the resisting factor. Figure 2.27b shows the condition when the load is increased and thus the stresses become imbalanced, where the driving variable is greater than the resisting variable. This will drive settlement as shown in Figure 2.27c. However, when the embankment settled, it will compress the foundation soil to a denser state and subsequently increase the mobilised shear strength. This is indicated by the bigger upward arrow representing the increased mobilised shear strength. When this resisting variable equates to the driving variable, then the state of stress equilibrium is reinstated, and the settlement will stop as illustrated in Figure 2.27c.

Figure 2.28a–c illustrates the mechanics of wetting collapse under the concept of effective stress and shear strength interaction. This phenomenon occurs under the reduction of the resisting variable, which is the reduction of mobilised shear strength due to inundation, which results in the loss of suction in the initially partially saturated soil. Figure 2.28a shows the initial state of stress equilibrium before the rise of the GWT. However, in Figure 2.28b, the GWT rises and causes the mobilised shear strength to decrease due to the loss of suction and thus results in the stress imbalance that will drive the settlement. However, Figure 2.28c illustrates the subsequent increase in the mobilised shear strength when the foundation soil is compressed. In addition, when the magnitude of the mobilised shear strength equates the driving variable, then the state of stress equilibrium is reinstated and the settlement stopped.

This application of this conceptual mechanics of loading and wetting collapse will be explained in the Rotational Multiple Yield Surface Framework. The framework is developed from the soil stress–strain behaviour, which is the most fundamental soil

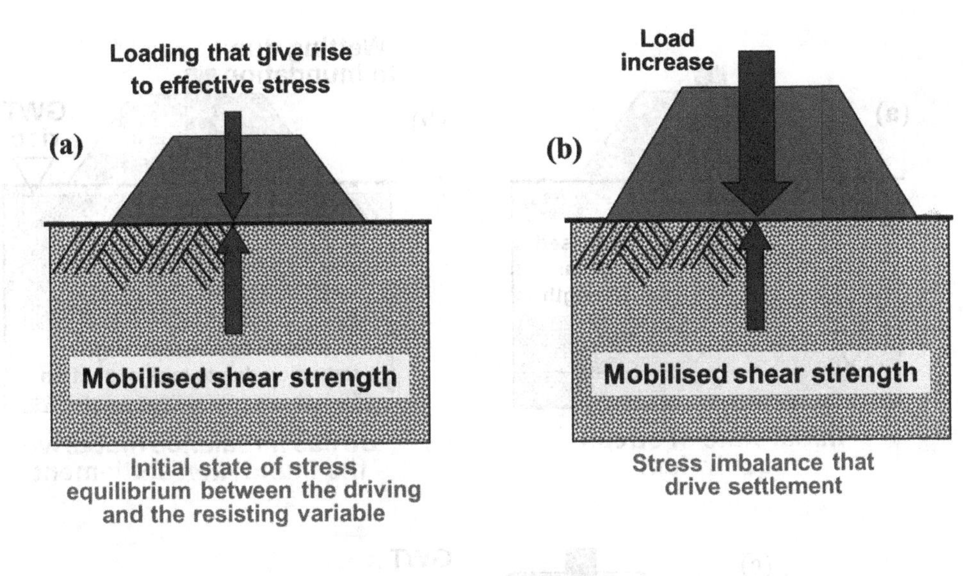

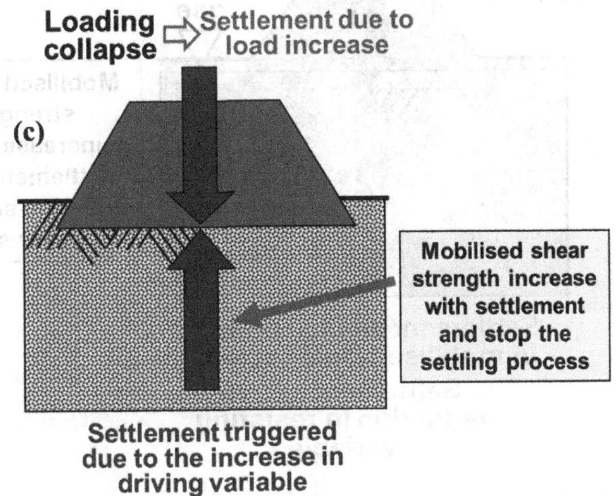

Figure 2.27 The mechanism of loading collapse. (a) Initial state of stress equilibrium between the driving and the resisting variables. (b) State of stress imbalance when the load is increase which drive the settlement. (c) Mobilised shear strength increase with settlement and reinstate the stress equilibrium.

characteristic that explains how the soil responds to stress. The framework utilises the set of mobilised shear strength envelope as the yield surface cum the soil-resisting variable, which is unique to the soil. Therefore, the framework uses the concept of multiple yield surfaces. Thence, if this Rotational Multiple Yield Surface Framework can model the weird soil settlement behaviour as discussed in Section 2.2, then it can be regarded as the first theoretical framework for soil settlement behaviour.

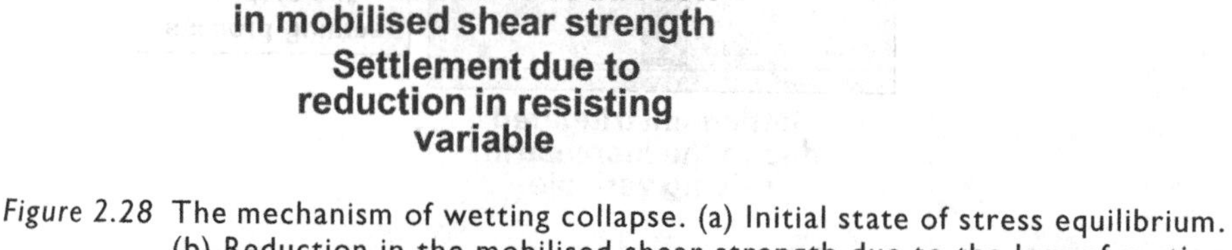

Figure 2.28 The mechanism of wetting collapse. (a) Initial state of stress equilibrium. (b) Reduction in the mobilised shear strength due to the loss of suction when wetted and resulted in stress imbalance that drives the wetting settlement. (c) Settlement increase the state of mobilised shear strength and reinstate stress equilibrium.

2.9 Interaction between the effective stress Mohr circle and the mobilised shear strength envelope during anisotropic settlement in sand and clay under loading collapse

This section is to demonstrate that the mobilised shear strength envelope is acting as the yield locus and representing the state of shear strength developed and stored in the soil. In addition, whenever the stress condition exceeds the yield locus, which is when the effective stress Mohr circle extends above the current state of mobilised shear strength, the soil will react by undergoing settlement to reinstate the state of equilibrium. The state of the applied stress condition is defined by the effective stress Mohr circle. When the state of stress is increased by increasing the effective vertical stress, the effective stress Mohr circle will enlarge and if the Mohr circle extends beyond the current mobilised shear strength envelope, then the soil will react by undergoing settlement. This type of settlement is referred to as loading collapse because the settlement is driven by the increase in the effective stress. When the particles are rearranged to a denser state subsequently, the mobilised shear strength envelope will rotate to indicate the increase in the mobilised shear strength.

Finally, the condition will reinstate the one-point contact between the effective stress Mohr circle and the mobilised shear strength envelope. This is the state of equilibrium between the applied effective stress and the developed mobilised shear strength. At this point, the soil stops settling.

Figure 2.29a shows the initial state of stress equilibrium between applied effective stress to the soil body represented by the effective stress Mohr circle and the state of mobilised shear strength stored in the soil indicated by the mobilised shear strength envelope by the one-point contact between the two stress state variables. This is considering the soil to be fully saturated where the effective stress Mohr circle is sitting directly on the net stress axis or effective stress axis. The effective stress Mohr circle is defined by the effective vertical stress σ_1' and effective lateral stress σ_3'. However, when the effective vertical stress is increased by applying a bigger load, then the effective stress Mohr circle will immediately be enlarged as the value of σ_1' has increased as shown in Figure 2.29b. The Mohr circle extends higher than the existing mobilised shear strength envelope. Then the soil will automatically react to this stress in-equilibrium. It will undergo settlement to increase the state of mobilised shear strength until the state of equilibrium is reinstated. Figure 2.29c shows the soil reaction by undergoing anisotropic settlement and rotated the mobilised shear strength envelope to the second position. However, the stress condition is still in-equilibrium. Thence, the soil needs to undergo further settlement and this will cause the mobilised shear strength envelope to further rotate to the third position as shown in Figure 2.29d. However, the stress condition is still in-equilibrium since the Mohr circle is still slightly extending higher than the current state of mobilised shear strength envelope. Thence, the soil needs to further react by continuing to be compressed and undergoes anisotropic particles rearrangement to achieve a denser state and thereby further rotates the mobilised shear strength envelope to the fourth position as shown in Figure 2.29e. At this condition, the mobilised shear strength envelope has reinstated the state of stress equilibrium by one-point contact with the effective stress Mohr circle. At this point, the settlement will stop. In sand, the soil reaction is immediate; in other words, the rotation of the mobilised shear strength envelope from position one to the fourth position is quick. However, in clay, the reaction is time-consuming because the rearrangement of the particles is a slow process.

Note that the settlement phenomenon discussed above is driven by load increase, which gives rise to the vertical effective stress increase. This type of settlement is called loading collapse. It is driven by the enlargement of the effective stress Mohr circle that extends beyond the current state of mobilised shear strength envelope. Figure 2.30 illustrates this loading collapse phenomenon in a three-dimensional space, i.e., $\tau:(\sigma-u_a):(u_a-u_w)$ when effective stress Mohr circle enlarged under partially saturated condition, i.e., at a certain suction. As the vertical net stress is

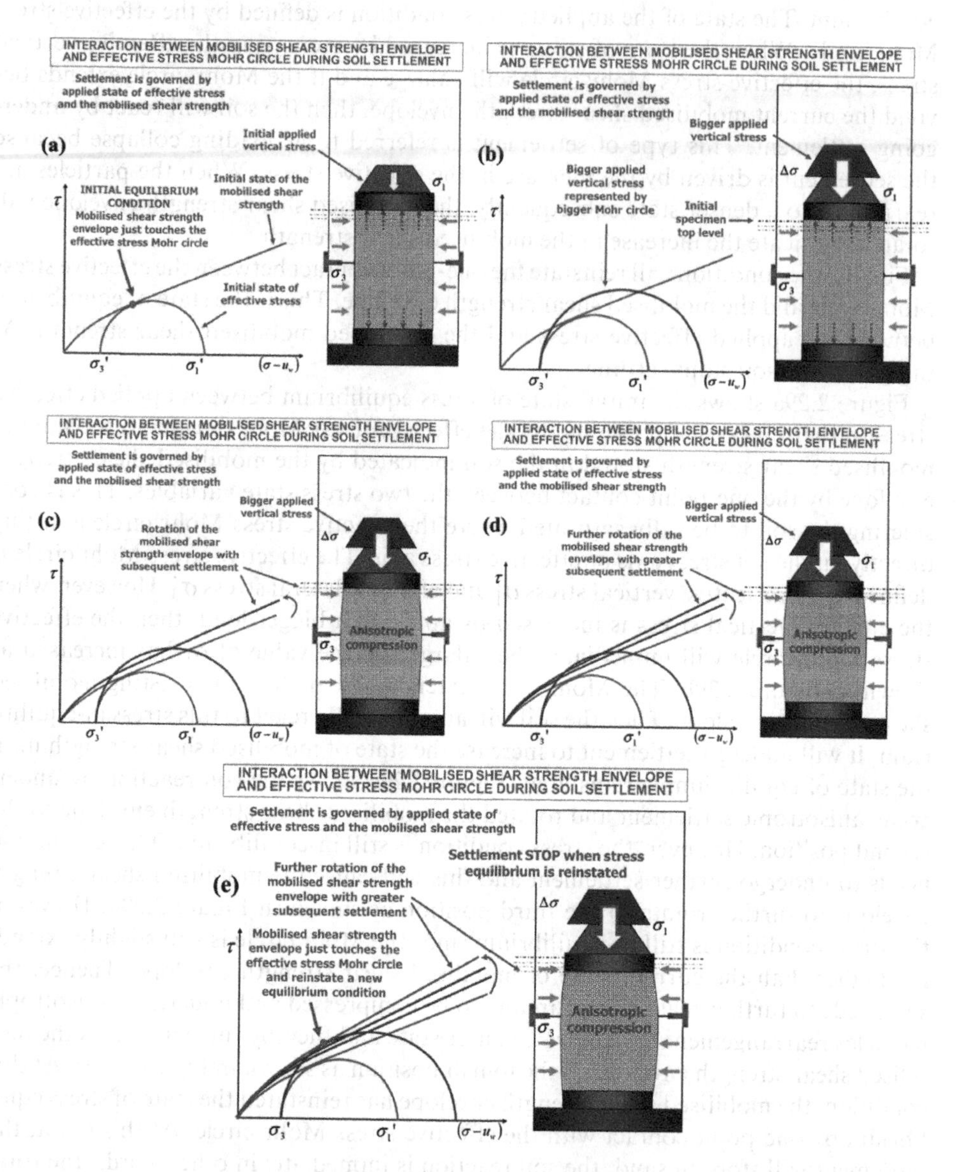

Figure 2.29 Concept of effective stress and shear strength interaction demonstrating the interaction between effective stress Mohr circle and mobilised shear strength envelope during soil anisotropic settlement. (a) Initial state of stress equilibrium with the mobilised shear strength envelope just touches the effective stress Mohr circle. (b) Load increase resulted in the enlargement of the effective stress Mohr circle and create stress imbalance. (c) Slight anticlockwise rotation of the mobilised shear strength envelope due to the soil underwent anisotropic settlement. (d) Further rotation of the mobilised shear strength envelope with further settlement. (e) The further rotation of the mobilised shear strength envelope reinstate the point contact between the envelope and the Mohr circle to mark the state of another stress equilibrium.

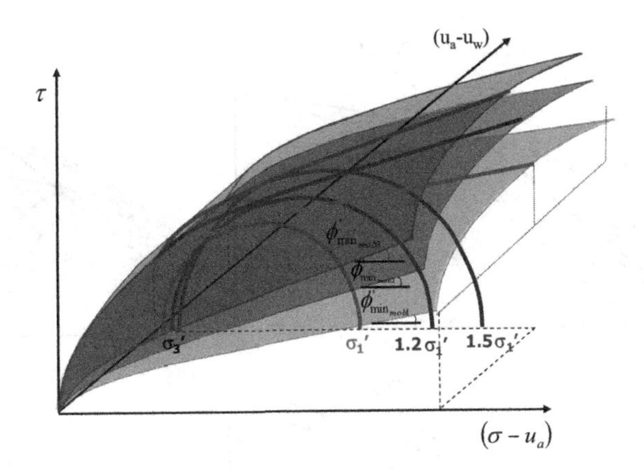

Figure 2.30 Rotation of the mobilised shear strength envelope due to an increase in the effective stress or net stress under unsaturated condition, i.e., under a certain state of suction.

increased from σ_1' to $1.2\sigma_1'$ and $1.5\sigma_1'$, the effective stress Mohr circle gets bigger and subsequently drives the mobilised shear strength surface envelope to rotate upwards until at the end of each stage the Mohr circle reinstates one-point contact with the mobilised shear strength surface envelope to indicate stress equilibrium condition and the rotation stopped. Correspondingly the compression of the soil also stopped. Note that the rotation of the mobilised shear strength surface envelope upwards is indicating soil compression and subsequent increase in the internal shear strength.

Note that the rotation of the mobilised shear strength surface envelope, in this case, is proportionate with the enlargement of the effective stress Mohr circle. This is indicating the linear behaviour of soil compression with respect to the applied vertical pressures as discussed in Section 2.2 item no. 3. Thence, this concept of effective stress and shear strength interaction can comply with this reported soil proportionate volume change behaviour with respect to the increase in the applied vertical pressure.

2.10 Interaction between the effective stress Mohr circle and the mobilised shear strength envelope during anisotropic settlement under wetting collapse

This section demonstrates that settlement can be triggered by suction reduction due to wetting. Figure 2.31a shows the initial state of stress equilibrium when the effective stress Mohr circle just touches the curved surface mobilised shear strength envelope. When suction is decreased, then the effective stress Mohr circle will move towards the frontal plane, i.e., the plane $\tau-(\sigma-u_a)$ as shown in Figure 2.31b. The diameter of the Mohr circle is maintained as there is no change in the loading where the major and minor principal stress, σ_1' and σ_3', remain the same and only the suction is reduced. The Mohr circle move towards the frontal plane, which is towards the lower side of the surface envelope, and thus it will extend higher than the current surface envelope. Note that the curved surface mobilised shear strength envelope is also acting as the

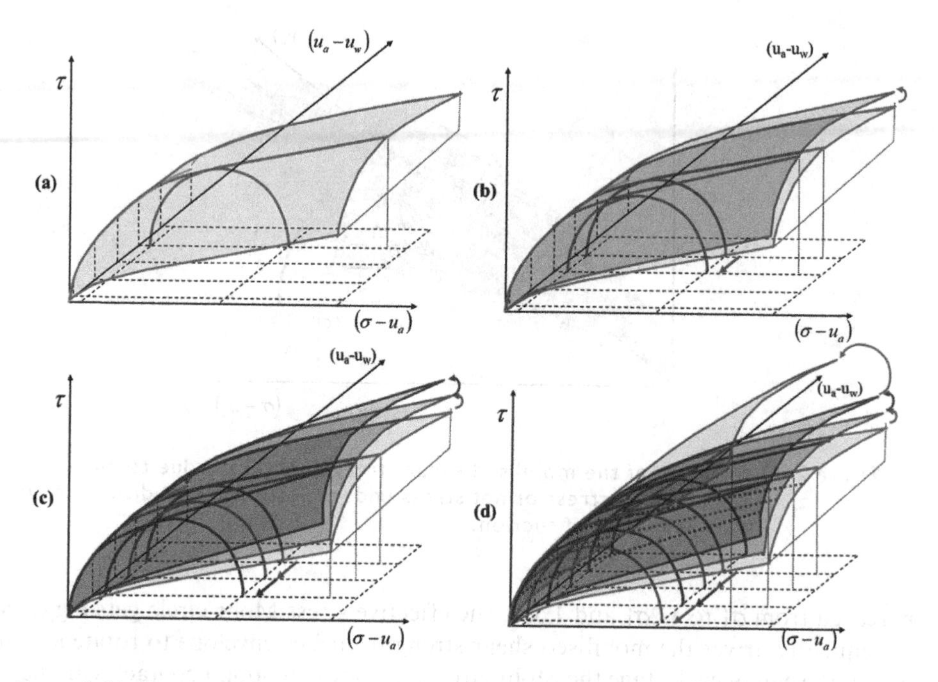

Figure 2.31 Rotation of the mobilised shear strength envelope triggered by the reduction in suction due to wetting. (a) Initial state of stress equilibrium where the surface shear strength envelope is resting on to the effective stress Mohr circle. (b) Slight rotation upwards of the surface shear strength envelope when the effective stress Mohr circle has extended above the surface envelope as it moves towards the $\tau - (\sigma - u_a)$ plane when suction is reduced due to wetting. (c) Further rotation of the surface envelope as the Mohr circle is further driven towards the $\tau - (\sigma - u_a)$ plane as suction is further reduced. (d) Massive rotation of the surface envelope as the top of the Mohr circle crosses the steepest curvature of the surface envelope as it moves towards zero suction.

yield surface. Whenever the effective stress Mohr circle extends higher than the yield surface, the soil is in the state of stress imbalance. The soil will respond by undergoing settlement to increase the state of the mobilised shear strength to reinstate another stress equilibrium condition, otherwise, the settlement will proceed indefinitely. This is achieved when the curved surface mobilised shear strength envelope rotates higher and reinstated the one-point contact between the effective stress Mohr circle and the curved surface mobilised shear strength envelope as shown in Figure 2.31b. Similar rotation of the curved surface envelope happens when the Mohr circle moves further due to the further reduction in suction as shown in Figure 2.31c. Thence, the rotation of the curved surface mobilised shear strength envelope is an indication that the soil is undergoing settlement and there is a subsequent increase in the mobilised shear strength within the soil mass.

This complex settlement behaviour is difficult to perceive where it is happening under effective stress decrease and this settlement behaviour that contradicts the effective stress concept of Terzaghi (1936). Thence, this concept of effective stress and shear strength interaction will reveal the mechanics that cause this weird behaviour. This is the advantage of applying the true soil shear strength behaviour with respect to suction where there is a gradual drop in the shear strength at a higher suction followed by a steep drop in the shear strength when the approaches zero or saturation. Thence, this concept manages to explain the occurrence of wetting collapse as reported in item no. 1 in Section 2.2.

2.11 Shear strength behaviour at low suctions and its role in governing massive settlement near saturation

Note that the rotation of the mobilised shear strength envelope is indicating that the soil undergoes settlement at the same time indicating that the mobilised shear strength has increased within the soil mass. The greater amount of rotation is indicating the greater amount of settlement. The settlement driven by the rise of the groundwater table or inundation is termed as inundation settlement or wetting collapse. It is triggered by the reduction in suction under constant net stress. Figure 2.32a shows the gradual rotation of the curved surface mobilised shear strength envelope from position a to b and to c as Mohr circle moves from position 1 to 2 and to 3 respectively. However, when the Mohr circle moves towards zero suction, i.e., from position 3 to 4, it is moving across the steep downturn of the surface envelope. Hence, the stress imbalance is greater when the Mohr circle sits on the net stress axis extending a lot higher than the previous move. This requires a greater rotation of the surface envelope to reinstate the one-point contact. This greater rotation is indicated by the rotation of the surface envelope from position c to d corresponding Mohr circle movement from position 3 to 4 as shown in Figure 2.32a. Besides, this greater rotation of the curved surface envelope as suction approaches zero can be seen clearly when the rotation is viewed from the arrow direction in Figure 2.32a and seen as illustrated in Figure 2.32b. Both figures indicate the position of the Mohr circles 1, 2, 3 and 4 and the Mohr circles as seen as vertical lines of equal height in Figure 2.32b. As the Mohr circle moves from position 3 to 4 in Figure 2.32b, there is massive push upwards on to the curved surface envelope from position c to d. This massive rotation of the curved surface mobilised shear strength envelope is due to the steep curving down of the surface envelope when it approaches the net stress axis. It can be seen here that this is a very important characteristic of the curved surface shear strength envelope that cannot be neglected if the weird massive soil volume change behaviour is to be understood. Thence, this concept manages to explain the occurrence of massive wetting collapse as reported in item no. 2 in Section 2.2

2.12 Summary

This chapter starts with the discussion on the application of the effective stress concept in characterising the soil shear strength and settlement behaviour. This concept has been the most fundamental in soil mechanics. However, the concept failed to explain some weird soil settlement behaviours like inundation settlement that contradicts the concept of effective stress. There are many complex soil settlement behaviours, which cannot be explained by the current concept. This thereby demands to improvise the existing fundamental concept of soil mechanics; otherwise, the weird soil settlement

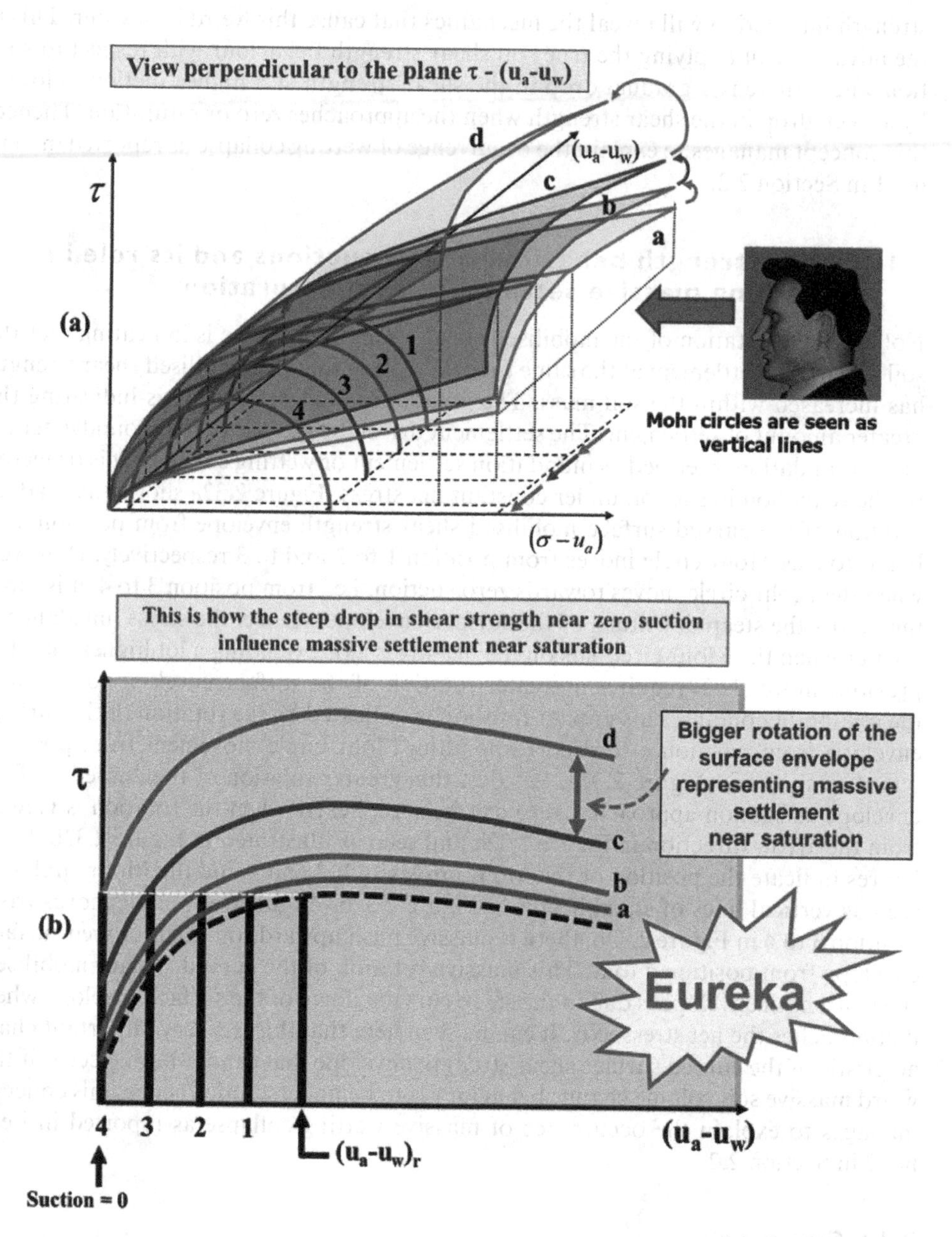

Figure 2.32 Greater rotation of the mobilised shear strength envelope as suction approaches zero to indicate massive settlement near saturation. (a) The overall view of the movement of the effective stress Mohr circle towards the $\tau-(\sigma-u_a)$ plane as the soil is being wetted. (b) The view of the Mohr circle movement during wetting effect in the $\tau-(u_a\text{-}u_w)$ plane and its association with the massive rotation of the surface envelope near zero suction.

behaviour cannot be understood and thence cannot be resolved. Then the chapter recalled the concept of strain hardening and mobilised shear strength according to the linear mobilised shear strength envelope whereas the truly mobilised shear strength envelope is non-linear. However, this concept is not incorporated in the current concept of effective stress and it is very difficult to be incorporated in the soil volume change model.

Then the true non-linear shear strength behaviour with respect to effective stress or net stress and suction is discussed. The non-linearity behaviour at low stress and low suction plays a very important effect on the soil mechanical behaviour like shallow landslide and wetting collapse. Thereby, this aspect of shear strength behaviour cannot afford to be neglected and need an emphasis on the development of the soil shear strength model. All these non-linearity attributes of shear strength have been incorporated in the curved surface soil shear strength model. Six equations are required to define the model and there are seven shear strength parameters involved. The brief procedure to determine the shear strength parameters is also being discussed.

This is followed by the presentation to prove the development of mobilised shear strength in the form of curvilinear mobilised shear strength envelopes when the soil undergoes anisotropic compression during the application of deviator stress in the triaxial test. The mobilised shear strength envelope rotates upwards to mark the increase in the mobilised shear strength when the soil undergoes particle rearrangement to a denser state. This demonstrates the process of strain hardening when the soil undergoes compression and this attribute is important to be incorporated in the soil volume change framework.

The chapter then introduced the concept of effective stress and shear strength interaction, which is the state-of-the-art concept extended from the existing concept of effective stress. This new concept is claimed as the first theoretical soil settlement model, whereas the existing models are empirical and semi-empirical. This new concept introduced the resisting variable in the form of the development of mobilised shear when the soil is compressed while the effective stress is maintained as the driving variable. The new concept utilises the interaction between the effective stress Mohr circle and the mobilised shear strength envelope during soil anisotropic settlement. In this manner, the concept incorporates the effect of strain hardening in governing the soil settlement behaviour. The mobilised shear strength envelope is taken as the yield surface and whenever the Mohr circle extends above the yield envelope, it will drive settlement. The new concept also manages to demonstrate conceptually the occurrences of loading and wetting collapse. Besides, the new concept can explain conceptually the occurrence of the massive settlement near saturation by utilising the steep drop in shear strength when suction approaches zero.

References

Alonso, E. E., Gens, A., and Josa, A. (1990). "A constitutive model for partially saturated soil." *Geotechnique*, 40(3), 405–430.

Barden, L., Madedor, A. O., and Sides, G. R. (1969). "Volume change characteristics of unsaturated clay." *Journal Soil Mechanics Foundation Engineering*, 95, 33–52.

Basma, A. A., and Tuncer, E. R. (1992). "Evaluation and control of collapsible soils." *Journal of Geotechnical Engineering*, 118(10), 1491–1504.

Bishop, A. W. (1959). "The principle of effective stress." *Teknisk Ukeblad*, 106(39), 859–863.
Bishop, A. W. (1966). "The strength of soils as engineering materials." *Geotechnique*, 16(2), 91–130.
Bishop, A. W., Alpan, I., Blight, G. E., and Donald, I. B. (1960). "Factors controlling the shear strength of partly saturated cohesive soils." *ASCE Research Conference Shear Strength of Cohesive Soils*, University of Colorado, Boulder, 503–532.
Bishop, A. W., and Blight, G. E. (1963). "Some aspects of effective stress in saturated and unsaturated soils." *Geotechnique*, 13(3), 177–197.
Blanchfield, R., and Anderson, W. F. (2000). "Wetting collapse in opencast coalmine backfill." *Proceedings of the ICE Geotechnical Engineering*, 143(3), 139–149.
Brackley, I. J. A. (1971). "Partial collapse in unsaturated expansive clay." *Proceedings of the 5th Regional Conference Soil Mechanics Foundation Engineering*, Durban, South Africa, 23–30.
Brandon, T. L., Duncan, J. M., and Gardner, W. S. (1990). "Hydrocompression settlement of deep fills." *Journal of Geotechnical Engineering*, 116(10), 1536–1548.
Burland, J. B. (1964). "Effective Stresses in partly saturated soils. Discussion of: Some aspects of effective stress in saturated and partly saturated soils, by G. E. Blight and A. W. Bishop." *Geotechnique*, 14, 65–68.
Burland, J. B. (1965). "Some aspects of the mechanical behaviour of partly saturated soils." Moisture Equilibria and Moisture Changes in the Soils Beneath Covered Areas, A Symposium in Print, Butterworth, Sydney, Australia, 270–278.
Charles, J. A., and Watts, K. S. (1980). "The influence of confining pressure on the shear strength of compacted rockfill." *Geotechnique*, 30(4), 353–367.
Coulomb, C. A. (1776). "Essai sur une application des regles de maximia et minimis a quelques problemes de statique relatifs a l'architechture." *Memoires de la Mathematique st de Physique, presentes a l'Academic Royale des Sciences, par divers savants, et lus dans ces Assemblees*, L'Imprimerie Royale, Paris, 3–8.
Cox, D. W. (1978). "Volume change of compacted clay fill." *Proceedings of the Conference on Clay Fills*, Institution of Civil Engineers, London, UK, 79–86.
De Beer, E. E., and Martens, A. (1951). "Method of computation of an upper limit for influence of heterogeneity of sand layers on the settlement of bridges." *Proceedings of the 4th International Conference on Soil Mechanics and Foundation Engineering*, London, England, 275–278.
Escario, V., and Juca, J. (1989). "Strength and deformation of partly saturated soils." Proceedings of the 12th International Conference on Soil Mechanics and Foundation Engineering, Rio de Janeiro, Vol. 3, 43–46.
Escario, V., and Saez, J. (1973). "Measurements of the properties of swelling and collapsing soils under controlled suctions." *Proceedings of the 3rd International Conference on Expansive Soils*, Haifa, Israel, Vol. 1, 195–200.
Escario, V. (1980). "Suction controlled penetration and shear tests." *Proceedings of the 4th International Conference on Expansive Soils, Denver, Colorado, June 16-18, 1980, Vol. 1.*
Escario, V., and Saez, J. (1986). "The shear strength of partly saturated soils." *Geotechnique*, 36(3), 453–456.
Feda, J. (1988). "Collapse of loess upon wetting." *Engineering Geology*, 25, 263–269.
Fredlund, D. G. (2000). "The 1999 R.M. Hardy Lecture: The implementation of unsaturated soil mechanics into geotechnical engineering." *Canadian Geotechnical Journal*, 37, 963–986.
Fredlund, D. G., and Morgenstern, N. R. (1976). "Constitutive relations for volume change in unsaturated soils." *Canadian Geotechnical Journal*, 13(3), 261–276.
Fredlund, D. G., Morgenstern, N. R., and Widger, R. A. (1978). "Shear strength of unsaturated soils." *Canadian Geotechnical Journal*, 15(3), 313–321.
Fredlund, D. G., Rahardjo, H., and Gan, J. K. M. (1987). "Non-linearity of strength envelope for unsaturated soils." *Proceedings of the 6th International Conference on Expansive Soils*, New Delhi, 49–54.

Fredlund, D. G., and Xing, A. (1994). "Equations for the soil-water characteristic curve." *Canadian Geotechnical Journal*, 31(4), 533–546.
Gan, J. K. M., and Fredlund, D. G. (1988). "Multistage direct shear testing of unsaturated soil." *ASTM, Geotechnical Testing Journal*, 11(2), 132–138.
Gan, J. K. M., and Fredlund, D. G. (1996). "Shear strength characteristics of two saprolitic soils." *Canadian Geotechnical Journal*, 33, 595–609.
Gan, J. K. M., Fredlund, D. G., and Rahardjo, H. (1988). "Determination of the shear strength parameters of an unsaturated soil using the direct shear test." *Canadian Geotechnical Journal*, 25(3), 500–510.
Goodwin, A. K. (1991). "One dimensional compression behaviour of unsaturated granular soils at low stress level." Ph.D. Thesis, University of Sheffield.
Henkel, D. J. (1958). "The correlation between deformation, pore-water pressure and strength characteristics of saturated clays." Ph.D. Thesis, University of London.
Ho, D. Y. F., and Fredlund, D. G. (1982). "A multistage triaxial test for unsaturated soils." *ASTM Geotechnical Testing Journal*, 5(1/2), 18–25.
Janbu N., Bjerrum, L., and Kjaernsli, B. (1956). "Veiledring ved losning av fundermentering-soppgaver." Norwegian Geotechnical Institute, Publication No. 16, Oslo.
Krahn, J., Fredlund, D. G., and Klassen, M. J. (1989). "Effect of soil suction on slope stability at Notch Hill, *Canadian Geotechnical Journal*, 26(2), 269–278
Lloret, A., and Alonso, E. E. (1980). "Consolidation of unsaturated soils including swelling and collapse behavior." *Geotechnique*, 30(4), 449–477.
Mahalinga-Iyer, U., and Williams, D. J. (1995). "Unsaturated strength behaviour of compacted lateritic soils." *Geotechnique*, 45(2), 317–320.
Maswoswe, J. (1985). "Stress paths for a compacted soil during collapse due to wetting." Ph.D. Thesis, Imperial College.
Matyas, E. L., and Radhakrishna, H. S. (1968). "Volume change characteristics of partially saturated soils." *Geotechnique*, 18(4), 432–448.
Md Noor, M. J. (2006). "Shear strength and volume change behaviour of saturated and unsaturated soils." Ph.D. Thesis, University of Sheffield.
Md Noor, M. J., and Anderson, W. F. (2006). "A comprehensive shear strength model for saturated and unsaturated soils." *Proceedings of the 4th International Conference Unsaturated Soils*, ASCE Geotechnical Special Publication No. 147, Carefree, Arizona, Vol. 2, 1992–2003.
Md Noor, M. J., Mat. Jidin, R., and Hafez, M. A. (2008). "Effective stress and complex soil settlement behaviour." *Electronic Journal Geotechnical Engineering*, 13, 1–13.
Nishimura, T., and Fredlund, D. G. (2000). "Relationship between shear strength and matric suction in an unsaturated silty soil." *Proceedings for the Unsaturated Soil for Asia*, Singapore, 563–568.
Rassam, D. W., and Williams, D. J. (1999). "Relationship describing the shear strength of unsaturated soils." *Canadian Geotechnical Journal*, 32(2), 363–368.
Salman, T. H. (1995). "Triaxial behaviour of partially saturated granular soils at low stress levels." Ph.D. Thesis, University of Sheffield.
Schmertmann, J. H., Hartman, J. P., and Brown, P. R. (1978). "Improved strain influence factor diagrams." *Journal of Geotechnical and Geoenvironmental Engineering*, 104(GT8), 1131–1135.
Steinbrenner, W. (1934). "Tafeln zur Setzungsberechnung." *Die Strasse*, 1, 121–124.
Tadepalli, R., Rahardjo, H., and Fredlund, D. G. (1992). "Measurement of matric suction and volume change during inundation of collapsible soil." *ASTM Geotechnical Testing Journal*, 15(2), 115–122.
Terzaghi, K. (1936). "The shear resistance of saturated soils." *Proceedings for the 1st International Conference on Soil Mechanics and Foundation Engineering*, Cambridge, MA, Vol. 1, 54–56.
Terzaghi, K. (1943). *Theoretical soil mechanics*, Wiley Publications, New York.

Toll, D. G., Ong, B. H., and Raharjo, H. (2000). "Triaxial testing of unsaturated samples of undisturbed residual soil from Singapore." *Proceedings of the Unsaturated Soil for Asia*, Singapore, 581–586.

Vanapalli, S. K., Fredlund, D. G., Pufahl, D. E., and Clifton, A. W. (1996). "Model for the prediction of shear strength with respect to soil suction." *Canadian Geotechnical Journal*, 33(3), 379–392.

Vanapalli, S. K., Sillers, W. S., and Fredlund, M. D. (1998). "The meaning and relevance of residual state to unsaturated soils." 51st *Canadian Geotechnical Conference*, Edmonton, Alberta, Canada, 1–8.

Vilar, O. M. (1995). "Suction control oedometer tests on compacted clay." *Proceedings of the 1st International Conference on Unsaturated Soils*, Paris, France, 201–206.

Satija, B. S. (1978). "Shear behaviour of partially saturated soils." Ph.D. thesis, Indian Institute of Technology, New Delhi, India.

Chapter 3

Rotational Multiple Yield Surface Framework

3.1 Previous unsaturated soil volume change models based on two independent stress state variables

Volume change in unsaturated soils is a very complex behaviour to characterise. It involves expansion and contraction. On wetting, expansive soil exhibits an increase in volume under low confining pressure, whilst collapsible soil always exhibits a decrease in volume. Wetting reduces suction and causes subsequent decreases in effective stress. The normal type of collapse is due to load increase, which subsequently increases the effective stress and, at a certain point, triggers collapse. The occurrence of collapse due to both increase and decrease in effective stress has made it difficult to formulate them under a single framework. This is the reason why volume change is the most difficult unsaturated soil behaviour to characterise (Fredlund, 2000). To make the situation even more complex, alternate wetting and drying have been reported to cause collapse (Alonso et al., 1995; Sharma, 1998). Earlier studies of volume change behaviour in unsaturated soils are solely based on the influence of net stress and suction and the influence of shear strength was not incorporated until the critical state approach began (Alonso et al., 1987). In the critical state model, three stress state variables are applied, which are the deviator stress, q, net mean stress, p', and specific volume, v. Note that one of the stress state variables applied is the deviator stress q and this is not the shear strength τ. Therefore, even the critical state model has not incorporated shear strength in its framework. Collapse produces closer packing of soil particles. The movement of particles is triggered when the friction between the particles is reduced as suction decreases upon wetting and the overburden pressure then mobilises the movement. Soils compacted dry of optimum have low densities and often exhibit collapse behaviour upon wetting (Barden et al., 1969).

The first attempt to explain the volume change behaviour of unsaturated soil based on two independent stress state variables was by Bishop (1959) with the widely quoted effective stress equation for unsaturated soil as in Equation 2.8. However, a unique relationship between the weighting parameter, χ, and volume change was not achieved. Then Jennings and Burland (1962) conducted oedometer one-dimensional compression tests on silt and clay soils and assessed the volume change behaviour using the Bishop (1959) effective stress equation, but no unique relationship between the effective stress and volume change was found. They thereby questioned the validity of the effective stress equation of Bishop (1959). Normally volume reduction occurs under increasing effective stress such as settlement in the event of groundwater table lowering, but then collapse is encountered under effective stress decreasing in the case of the soil being inundated. The proposed effective stress equation thus failed to explain theoretically the mode of wetting collapse and the massive volume change behaviour near saturation.

Bishop and Blight (1963) described the volume change behaviour of unsaturated soil under isotropic loading and expressed its path in void ratio, e, net stress, $(\sigma - u_a)$, and suction, $(u_a - u_w)$, space as shown in Figure 2.5. The figure indicates that wetting the soil at a low value of $(\sigma - u_a)$ caused swelling of the soil along the path AD, whereas wetting the soil at a high value $(\sigma - u_a)$ caused collapse of the soil along the paths BE and CF. For an unsaturated soil loaded at constant water content starting at point A, the volume decreased and so did the suction as represented by path ABC. The path is moving towards full saturation as void ratio decreases when the line ABC is approaching the plane e versus $(\sigma - u_a)$.

Matyas and Radhakrishna (1968) carried out triaxial compression tests on a mixture of 80% flint and 20% kaolin. The tests were carried out in isotropic compression. Wetting tests were carried out by allowing the specimen to imbibe water, either under a constant volume or under constant applied stress. The results were plotted in a three-dimensional space of void ratio against net stress and suction as shown in Figure 2.6. The results showed a steep drop in volume near saturation indicating collapse. Furthermore, they noted that the collapse behaviour caused by suction reduction at high values of $(\sigma - u_a)$ could not be explained by the principle of effective stress as suggested by Bishop (1959), as volume decreases (normally it swells) when effective stress decreases.

Barden et al. (1969) conducted a series of anisotropic consolidation tests in a modified Rowe cell using 150 mm diameter by 25 mm thick samples of compacted Westwater and Derwent clays. The specimens were initially consolidated at given values of suction and net stress and then taken through three different stress paths involving wetting at constant net stress, consolidation at constant suction and both consolidation and wetting at the same time. The results indicated that for specimens compacted dry of optimum, collapse occurred on wetting to zero suction under large p'. However, the higher the clay content, the smaller the collapse. The smaller collapse is most probably caused by volume compensation due to the simultaneous swelling of the clay microstructure. However, specimens compacted close to optimum moisture content showed no collapse. This gives an indication that the degree of collapse depends on the availability of space for the soil particles to move. Therefore, for coarse-grained soil, the limit of collapse is when the particles have come to a fully interlocking position. It was also realised that the volume change behaviour of unsaturated soil is best analysed in terms of the two independent stress tensors, $(\sigma - u_a)$ and $(u_a - u_w)$.

Escario and Saez (1973) have noted that in one-dimensional compression under controlled suction, as negative pore water pressure decreases, settlement increases and bigger settlement is achieved at a higher vertical pressure as illustrated in Figure 3.1. Compression settlement was thought to be strongly governed by the total net vertical stress $(\sigma - u_a)$ or the deviator stress $(\sigma_1 - \sigma_3)$ rather than the effective stress, because effective stress decreases when negative pore water pressure approaches zero, whereas settlement increases. Furthermore, the settlement induced by the suction decrease process was not large until a relatively low suction value was attained, which led to a bigger collapse. This is an important characteristic of collapse settlement, which ought

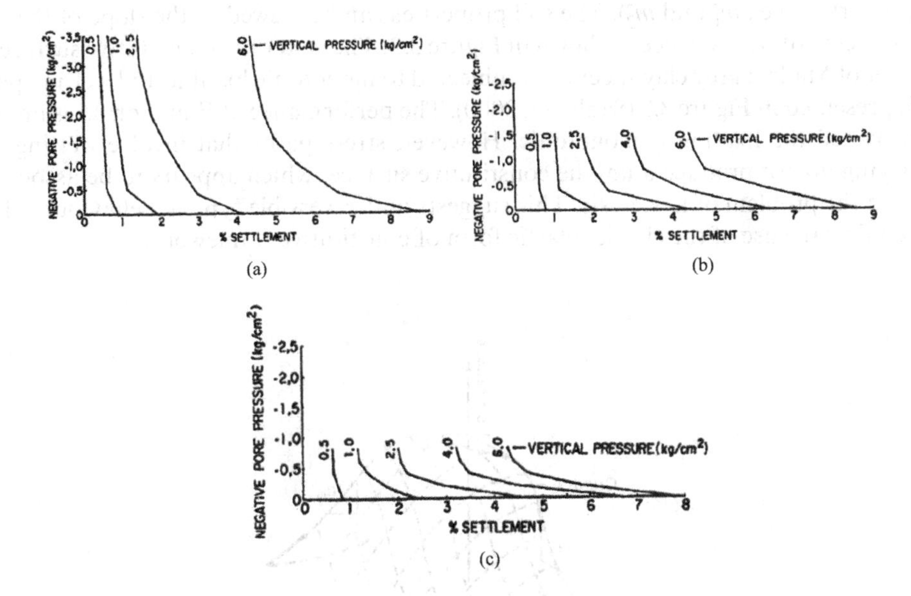

Figure 3.1 Variation of settlements with various constant vertical pressures and decreasing negative pore water pressures at different degrees of compaction (Escario and Saez, 1973) with a Proctor optimum moisture content of 11%. (a) Compaction m.c. 3%. (b) Compaction m.c. 6%. (c) Compaction m.c. 8%.

not to be neglected in the development of a new shear strength–volume change framework for unsaturated soil. In addition, the collapse is plastic in behaviour, because deformation is irrecoverable.

Fredlund and Morgenstern (1976) proposed a semi-empirical constitutive relation for unsaturated soil using the two independent stress state variables, $(\sigma - u_a)$ and $(u_a - u_w)$, as in Equation 3.1. However, physically the model is identical to the constitutive void ratio warped surface presented by Matyas and Radhakrishna (1968). This model quantified the rate of soil volume change with respect to the initial volume based on the change in net stress and suction by associating each term with the coefficient of soil volume change. Besides, the model does not incorporate the role of soil shear strength; thence, there is no effect of strain hardening involved.

$$\frac{dV_v}{V_o} = m_1^s d(\sigma - u_a) + m_2^s d(u_a - u_w) \tag{3.1}$$

where

V_v=volume of voids

V_o=initial total volume of soil

m_1^s=coefficient of soil volume change with respect to a change in net normal stress

m_2^s=coefficient of soil volume change with respect to a change in suction

Equation 3.1 can be viewed as comprising of two parts: a part that is designated to the stress states (i.e., $(\sigma - u_a)$ and $(u_a - u_w)$) and a part that is designated to the soil properties (i.e., m_1^s and m_2^s). The soil properties can be viewed as the slope of the void ratio constitutive surface as shown in Figure 3.2. The void ratio constitutive surface for a set of Madrid grey clay specimens subjected to monotonic loading and wetting paths is presented in Figure 3.3 (Sabbagh, 2000). The performance of Equation 3.1 is unique, provided the loading is monotonic. However, stress paths that involve wetting and drying do not produce a unique constitutive surface, which appears to be associated with the problem of hysteresis. This suggests an irreversible type of behaviour, which requires the use of the elastic–plastic form of constitutive framework.

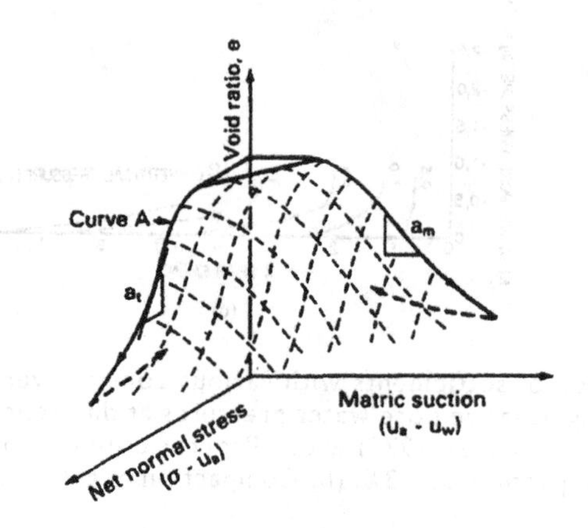

Figure 3.2 Three-dimensional void ratio constitutive surface for unsaturated soil based on the two independent stress state variables, $(\sigma - u_a)$ and $(u_a - u_w)$ (Fredlund and Morgenstern, 1976).

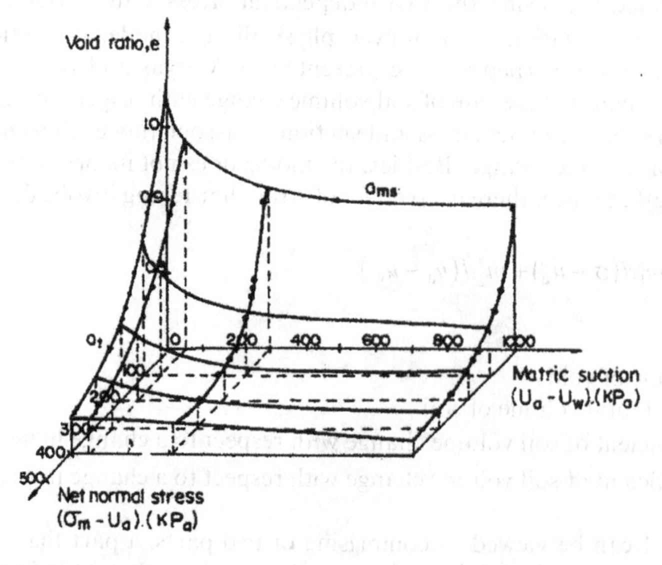

Figure 3.3 Void ratio constitutive surface for Madrid grey clay (Sabbagh, 2000).

The application of an elastic–plastic critical state framework to explain the different modes of volume change behaviour in fine-grained unsaturated soils has limitations. This is as noted by Wheeler et al. (2003), because the framework cannot model the irreversible volume change during cycles of wetting and drying as reported by Alonso et al. (1995) and Sharma (1998). The limitation faced by the critical state framework is suspected due to the inadequacy of the definition of failure by suction changes.

Collapse due to drying is triggered whenever suction exceeds the past maximum suction (Alonso et al., 1990; Wheeler and Sivakumar, 1995; Tang and Graham, 2002). But later, Wheeler et al. (2003) have noticed that this definition has failed to explain the collapse behaviour due to alternate wetting and drying based on the reports by Alonso et al. (1995) and Sharma (1998), as shown in Figure 3.4a and b, respectively. This includes the "closed"-type framework proposed by Tang and Graham (2002), because it applies the same definition of failure due to suction. In the first and second drying curves in Figure 3.4a and b, the collapse takes place even at a suction lower than the past maximum suction indicated by point A in both figures. Furthermore, the critical state framework also claims that whenever the stress path is within the elastic zone (i.e., underneath the yield surface envelope) of the framework, the collapse will not be triggered. However, the corresponding alternate wetting and drying paths in Figure 3.4a and b are indicating otherwise. The path ABC in Figure 3.4a and b is correspondingly illustrated in the p-q-s space in Figure 3.5, which shows the expansion of the yield envelope when the wetting path AB hits the inner loading collapse line LC along the way. Then the subsequent drying path BC is within the elastic zone and there should not be any collapse failure along this path. However, the corresponding path shown in Figure 3.4a and b indicates that plastic straining is taking place. This could be a problem of defining yield purely based on suction without considering the combined effect that it has with the major and minor principal stresses on the mobilised shear strength and the shear stress. Therefore, the definition of failure purely based on suction increase through the SI yield line is not adequate. That is why some authors like

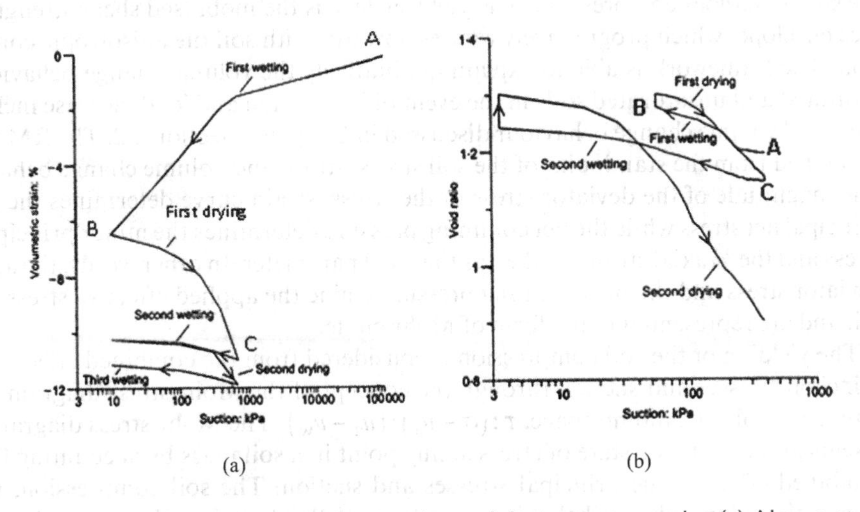

Figure 3.4 Volume reductions under wetting and drying stress paths. (a) Alonso et al. (1995). (b) Sharma (1998).

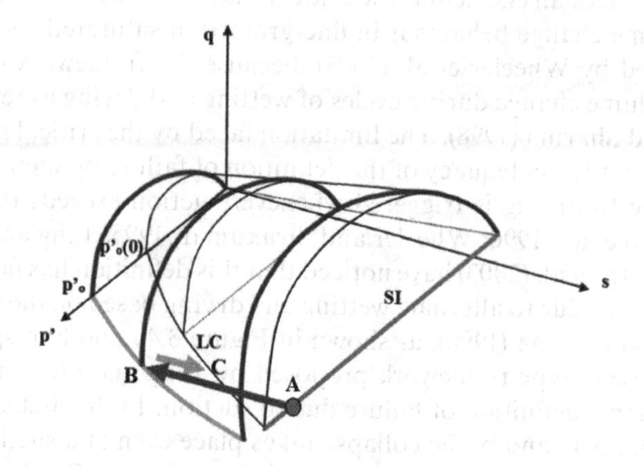

Figure 3.5 Superimposition of the wetting path followed by the drying path onto the elasto-plastic framework of Alonso et al. (1990) as reported by Alonso et al. (1995) in Figure 3.4a.

Gallipoli et al. (2003) and Wheeler et al. (2003) have already switched from this type of yield definition, which is based solely on the limiting value of q, p' and s, to defining yield based on inter-particle stress.

3.2 Rotational Multiple Yield Surface Framework (RMYSF) soil volume change model by the concept of effective stress and shear strength interaction

This section discusses the soil volume change framework named as Rotational Multiple Yield Surface Framework (RMYSF) introduced by Md Noor and Anderson (2007). It is named as multiple yield surface, because it utilises the yield surface progressively as the soil undergoes compression. The yield surface is the mobilised shear strength surface envelope, which progressively rotates upwards with soil the anisotropic compression. The framework is able to explain qualitatively the volume change behaviour of saturated and unsaturated soils in the event of inundation and load increase including the weird volume change behaviour discussed in Chapter 2 Section 2.2. The RMYSF is developed from the standpoint of the soil stress–strain and volume change behaviour. The magnitude of the deviator stress in the stress–strain curve determines the major principal net stress while the net confining pressure determines the minor principal net stress and the % axial strain is taken as the yield parameter. In other words, the applied deviator stress and the net confining pressure define the applied effective stress to the soil and are represented in the form of Mohr circle.

The yielding or the soil compression is considered from the combined effects of the principal stresses and suction through the concept of the Mohr stress diagram in the extended Mohr–Coulomb space, $\tau:(\sigma - u_a):(u_a - u_w)$. The Mohr stress diagram represents graphically the state of stress at any point in a soil mass by accounting for the combined effect of the principal stresses and suction. The soil compression, which defines the volume change behaviour, can be modelled based on the interaction of the state of stress and the state of the mobilised shear strength. Mobilised shear strength increases with the mobilised friction angle ϕ'_{mob}. Thence, the curvilinear shape of the mobilised shear strength envelope rotates about the suction axis towards the soil shear strength envelope at failure as the soil structure is compressed. The uniqueness of this framework is that the mobilised shear strength envelope is also acting as the yield locus, because the axial strain along the locus is considered constant irrespective of the net confining stress (i.e., effective stress or the net stress) or the soil suction. Soil compression is triggered when the state of stress exceeds the mobilised shear strength or, in other words, when the state of stress exceeds the yield surface. Thus to verify the applicability of this volume change framework (i.e., RMYSF), the data from the triaxial compression tests for both saturated and unsaturated specimens must indicate that the yield surface represents identical axial strains.

Because the mobilised shear strength envelope is a curvilinear surface, as shown in Figure 3.6, similar to the shape of the soil failure shear strength envelope, the axial strain needs to be correlated to a parameter, which is a constant to the yield envelope. This has to be the effective mobilised minimum friction angle $\phi'_{min_{mob}}$, which is the inclination of the linear section of the envelope to the horizontal (refer Figure 3.6). The changing of the effective mobilised minimum friction angle during soil compression is correlated to the axial strain and a unique relationship is sought. If this unique relationship exists at saturation and at different values of suction, this would mean that the rotation of the yield surface envelope about the suction axis is like a rigid surface, which retains its shape when identical soils at different suctions are compressed, i.e., the rotation is uniform. Besides, the unique relationship, which indicates that the increase in the axial strain relative to the increase in the effective mobilised minimum friction angle would also imply greater rotation of the yield surface envelope, corresponds to a greater volume change. This is significant because the greater rotation of the yield envelope obtained during the simulation of wetting collapse when suction approaches zero (refer Chapter 2, Section 2.11) would correspond to the massive volume change.

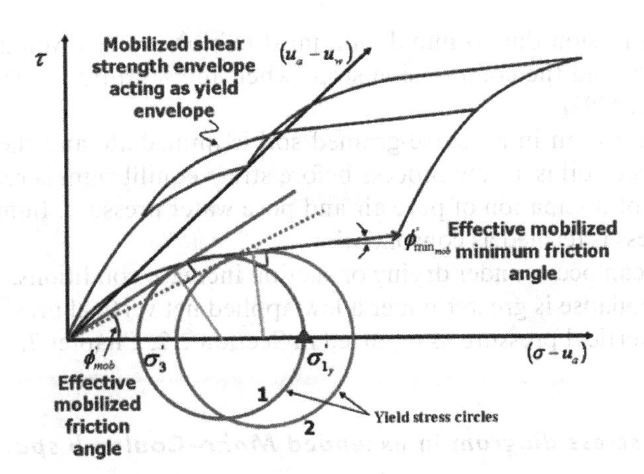

Figure 3.6 Mobilised minimum friction angle, $\phi'_{min_{mob}}$, describes the position of the current yield envelope. All the points on the yield envelope represent the same axial strain, irrespective of the net confining stress or suction (Md Noor, 2006).

3.2.1 General characteristics of RMYSF

The review of the overall volume change behaviour for saturated and unsaturated soils has revealed the general characteristics of the soil volume change behaviour. The RMYSF must be in compliance with all of these characteristics for it to be comprehensively acceptable. The characteristics of the volume change behaviour of saturated and unsaturated soils are summarised as follows.

1. First and foremost, the soil compression must be in accordance with its stress–strain behaviour, which is elastic–plastic. The settlement due to particle rearrangement is irrecoverable and therefore, the simulation of the collapse compression by the RMYSF must be elastic–plastic, provided the applied effective stress is the maximum in the soil history.
2. The changes in the applied stress are driving the stress state to the yield limit or the mobilised shear strength irrespective of whether it is effective stress increase (loading collapse) or decrease (wetting collapse).
3. Collapse failure does not occur at the critical state; it occurs under stress changes. Note that critical state is the straining at constant stress. Besides, the collapse occurs due to the availability of the space for the particles to move into (Barden et al., 1969), which causes the reduction in the overall volume, and it is not straining under constant volume, which is a characteristic of a critical state.
4. The framework involves a multiple yield surface because under a specific constant loading condition, the soil compression ceases at one point, and changing the loading condition produces further compression, which also ceases at another point (Rahardjo and Fredlund, 2003). It is suspected that the soil compression ceases when stress equilibrium is achieved between the state of stress in the soil and the mobilised shear strength developed in the soil mass.
5. Therefore, the soil compression mechanism must be derived from the concept of stress equilibrium between the state of stress in the soil and the mobilised shear strength.
6. Soil compression due to inundation must exhibit massive volume decrease near saturation, and the compression stops when 100% saturation is achieved (Tadepalli et al., 1992).
7. Soil compression in a coarse-grained soil is immediate and the compression in fine-grained soil is a slow process before stress equilibrium is reached due to the slow rate of dissipation of pore air and pore water pressure. In fine-grained soils, this process is termed as consolidation.
8. Collapse can occur under drying or suction increase conditions.
9. Wetting collapse is greater under a low applied net vertical pressure compared to high net vertical pressure as reported in Section 2.2, Chapter 2.

3.2.2 Mohr stress diagram in extended Mohr–Coulomb space

RMYSF applied the extended Mohr–Coulomb space of $\tau:(\sigma-u_a):(u_a-u_w)$, introduced by Fredlund et al. (1978); the influence of the two independent stress state variables on shear strength is considered separately through the net stress and suction axes. The same concept of Mohr stress diagram in the three-dimensional extended

Mohr–Coulomb space will be applied for this interpretation of the curvilinear shear strength behaviour and volume change behaviour with respect to net stress and suction.

Consider an infinitely small element of soil within a soil mass subjected to a triaxial stress state as shown in Figure 3.7a. The element is subjected to major and minor principal net stresses (i.e., $(\sigma_1 - u_a)$ and $(\sigma_3 - u_a)$) in the vertical and the horizontal directions, respectively. The suction $(u_a - u_w)_1$ is considered as an isotropic tensor and thus it acts equally in all directions (Fredlund and Rahardjo, 1993). Let the net normal stress and shear stress on an inclined plane AC at an angle α from the major principal plane AB be $(\sigma_\alpha - u_a)$ and τ_α, respectively. The expressions for the net normal stress and shear stress on the inclined plane are given in Equations 3.2 and 3.3.

$$(\sigma_\alpha - u_a) = \left(\frac{\sigma_1 + \sigma_3}{2} - u_a\right) + \left(\frac{\sigma_1 - \sigma_3}{2}\right)\cos 2\alpha \tag{3.2}$$

$$\tau_\alpha = \left(\frac{\sigma_1 - \sigma_3}{2}\right)\sin 2\alpha \tag{3.3}$$

When Equations 3.2 and 3.3 are squared and added, the resulting equation becomes the equation of a circle (Equation 3.4), which is known as the Mohr stress circle (Fredlund and Rahardjo, 1993).

$$\left[(\sigma_\alpha - u_a) - \left(\frac{\sigma_1 + \sigma_3}{2} - u_a\right)\right]^2 + \tau_\alpha^2 = \left(\frac{\sigma_1 - \sigma_3}{2}\right)^2 \tag{3.4}$$

The Mohr stress circle forms a graphical method of presenting the state of net normal stress and shear stress at a point in a soil mass, provided the inclination of the point from the principal planes is known. The Mohr stress circle is drawn on a plane with the net normal stress, $(\sigma - u_a)$, as the abscissa and the shear stress, τ, as the ordinate. The centre of the circle has the coordinate $\left[\left(\frac{\sigma_1 + \sigma_3}{2} - u_a\right), 0\right]$, and the radius is $\left(\frac{\sigma_1 - \sigma_3}{2}\right)$.

The stress condition on a plane at an angle α from the major principal plane AB on the soil element as shown in Figure 3.7a is represented by the point Z at double the angle (i.e., 2α) in the Mohr stress circle through the same course of rotation from the referred major principal plane OY as shown in Figure 3.7c. The coordinate of the resultant point Z on the Mohr stress circle represents the stress state along the inclined plane in the element. For unsaturated soils, the Mohr stress diagram uses a third orthogonal axis to represent suction, because it must be included in the description of the stress state. The position of the circle along the suction axis is determined by the magnitude of the suction.

The corresponding extended Mohr diagram for the triaxial stress state shown in Figure 3.7a is presented at position 1 in Figure 3.7c. In this case, the plane of the Mohr circle is located at distance $(u_a - u_w)_1$ from the net stress axis. When the condition becomes saturated and the suction becomes zero, while the net principal stresses remain constant like the triaxial stress state represented in Figure 3.7b, the Mohr circle is shifted to position 2 in Figure 3.7c, which sits exactly on the net stress axis. This is

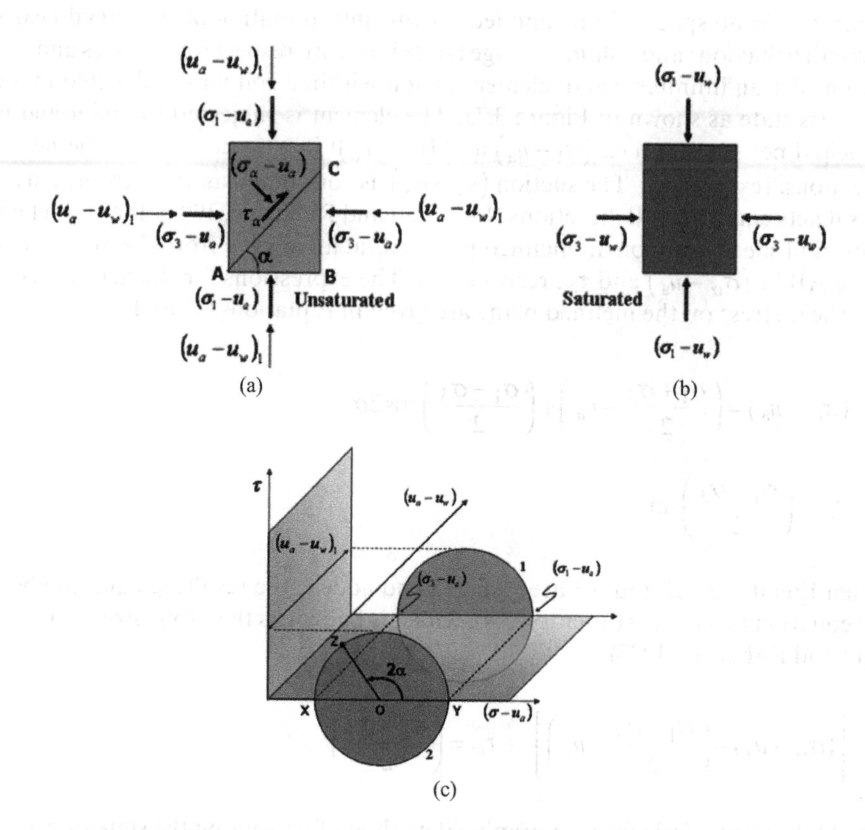

Figure 3.7 Saturated and unsaturated triaxial stress states and the corresponding Mohr stress circles. (a) Stress state in an unsaturated triaxial specimen. (b) Stress state in a saturated triaxial specimen. (c) Representation of unsaturated stress state by Mohr circle 1 and the saturated stress state by Mohr circle 2 in extended Mohr–Coulomb space.

the way that the change in the stress state due to the change in suction is presented in the extended Mohr–Coulomb space. The Mohr stress circles from triaxial tests on unsaturated specimens at different suctions are plotted in this manner to obtain the curved-surface extended Mohr–Coulomb shear strength envelope.

3.2.3 *Principle concept of RMYSF*

Investigating the changes on the shear strength parameters when the soil structure is compressed has essentially revealed an increase in the mobilised friction angle, ϕ'_{mob}, as discussed in Section 2.3, Chapter 2. This is indicated through the rotation of the mobilised shear strength envelope in the extended Mohr–Coulomb space about the suction axis. Consider the compression of three triaxial test specimens subjected to three different effective stresses. They would exhibit three different magnitudes of peak stress at identical axial strain, provided there is no particle breakage in the specimens. This is based on the report that particle breakage delays the arrival of peak strength until higher axial strains (Bishop, 1966, 1972; Billam, 1972). Reaching peak strengths at the same axial strain implies that the specimens have undergone the same volume change before achieving the peak strength, irrespective of the applied effective stresses.

Therefore, the shear strength surface envelope represents identical axial strains experienced by the specimens even under different effective stresses, provided there is no particle breakage. Similarly, it is anticipated that the mobilised shear strength envelope or the yield surface at a certain stage of the compression represents the same axial strain. In other words, the same rotation angle of the mobilised shear strength envelope is anticipated to cause the same axial strain, irrespective of the magnitude of the effective stress. This holds as long as it is assumed that at low-stress levels that particle breakage does not occur. Owing to the rotation of the mobilised shear strength envelope, or the yield surface envelope, when the soil is compressed, the proposed soil volume change model is called the "Rotational Multiple Yield Surface Framework" (Md Noor, 2006). High stress levels would cause particle breakage, would cause a variable axial strain along the mobilised shear strength envelope and, therefore, could not be regarded as the yield surface. This limits the application of the framework.

In the field with a horizontal ground surface, as horizontal stress increases with vertical stress, the soil must still develop inclined planes of maximum shear stress. Thus, it is anticipated that soil deformation is due to the combined action of shear and normal stress on the inclined planes. Initially, in a loose state, the soil particles have room to adjust themselves in the direction of shear as they are being displaced vertically downwards. However, when they reach a tightly packed arrangement, they have less freedom of movement and are said to be in the state of interlocking. At this state, increase in the vertical stress can only move the particles vertically downwards, because lateral displacement has been restrained by the surrounding soil (Powrie, 1997). Therefore, in this case, soil deformation is assumed to be vertically one-dimensional. This mode of deformation is approximately appropriate for soil subjected to vertical loads like the pad, strip and raft foundations. This means that the action of shear and normal stress is the driving factor that governs the field soil compression, and graphically, it can be represented by the Mohr stress diagram. The mobilised shear strength, τ_{mob}, is resisting the deformation and graphically it can be represented by the mobilised shear strength envelope. The deformation ceases when the mobilised shear strength equals or exceeds the stress state in the soil mass. Therefore, soil deformation needs to be assessed from the standpoint of shear stress, normal stress, suction and the mobilised shear strength in the soil.

Triaxial testing is the common laboratory method for evaluating the fundamental strength and deformation properties of soil based on the stress–strain behaviour. A close investigation of the stress–strain curve indicates that at one point on approaching the peak strength, the soil starts to undergo a significant compression, which corresponds to a small increase in the deviator stress. This is the point where the soil starts to undergo dilation and this point can be recognised as the point where the stress–strain curve starts to divert from a linear or an almost linear behaviour. However, in contrast to this stress–strain curve, in the field, the soil is suppressed from dilating by the vertical net stress as well as being constrained laterally by the surrounding soil. The utmost volumetric state that the soil can achieve in the field is a full interlocking state, which corresponds to that point on the stress–strain curve at which the soil starts to dilate. Further compression beyond this point in the field (i.e., one-dimensional compression) is attributed to particle breakage. Nonetheless, the peak strength is still attained in the field but not through dilation as in the triaxial specimen. It is achieved by particles squeezing between each other as they try to deform under the resultant net normal and shear stresses along the plane of maximum shear under the interlocking condition. Thence, the stress–strain behaviour that ranges up to the peak strength only will be applied in the development of this shear strength–volume change framework. However, the corresponding maximum axial strain is taken as the abscissa of the intersection point between the extrapolated line from the linear section of the stress–strain curve and the horizontal line that passes through the peak deviator stress.

The maximum shear and normal stresses acting on an inclined shear plane are derived from the combination of major principal net stress, $(\sigma_1 - u_a)$, minor principal net stress, $(\sigma_3 - u_a)$, and suction, $(u_a - u_w)$, as discussed in Section 3.2.2. The mobilised shear strength, τ_{mob}, depends on the maximum attained net normal stress and the maximum achieved effective mobilised friction angle, ϕ'_{mob}, or the maximum value of the effective mobilised minimum friction angle, $\phi'_{min_{mob}}$, achieved as illustrated in Figure 3.6.

During inundation, the normal stress, σ_n, and the shear stress, τ_D, along a shear plane remain constant even though there is a reduction in suction. This is because the principal net stresses (i.e., $(\sigma_1 - u_a)$ and $(\sigma_3 - u_a)$), which determine the circle diameter, remain constant as the Mohr circle retreats towards the net stress axis. However, the maximum mobilised shear strength, τ_{mob}, defined by the mobilised shear strength envelope is still constant. Nonetheless, the maximum mobilised shear strength at a specific suction decreases with suction according to the envelope. Thence, at one point, the soil state of stress as defined by the effective stress Mohr circle will touch the mobilised shear strength envelope. At this point, the soil compression will be triggered. During compression, the effective stress Mohr circle is growing in diameter and always at the point of touching the mobilised shear strength envelope, which rotates with it. The point at which the effective Mohr circle stops enlarging, which is when the dissipation of the pore water pressure has ceased the circle, is located slightly below the envelope. In addition, at this point, the settlement is completely stopped unless there is another change in the effective stress condition. Soil compression is an irrecoverable volume change, and therefore, it must be taking place after the yield point in the stress–strain curve. As the mobilised friction angle, ϕ'_{mob}, increases with soil compression when the soil particles move to a denser arrangement, the mobilised shear strength, τ_{mob}, also increases with the collapse.

If the deviator stress is released at one stage of the shearing, like at point A in Figure 2.13a, the mobilised friction angle stays constant even though the shear stress may decrease thereafter. Simultaneously, the mobilised shear strength decreases as the net normal stress on the inclined shear plane decreases. As the deviator stress is released, the soil will recover part of the total strain that is elastic while the plastic strain is irrecoverable as discussed in Section 2.3, Chapter 2. Upon reapplication of the deviator stress, the soil will undergo elastic compression until the mobilised shear strength is exceeded. From this point onwards, the combination of plastic and elastic compression resumes until the peak strength is reached like at point P in Figure 2.13a.

3.2.4 General structure of RMYSF

When a soil undergoes compression, its mobilised shear strength increases from zero up to the soil shear strength at failure, which corresponds to the peak deviator stress in the triaxial compression test on a dense specimen. The increase in the mobilised shear strength is marked by the increase in the effective mobilised friction angle, ϕ'_{mob}, during the application of the deviator stress. As the soil yield envelope is a curvilinear surface envelope as shown in Figure 3.6, soil yielding cannot be correlated to the mobilised friction angle, ϕ'_{mob}, because it varies with effective stress unless the yield envelope is a plane surface. Referring to Figure 3.6, effective stress for Mohr stress circle 2 is higher than for Mohr stress circle 1. Therefore, for consistency in the relationship, the axial strain during soil compression needs to be correlated to the effective mobilised minimum friction angle, $\phi'_{min_{mob}}$, associated to a yield surface envelope, which is shown in Figure 3.6, rather than the mobilised friction angle, ϕ'_{mob}.

During a triaxial compression test on a saturated specimen, the confining pressure, σ_3, which determines the minor principal effective stress, $\sigma'_3 = \sigma_3 - u_w$ is maintained constant when the deviator stress is increased from q_{mob_1} to q_{mob_2} as shown in Figure 3.8. The Mohr stress circle grows from y1 to y2 and correspondingly the yield surface envelope is rotated from position 1 to position 2 as shown in Figure 3.9. The major principal effective stresses of the Mohr stress circles are given by Equations 3.5 and 3.6, respectively.

$$\sigma'_{1y1} = \sigma_3 + q_{mob1} - u \tag{3.5}$$

$$\sigma'_{1y2} = \sigma_3 + q_{mob2} - u \tag{3.6}$$

The change in the effective mobilised minimum friction angle, $\Delta\phi'_{min_{mob}}$, is assumed to be equal to the change in the effective mobilised friction angle, $\Delta\phi'_{mob}$, as illustrated in Figure 3.9. Consider a line connecting the origin to any point on the surface envelope.

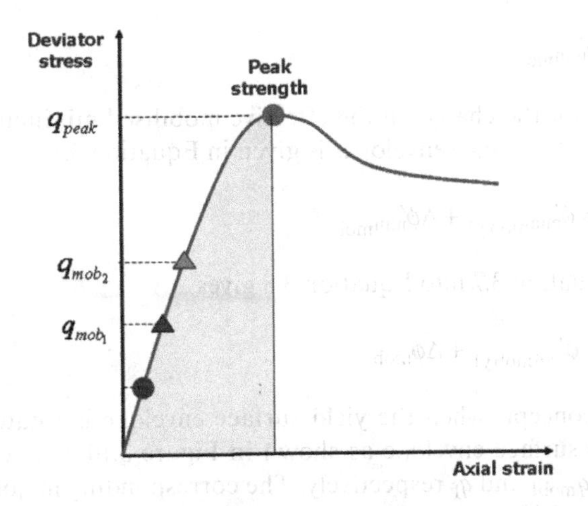

Figure 3.8 Typical stress–strain behaviour from the triaxial compression test.

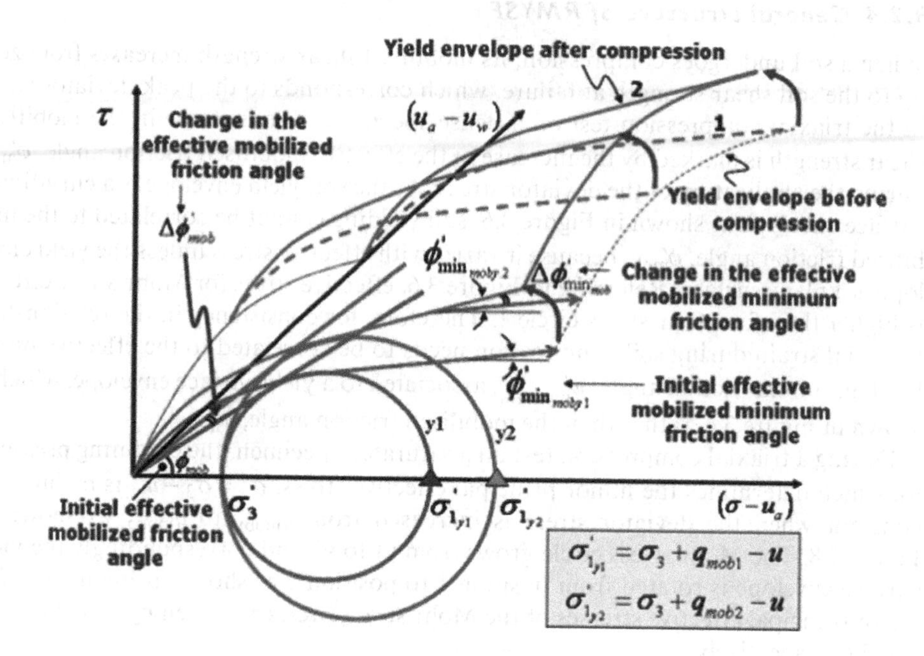

Figure 3.9 The rotation of the yield surface envelope as the deviator stress increases from q_{mob1} to q_{mob2} and the assumption that $\Delta\phi'_{\mathrm{min_{mob}}} = \Delta\phi'_{\mathrm{mob}}$ (Md Noor, 2006).

By considering the envelope to rotate as a rigid body, the change in the inclination of this line from the horizontal must be equal to the change in the inclination of the linear section of the surface envelope from the horizontal. The referred point on the surface envelope is taken as its intersection with the tangent line from the origin to the Mohr circle. This intersection point is assumed to remain the same when the Mohr circle enlarged, provided their values of σ'_3 are the same.

$$\Delta\phi'_{\mathrm{mob}} = \Delta\phi'_{\mathrm{min_{mob}}} \tag{3.7}$$

The expression for the change in the effective mobilised minimum friction angle between the two yield surface envelopes is given in Equation 3.8.

$$\phi'_{\mathrm{min_{mob\,y2}}} = \phi'_{\mathrm{min_{mob\,y1}}} + \Delta\phi'_{\mathrm{min_{mob}}} \tag{3.8}$$

Substituting Equation 3.7 into Equation 3.8 gives

$$\phi'_{\mathrm{min_{moby2}}} = \phi'_{\mathrm{min_{moby1}}} + \Delta\phi'_{\mathrm{mob}} \tag{3.9}$$

Based on this concept, when the yield surface envelope is rotated to the soil shear strength failure surface envelope as shown in Figure 3.10, the corresponding deviator stresses are q_{mob1} and q_{f} respectively. The corresponding major principal effective stresses in Figure 3.10 are calculated as follows:

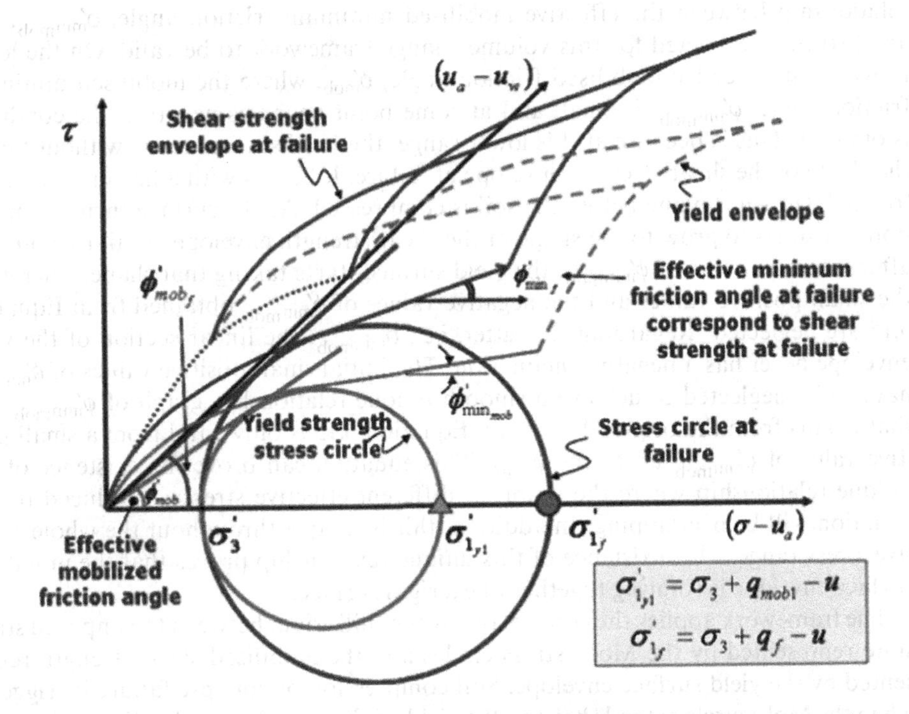

Figure 3.10 Rotation of the yield envelope to the shear strength envelope at failure and the assumption that $\phi'_{mobf} - \phi'_{mob} = \phi'_{minf} - \phi'_{minmob}$ (Md Noor, 2006).

$$\sigma'_{1y1} = \sigma_3 + q_{mob1} - u \tag{3.10}$$

$$\sigma'_{1f} = \sigma_3 + q_f - u \tag{3.11}$$

The effective mobilised minimum friction angle of the yield surface envelope, ϕ'_{minmob}, with reference to the effective mobilised minimum friction angle at failure, ϕ'_{minf}, (i.e., corresponding to the failure shear strength envelope) can be obtained as follows:

$$\phi'_{minf} = \phi'_{minmob} + \Delta\phi'_{minmob} \tag{3.12}$$

$$\Delta\phi'_{minmob} = \left(\phi'_{minf} - \phi'_{min}\right) \tag{3.13}$$

$$\phi'_{minf} = \phi'_{minmob} + \left(\phi'_{mobf} - \phi'_{mob}\right) \tag{3.14}$$

Rearranging,

$$\phi'_{minmob} = \phi'_{minf} - \left(\phi'_{mobf} - \phi'_{mob}\right) \tag{3.15}$$

This increase in the effective mobilised minimum friction angle, $\Delta\phi'_{minmob}$, with respect to axial strain as the soil structure is compressed is the basic principle used in the proposed shear strength–volume change framework, i.e., RMYSF. A unique relationship between the effective mobilised minimum friction angle, $\phi'_{min_{mob}}$, and axial strain is expected for this volume change framework to be valid. On the lower range of the effective mobilised friction angle, ϕ'_{mob}, where the mobilised minimum friction angle, $\phi'_{min_{mob}}$, is small and at some point becomes negative, the condition is omitted. This is because at this lower range, the yield surface turns without taking the shape of the shear strength envelope at failure. It begins with a flat surface on the $(\sigma - u_a):(u_a - u_w)$ plane before the soil is compressed. As the surface turns from this point, it starts to grow to the shape of the shear strength envelope at failure and only after a certain value of $\phi'_{min_{mob}}$, the yield surface starts taking that shape. Therefore, the small positive values and the negative values of $\phi'_{min_{mob}}$ obtained from Equation 3.15 are neglected. Regarding the latter, i.e., $\phi'_{min_{mob}}$, the linear section of the yield envelope never has a negative inclination. The initial small positive values of $\phi'_{min_{mob}}$ have to be neglected to achieve a smooth unique relationship graph of $\phi'_{min_{mob}} - \varepsilon_a$ that begins from the origin. Therefore, Equation 3.15 is only valid from a small positive value of $\phi'_{min_{mob}}$ up to the ϕ'_{minf}. This equation can prove the existence of the unique relationship where the graph of different effective stresses produced by the equation will be overlapping. In addition, this is unique throughout the whole effective stress range. This existence of this unique relationship proves that the mobilised surface envelope is rotating together like a rigid surface.

The framework applies the concept of stress equilibrium between the imposed stress state represented by the Mohr stress circles and the mobilised shear strength represented by the yield surface envelope. Soil compression or collapse failure is triggered when the Mohr circle extends beyond the yield surface envelope indicating an imposed loading stress state that has exceeded the yield strength or the mobilised shear strength of the soil. Once this has happened, the yield surface envelope will rotate about the suction axis until it returns to one-point contact with the Mohr circle. At this instance, stress equilibrium is reinstated and the soil compression stops. As long as the change in stress state makes the Mohr circle move underneath the current yield envelope without touching or extending above it, no failure will be triggered. This is equivalent to the condition that the deviator stress changes below the point of maximum achieved deviator stress during the unloading and reloading process discussed in Section 2.3, where there will be no plastic straining.

The collapse failure can be triggered by the Mohr circle expanding beyond the surface envelope (i.e., loading collapse), or moving laterally parallel to the suction axis towards a lower suction where there is a steep drop in the mobilised shear strength with respect to suction (i.e., wetting collapse) or moving towards a higher suction where the surface envelope is curving down gradually (i.e., drying collapse). The rate of rotation of the yield surface envelope or the rate of increase in the effective mobilised friction angle depends on the compression rate of the soil structure. This in turn depends on the rate of dissipation of pore water pressure in the case of a saturated soil and the dissipation of pore air and pore water pressure in the case of an unsaturated soil.

Consider the case of inundation where this is equivalent to the Mohr circle moving parallel to the suction axis towards a lower suction. The turning of the shear surface envelope about the suction axis is immediate whenever the Mohr circle extends above the envelope. As the Mohr circle moves parallel to the suction axis towards the net stress axis in the case of wetting collapse, the yield surface envelope rotates and when it is close to zero suction, there is an abrupt greater turning of the yield envelope, because the envelope bends steeply downward (described in Section 2.11 and illustrated in Figure 2.32). If the turning angle of the yield envelope represents the volume change, this would then correspond to the massive volume change near saturation. This is a sign of the potential to explain loading, wetting and drying collapse within the same framework. However, this needs justification by the experimental data and if this is true, then the framework can model all the three types of collapse failure including the massive volume reduction near saturation.

Generally, the overall structure of the RMYSF, which incorporates the growing of the mobilised shear strength with the soil compression, can be summarised as follows:

1. The framework is derived from the stress–strain behaviour of the soils obtained from the triaxial test anisotropic compression.
2. The yield surface is taken as the mobilised shear strength envelope.
3. The yield surface represents the same axial strain irrespective of effective stress and suction.
4. Particle breakage is the limit for the application of the framework; therefore, the framework is limited to relatively low-stress levels.
5. The concept of a Mohr stress diagram in the extended Mohr–Coulomb space defining the soil state of stress at a point in the soil mass has already accounted for the combined effect of principal stresses and suction.
6. The state of soil current shear strength is taken as the mobilised shear strength envelope.
7. Soil compression or collapse failure is triggered whenever the state of stress in the soil is greater than the mobilised shear strength. Graphically, this can be represented by the Mohr stress circle extending above the mobilised shear strength envelope.
8. When collapse failure is triggered, the soil will be compressed, the mobilised friction angle will be increased and, in turn, the mobilised shear strength will also increase. Graphically, this is represented by the rotation of the yield surface envelope about the suction axis until the envelope is at the one-point contact with the Mohr stress circle. At this point, the soil is in a state of stress equilibrium and the compression of the soil has stopped.
9. Extension of the Mohr circle above the mobilised shear strength envelope can be brought about either by a Mohr stress circle enlargement, which is referred to as loading collapse, or by the Mohr stress circle moving towards a lower suction, which is referred to as wetting collapse.
10. Loading collapse is driven by the increase in the vertical load, which results in an increase in the major principal net stress, and wetting collapse is caused by inundation, which decreases the suction in the soil.
11. The framework incorporates the effect of strain hardening when it considers the mobilised shear strength surface envelope to rotate with soil compression.
12. The framework also incorporates the soil elastic–plastic response, because it is being developed from the soil stress–strain behaviour.
13. The framework is developed from the soil anisotropic compression behaviour obtained from conducting consolidated drained or consolidated undrained triaxial tests.

3.2.5 Unique relationship between mobilised minimum friction angle and axial strain

The normal form of the constitutive equations for volume change relates deformation state variables to the stress state variables. This relationship requires the soil properties, which are usually evaluated experimentally and appear as a volumetric deformation coefficient. However, in this volume change framework, there is a slight change in the approach where the influence of soil properties and the imposed stress are already incorporated into the shear strength–volume change framework. The modelling of soil compression in the framework produces an increase in the mobilised minimum friction angle, $\phi'_{min_{mob}}$. Therefore, it is anticipated that there is a unique relationship between the mobilised minimum friction angle, $\phi'_{min_{mob}}$, and the vertical or % axial strain, $\%\varepsilon_a$, which is taken as the deformation state variable.

This unique relationship, $\phi'_{min_{mob}} - \%\varepsilon_a$, during the specimen compression, will be inspected in triaxial tests on saturated and unsaturated specimens in the strain range up to the point of failure. This will only cover the case of loading collapse at different effective stress and suction values. There were no triaxial tests involving inundation of unsaturated specimens in the test programme and therefore, the soil compression due to suction decrease cannot be validated here. Likewise, there were no tests that involved specimen drying.

This unique relationship can be obtained using two techniques:

1. Technique A is doing it manually. First, by selecting the axial strains, the mobilised shear strength envelopes are to be drawn. From each axial strain selected determine the corresponding deviator stress for the respective stress–strain curves. Then draw the Mohr circles related to each axial strain and subsequently draw the mobilised shear strength envelope using Equations 2.16 and 2.17. Note the mobilised minimum friction angle, $\phi'_{min_{mob}}$, and the corresponding % axial strain for each envelope. Thence, draw the unique relationship, $\phi'_{min_{mob}} - \%\varepsilon_a$.
2. Technique B is the determination of $\phi'_{min_{mob}} - \%\varepsilon_a$ by applying the developed formulae. This is started by applying Equation 2.15 to determine effective mobilised friction angle, ϕ'_{mob}, for every deviator stress and for every stress–strain. Then, the effective mobilised friction angle, ϕ'_{mob_f}, at peak deviator stress or at failure point is calculated using Equation 2.15 for every stress–strain curve, i.e., using q_A equal to maximum deviator stress. This is followed by drawing of the curvilinear shear strength envelope at failure using Equations 2.16 and 2.17 and determining the ϕ'_{min_f}. Determine $\phi'_{min_{mob}}$ for each deviator stress according to Equation 3.15 for a stress–strain curve. Then, finally, plot $\phi'_{min_{mob}}$ versus axial strain, ε_a, and neglect the section of the graph that has a negative value.

3.2.6 Defining yield surface envelopes according to the RMYSF

The RMYSF is based on a unique relationship between the mobilised minimum friction angle, $\phi'_{min_{mob}}$, and the axial strain, ε_a, obtained from triaxial tests. This is because the existence of this unique relationship for different net confining pressures (i.e., net normal stress for unsaturated conditions and effective stress for saturated conditions) will allow the mobilised shear strength envelope to be considered as the yield surface. This yield surface represents identical axial strains irrespective of the different values of net confining stress as long as they are in the range of low-stress levels where the occurrence of particle breakage is disregarded. The absence of particle breakage during shearing at different net confining pressure is assumed to produce failure at identical axial strains. This is the primary requirement to obtain a unique relationship of $\phi'_{\text{minmob}} - \varepsilon_a$. Failure deviator stress attained at a higher axial strain indicates the occurrence of particle breakage during shearing (Bishop, 1966, 1972; Billam, 1972), and particle breakage will invalidate the applicability of the framework. Also, the graph of ε_v versus ε_a must show the occurrence of dilation during shear, because particle breakage will suppress the specimen from dilating. Then the volume change behaviour is obtained from the relationship between axial strain, ε_a, and volumetric strain, ε_v.

The following procedure describes how, according to the hypothetical volume change framework, the yield envelopes at various stages of the shearing stage in triaxial tests at different net confining pressures can be obtained. This will demonstrate that the yield envelope is rotating about the suction axis towards the soil shear strength envelope at failure as the specimens are being compressed by the increasing deviator stress. The position of the yield envelope can be recognised by the value of the mobilised minimum friction angle, ϕ'_{minmob}. The corresponding axial strain, ε_a, represented by the yield envelope can be obtained from the unique relationship of $\phi'_{\text{minmob}} - \varepsilon_a$.

This is the step-by-step procedure to determine mobilised shear strength envelopes and the unique relationship $\phi'_{\text{minmob}} - \%\varepsilon_a$.

Technique A

1. Decide the values (i.e., at certain increments) of % axial strain, $\%\varepsilon_a$, that the mobilised shear strength envelopes are to be drawn. Note that the mobilised shear strength envelopes are the yield surfaces and they are representing a specific $\%\varepsilon_a$.
2. Determine the corresponding deviator stresses, q for each $\%\varepsilon_a$ from the stress–strain curves, i.e., for every stress–strain curve that is representing a certain effective stress.
3. From the deviator stresses, q, draw the Mohr circles representing one specific axial strain, $\%\varepsilon_a$ at different effective stresses.
4. From the drawn Mohr circles, draw the best-fit curvilinear mobilised shear strength envelope using Equations 2.16 and 2.17.
5. Note the value of the ϕ'_{minmob} for the drawn mobilised shear strength envelope.
6. Note the $\%\varepsilon_a$ that the mobilised shear strength envelope represents.
7. Plot the unique relationship, $\phi'_{\text{minmob}} - \%\varepsilon_a$.

Technique B

1. Plot deviator stress versus axial strain for the tests at different net confining pressures, i.e., the stress–strain curves from CD or CIU triaxial tests.
2. Calculate the effective mobilised friction angle, ϕ'_{mob}, for every deviator stress according to Equation 2.15 and for every stress–strain curve.
3. From the graph of deviator stress versus % axial strain, determine the deviator stress at failure and the corresponding % axial strain.
4. Plot the graph of ϕ'_{mob} against % axial strain for every stress–strain curve.

6. Determine the effective mobilised friction angle, ϕ'_{mobf}, at peak deviator stress or failure point using Equation 2.15 for every stress–strain curve, i.e., using q_A equal to maximum deviator stress.
7. From the triaxial tests at different net confining pressures, draw the curvilinear shear strength envelope at failure using Equations 2.16 and 2.17 and determine the ϕ'_{minf}, which is the inclination of the linear section of the envelope from horizontal, for the soil.
8. Determine the value of the $\phi'_{min_{mob}}$ corresponding to each computed deviator stress according to Equation 3.15 for every test.
9. To draw the yield surface corresponding to a value of $\phi'_{min_{mob}}$, determine the corresponding major principal stress for every case of net confining pressure and draw the Mohr stress circles on the graph of shear stress versus net stress (i.e., unsaturated conditions) or the effective stress (i.e., saturated conditions). Then draw the corresponding curvilinear yield surface envelope using Equations 2.16 and 2.17 by trying the right values for the $(\sigma - u_w)_t$ and τ_t. Repeat this procedure for a higher value of $\phi'_{min_{mob}}$ and this will demonstrate that the yield surface is turning towards the shear strength surface envelope that has $\phi'_{min\,mob} = \phi'_{minf}$.
10. This is to determine the unique relationship, $\phi'_{min_{mob}} - \%\varepsilon_a$. Plot $\phi'_{min_{mob}}$ versus axial strain, ε_a, and neglect the section of the graph that has a negative value of $\phi'_{min_{mob}}$ (i.e., at that stage, the value of $\phi'_{min_{mob}}$ is assumed to be always zero and the mobilised shear strength envelope is flattening as it approaches the plane $(\sigma - u_a):(u_a - u_w)$ in the extended Mohr–Coulomb space). The graphs of $\phi'_{min_{mob}}$ versus ε_a for different net stresses or effective stresses will be overlapping due to their unique relationship. This will prove that the mobilised shear strength envelope is a yield surface, because it represents identical axial strains. The increase of the value of $\phi'_{min_{mob}}$ relative to % axial strain in the graph indicates the rotation of the curvilinear surface of the mobilised shear strength envelope or the yield envelope as the specimen is compressed and the corresponding yield, which is the axial strain, ε_a, can be determined.

However, the framework only considers all the plotted graphs up to the axial strain at failure. This is because, in the field, the soil cannot dilate due to the suppression by the net vertical stress and being constrained from lateral movement.

In other words, the framework is considering that if the soil shear strength is fully mobilised, then its state of shear strength is represented by the shear strength envelope at failure. In this state, the framework assumes that the particles are fully interlocked and there is no room for vertical compression except when there is particle breakage. The advantage of this framework is that it can explain the soil compression or the volume change behaviour due to load increase or inundation within the same framework. Furthermore, this hypothetical RMYSF can be used to predict the stress–strain curve from a triaxial compression test, which will be discussed in the following section.

3.3 Prediction of stress–strain curves by RMYSF

This section describes the procedure to predict the soil stress–strain curves for a triaxial compression test at any net confining stress by the proposed volume change framework. The primary requirements are the unique relationship $\phi'_{\text{minmob}} - \varepsilon_a$ and the shape of the shear strength envelope at failure of the soil. The latter is important because the shape of the mobilised shear strength envelope above a certain minimum angle of rotation (i.e., $\phi'_{\text{minmob}} \geq 0$) is assumed to take its shape. Lower than the effective mobilised minimum friction angle, ϕ'_{minmob}, of zero, the mobilised shear strength envelope is assumed to flatten by maintaining ϕ'_{minmob} equal to zero as it approaches the plane $(\sigma - u_a):(u_a - u_w)$.

The procedure for obtaining the predicted stress–strain curve is as follows:

1. Determine the stress–strain curves for various effective confining stresses for the soil.
2. Determine the shear strength envelope for the soil at failure.
3. Decide the values of the axial strains where the mobilised shear strength envelope is to be drawn.
4. Plot the mobilised shear strength envelopes for various axial strains and this is the intrinsic property of the soil.
5. From the plot of the mobilised shear strength envelopes, the unique relationship of $\phi'_{\text{minmob}} - \varepsilon_a$ for the soil can be determined.
6. From the unique relationship of $\phi'_{\text{minmob}} - \varepsilon_a$, select the effective mobilised minimum friction angles at certain increments and determine the corresponding axial strains, e.g., ϕ'_{minmob} equals 10°, 20°, 30° and 34°, and the corresponding axial strains are $\varepsilon_{a10°}$, $\varepsilon_{a20°}$, $\varepsilon_{a30°}$ and $\varepsilon_{a34°}$. This is illustrated in Figure 3.11a.
7. In the $\tau:(\sigma - u_a)$ space for unsaturated conditions and $\tau:(\sigma - u_w)$ for saturated conditions draw the shear strength envelope at failure and the yield surfaces at the values of ϕ'_{minmob} selected in step 1. Note that the yield surfaces and the shear strength envelope at failure for either saturated or unsaturated conditions represent the axial strains $\varepsilon_{a10°}$, $\varepsilon_{a20°}$, $\varepsilon_{a30°}$ and $\varepsilon_{a34°}$ respectively as illustrated in Figure 3.11b and c. For the saturated and the unsaturated conditions, the axis of rotation of the yield envelope is perpendicular to the page and passes through the origin and at distance c_s above the origin, respectively. Note that c_s is the apparent cohesion corresponding to the considered suction.
8. Identify the net confining pressure, $(\sigma_3 - u_w)$ or $(\sigma_3 - u_a)$, for which the stress–strain behaviour is to be predicted and mark it on the effective stress or the net stress axes as shown in Figure 3.11b and c for saturated and unsaturated conditions respectively.
9. Draw the predicted Mohr circles corresponding to the yield surfaces and the shear strength envelope at failure. The Mohr stress circles should be touching their respective yield surfaces as illustrated in Figure 3.11b and c.
10. Determine the magnitude of the net vertical stress corresponding to the Mohr circles drawn in step 4.
11. Deduce the corresponding deviator stress based on the equation $q = \sigma'_1 - \sigma_3 - u_a$ for unsaturated conditions and $q = \sigma'_1 - \sigma_3 - u_w$ for saturated conditions. Assume that the corresponding deviator stresses are $q_{10°}$, $q_{20°}$, $q_{30°}$ and $q_{34°}$.
12. Draw the predicted stress–strain curve using the deviator stresses of $q_{10°}$, $q_{20°}$, $q_{30°}$ and $q_{34°}$ against the corresponding axial strains $\varepsilon_{a10°}$, $\varepsilon_{a20°}$, $\varepsilon_{a30°}$ and $\varepsilon_{a34°}$ as illustrated in Figure 3.11d.

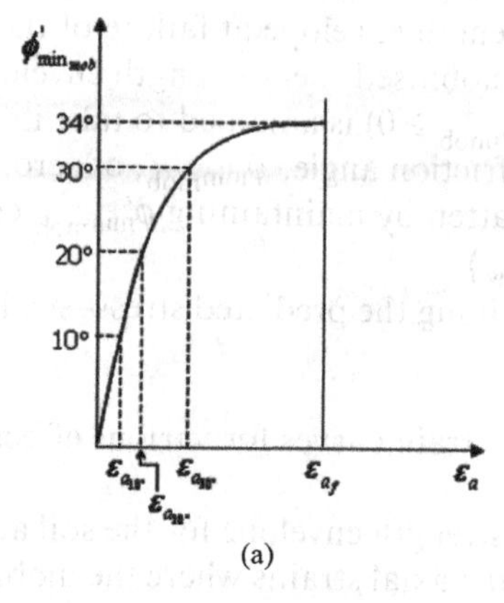

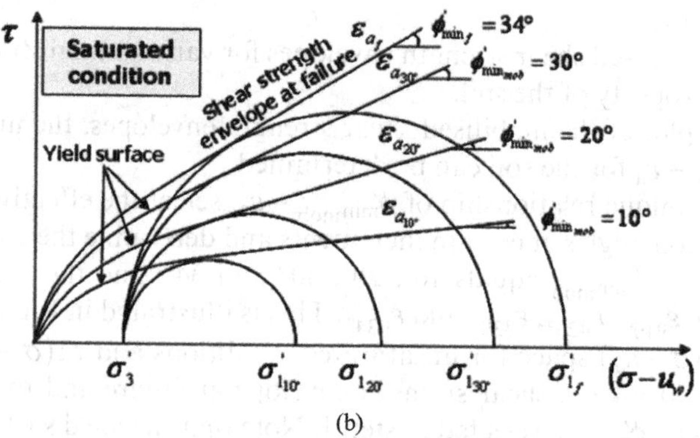

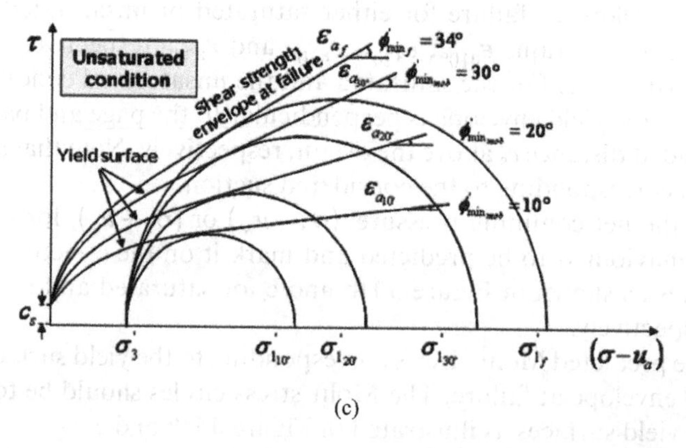

Figure 3.11 Process of predicting the stress–strain curve by the Rotational Multiple Yield Surface Framework. (a) The soil unique relationship of $\phi'_{\min_{\text{mob}}} - \varepsilon_a$ from the plotted curvilinear mobilised shear strength envelope at various % axial strains. (b) Construction of the different inclination of the yield surfaces based on the shape of the shear strength envelope at failure and the drawing of the corresponding Mohr stress circles for saturated conditions. (c) Construction of the different inclination of the yield surfaces based on the shape of the shear strength envelope at failure and the drawing of the corresponding Mohr stress circles for unsaturated conditions.

(Continued)

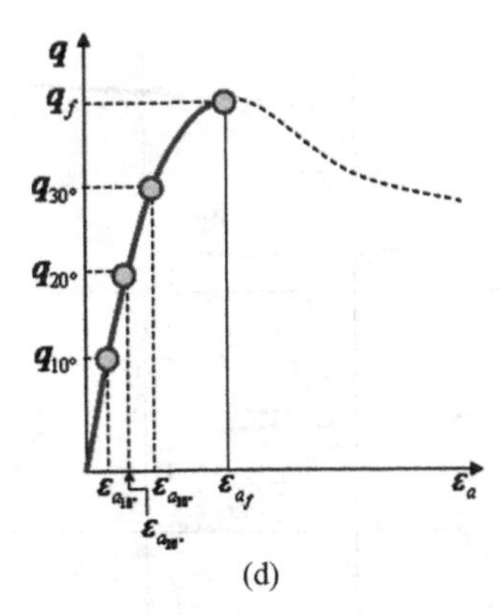

Figure 3.11 (Continued) (d) Predicted stress–strain curve of the soil at a net confining pressure.

From the above procedure, the stress–strain curve up to the failure point at any net confining pressure can be predicted by changing the value applied in step 3. This procedure will be applied for the prediction of the stress–strain behaviour by the RMYSF.

3.3.1 RMYSF prediction of the stress–strain curves for the consolidated drained triaxial test on saturated specimens of limestone gravel

This section is to demonstrate that RMYSF (Md Noor, 2006) can make a good prediction of stress–strain curves at any effective confining pressure. It starts with the derivation of the mobilised shear strength envelopes, and then, it is followed by the prediction of the stress–strain curves at a specific effective stress obtained from laboratory triaxial tests. They are the stress–strain curves obtained from conducting consolidated drained triaxail tests on limestone aggregate of 5 mm nominal diameter obtained from Mootlaw Quarry, near Matfan, Northumberland, United Kingdom. The tests were conducted at effective stresses of 100, 200 and 300 kPa. Figure 3.12 shows the stress–strain curves obtained from conducting consolidated drained triaxial tests. Essentially, the axial strains at failure are 1.1%, 1.2% and 1.4%, respectively. However, RMYSF in this case considers the prediction of deviator stresses at a maximum axial strain of 1.0% only.

The mobilised shear strength envelopes, as in Figure 3.13, were obtained using Technique B in Section 3.2.6. The effective mobilised friction angles, ϕ'_{mob}, for every deviator stress according to Equation 2.15 and for every stress–strain curve were determined and consequently, the graphs of ϕ'_{mob} against % axial strain for every stress–strain curve were plot as shown in Figure 3.14. Then the effective mobilised friction angles, ϕ'_{mobf}, at peak deviator stress were determined using Equation 2.15 for every stress–strain curve (i.e., using q_A equal to maximum deviator stress in the equation). Then, the values of the $\phi'_{min_{mob}}$ corresponding to each computed deviator stress according to Equation 3.15 for every test are determined. Then the angle of $\phi'_{min_{mob}}$ was selected as 10°, 20° and 30° while the $\phi'_{minf} = 34°$. Then in order to draw the yield surface corresponding to a value of the selected $\phi'_{min_{mob}}$, the corresponding deviator

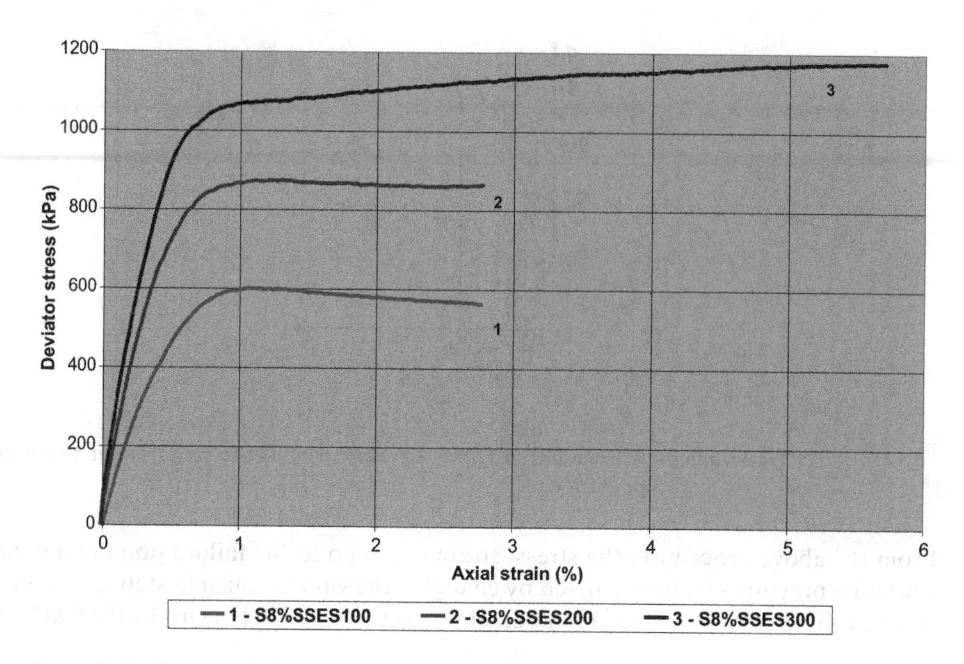

Figure 3.12 Curves of deviator stress versus axial strain for specimens compacted at 8% moisture content (Test series B).

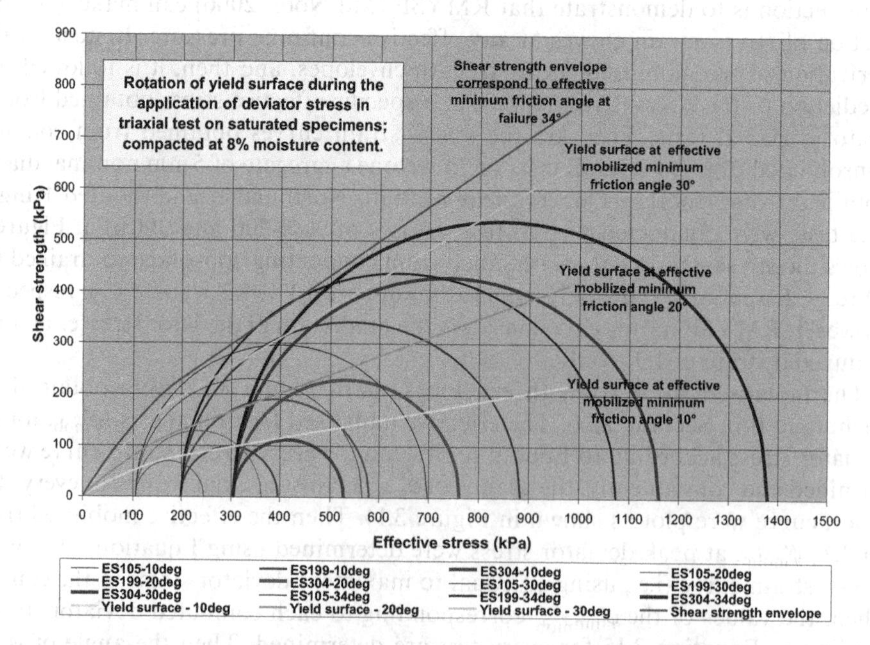

Figure 3.13 Rotation of the mobilised shear strength envelope during the triaxial compression test on saturated specimens compacted at 8% moisture content (Md Noor, 2006).

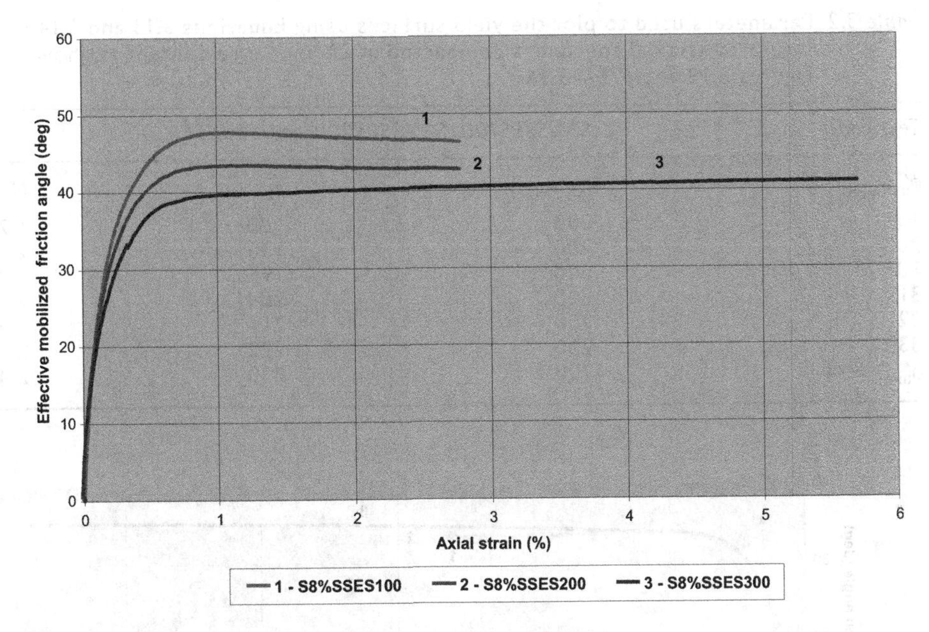

Figure 3.14 Plots of effective mobilised friction angle, ϕ'_{mob}, versus axial strain for triaxial tests on saturated specimens compacted at 8% moisture content using Equation 2.15 (Md Noor, 2006).

stress, q, and the major principal stress, σ'_1 for every case of net confining pressure are determined and the values are presented in Table 3.1. Then, with those values, the Mohr stress circles are drawn on the graph of shear stress versus net stress (i.e., unsaturated conditions) or the effective stress (i.e., saturated conditions). This is followed by drawing the corresponding curvilinear yield surface envelope using Equations 2.16 and 2.17 by trying the right values for the $(\sigma - u_w)_t$ and τ_t. Repeat this procedure for

Table 3.1 Parameters used to plot the Mohr circles corresponding to $\phi'_{min_{mob}}$ of 10°, 20°, 30° and 34° for saturated triaxial specimens compacted at 8% moisture content as shown in Figure 6.18

Test code	*S8%SSES100*			*S8%SSES200*			*S8%SSES300*		
Effective stress (kPa)	100			200			300		
Cell pressure (kPa)	590			589			690		
Back pressure (kPa)	490			390			390		
ϕ_{mob_f}	47.7			43.4			39.7		
$\%\varepsilon_a$ at failure	1.0%			1.0%			1.0%		
$\phi'_{min_{mob}}$	ϕ'_{mob}	q	σ'_1	ϕ'_{mob}	q	σ'_1	ϕ'_{mob}	q	σ'_1
10°	23.8	140.0	243.2	19.4	197.3	395.1	15.7	223.8	524.3
20°	33.8	257.1	359.5	29.4	381.3	578.6	25.7	457.5	756.2
30°	43.8	461.1	563.5	39.4	684.2	881.1	35.7	857.4	1163.3
$\phi'_{min_f} = 34°$	47.7	600.0	705.4	43.4	871.0	1064.8	39.7	1074.0	1376.1

Table 3.2 Parameters used to plot the yield surfaces using Equations 3.13 and 3.14 for saturated triaxial specimens compacted at 8% moisture content as shown in Figures 6.18 and 6.34–6.36

Test code	*S8%SSES100, S8%SSES200 and S8%SSES300*		
$\phi'_{\min_{mob}}$	$(\sigma - u_w)_t$	τ_t	N
10°	200	80	1.788
20°	200	130	2.272
30°	200	195	2.452
31°	200	204	2.43
32°	200	213	2.42
33°	200	222	2.41
$\phi'_{\min f} = 34°$	200	230	2.419

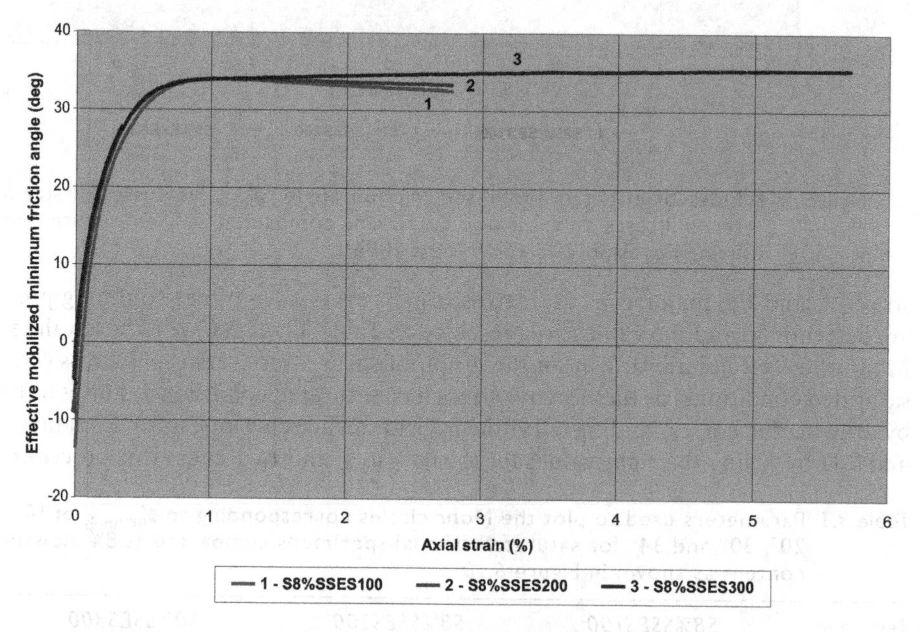

Figure 3.15 Unique graphs of $\phi'_{\min_{mob}} - \varepsilon_a$ for triaxial tests on saturated specimens compacted at 8% moisture content and sheared at effective stresses of 105, 201 and 304 kPa (Md Noor, 2006).

a higher value of $\phi'_{\min_{mob}}$ and this will demonstrate that the yield surface is turning towards the shear strength surface envelope at failure that has $\phi'_{\min_{mob}} = \phi'_{\min f}$. The mobilised shear strength envelopes are drawn in Figure 3.13 and the applied parameters used are shown in Table 3.2.

Then, to determine the unique relationship, $\phi'_{\min_{mob}} - \%\varepsilon_a$, the graphs of $\phi'_{\min_{mob}}$ versus % axial strain (i.e., $\%\varepsilon_a$) for every stress–strain curve are plotted, and the section of the graph that has a negative value is neglected. The graphs are shown in Figure 3.15. Essentially, the graphs of $\phi'_{\min_{mob}} - \%\varepsilon_a$ for every stress–strain curve overlapped to prove that the existence of this unique relationship is irrespective of

Table 3.3 The deduced values for the unique relationship of $\phi'_{min_{mob}} - \varepsilon_a$ for triaxial tests on saturated specimens compacted at 8% moisture content plotted in Figure 3.18

$\%\varepsilon_a$	$\phi'_{min_{mob}}$
0	0
0.08	10.0
0.10	12.0
0.20	20.0
0.30	25.5
0.40	29.0
0.43	30.0
0.48	31.0
0.50	31.4
0.54	32.0
0.60	32.7
0.65	33.0
0.70	33.2
0.80	33.7
0.90	33.8
1.00	34.0

the values of effective stress. This is proving that the curved surface mobilised shear strength envelope is rotating upwards monotonously irrespective of the effective stress values like a single body as the axial strain increases. Then the values representing the unique relationship are selected along the overlapping graphs and deduced as in Table 3.3 and the unique relationship graph, $\phi'_{min_{mob}} - \%\varepsilon_a$, is drawn in Figure 3.16.

The next step is to prove that RMYSF can make a good prediction of the soil stress–strain behaviour by comparing to the stress–strain curves obtained from conducting CD triaxial tests. Figures 3.17–3.19 show the plot of the predicted Mohr circles for effective stresses of 100, 200 and 300kPa, respectively. The Mohr circles are drawn to have one-point contact with the mobilised shear strength envelopes. Its diameter represents the deviator stress q, and the corresponding % axial strain is what the mobilised envelope represents. The predicted deviator stresses are presented in Table 3.4. The values are plotted superimposed on to the laboratory stress–strain curves as shown in Figure 3.20. Apparently, the prediction is excellent. This thence substantiates the applicability and the accuracy of the RMYSF in predicting the soil stress–strain response. In addition, the mobilised shear strength envelopes are the intrinsic property of the soil and are valid across the whole effective stress, and therefore, their prediction at any effective stress would be of similar accuracy.

3.3.2 *RMYSF prediction of the stress–strain curves for the consolidated drained triaxial test on unsaturated specimens of limestone gravel*

This section proved that the RMYSF has the same capability to predict the stress–strain curves for unsaturated soils. Consolidated drained triaxial tests were conducted

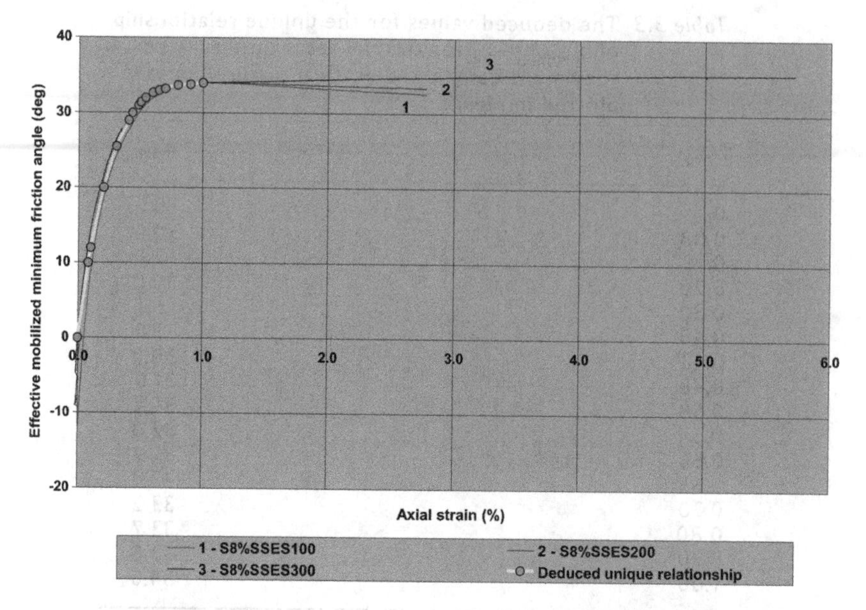

Figure 3.16 Deduced unique relationship of $\phi'_{\text{min}_{\text{mob}}} - \varepsilon_a$ for triaxial compression tests on saturated specimens compacted at 8% moisture content, i.e., note that $\phi'_{\text{min}_{\text{mob}}}$ is determined using Equation 3.15 (Md Noor, 2006).

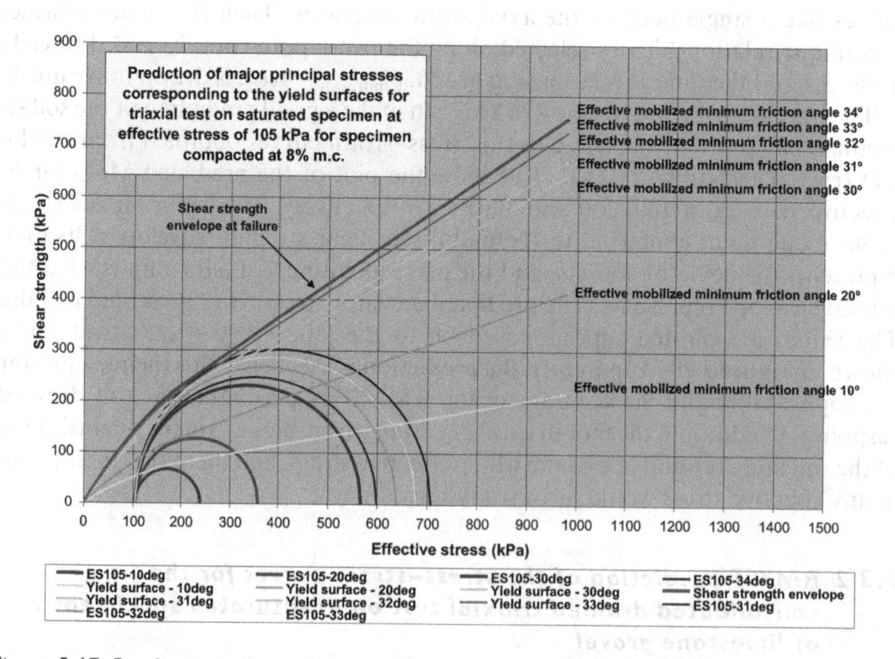

Figure 3.17 Prediction of major principal stresses during yielding in a triaxial compression test at an effective stress of 105 kPa for a specimen compacted at 8% moisture content (Md Noor, 2006).

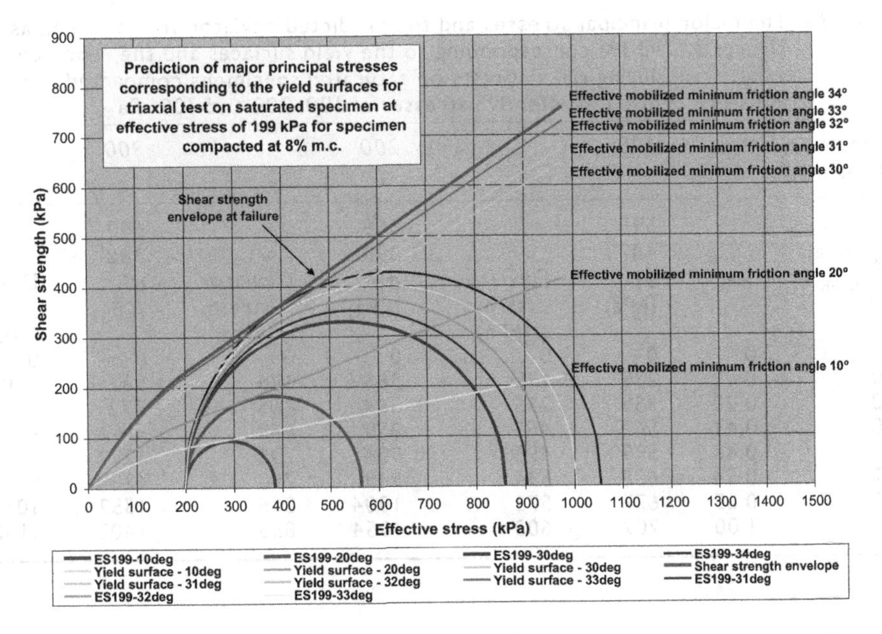

Figure 3.18 Prediction of major principal stresses during yielding in a triaxial compression test at an effective stress of 199 kPa for a specimen compacted at 8% moisture content (Md Noor, 2006).

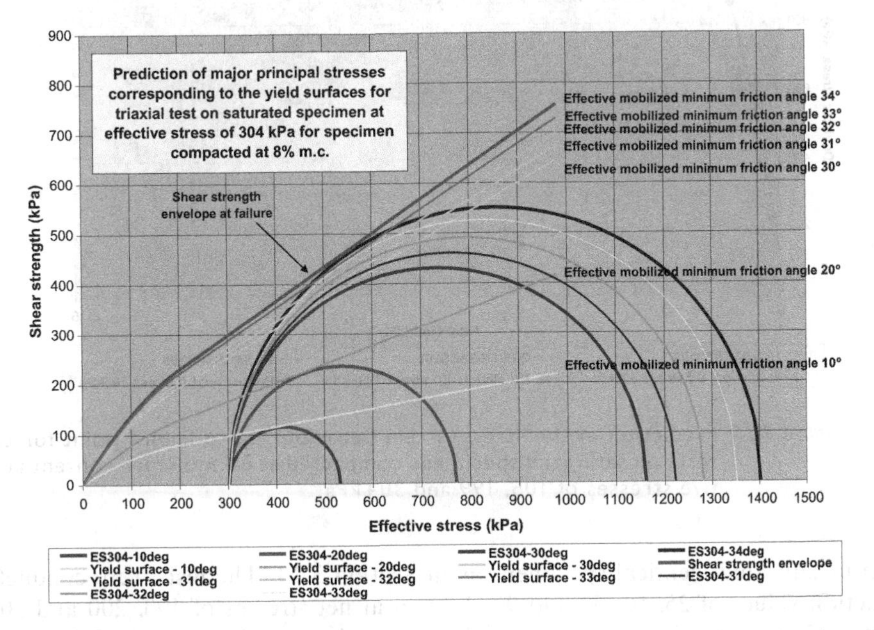

Figure 3.19 Prediction of major principal stresses during yielding in a triaxial compression test at an effective stress of 304 kPa for a specimen compacted at 8% moisture content (Md Noor, 2006).

Table 3.4 The major principal stresses and the predicted deviator stresses (i.e., as in Figures 3.17–3.19) corresponding to the yield surfaces and the represented axial strain during triaxial tests on saturated specimens compacted at 8% moisture content at effective stresses of 100, 200 and 300 kPa

Targeted effective stress (kPa)		*100*		*200*		*300*	
σ_3 (kPa)		591		589		690	
u_w (kPa)		487		390		388	
ϕ'_{minmob} (°)	$\%\varepsilon_a$	σ'_1 (kPa)	Deviator stress q (kPa)	σ'_1 (kPa)	Deviator stress q (kPa)	σ'_1 (kPa)	Deviator stress q (kPa)
0	0	0	0	0	0	0	0
10	0.08	239	135	385	186	532	230
20	0.20	359	255	564	365	772	470
30	0.43	567	463	859	660	1162	860
31	0.48	599	495	904	705	1222	920
32	0.54	638	534	949	750	1282	980
33	0.65	677	573	1004	805	1352	1050
34	1.00	707	603	1054	855	1402	1100

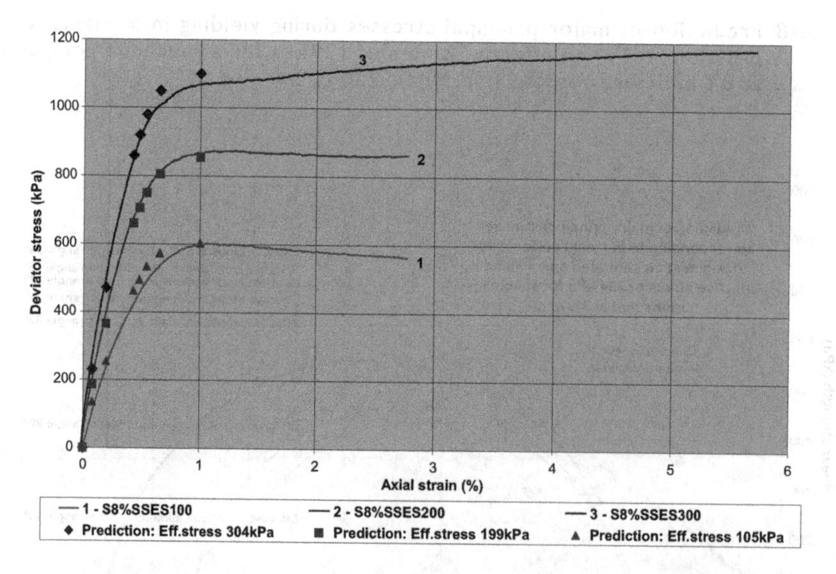

Figure 3.20 Prediction of the stress–strain behaviour up to failure point for triaxial tests on saturated specimens compacted at 8% moisture content at effective stresses of 105, 199 and 304 kPa.

on unsaturated identical specimens as in Section 3.2.1. The tests were conducted at suction values of 25, 50, 60 and 90 kPa and at net stresses of 100, 200 and 300 kPa for each suction using a multi-stage shearing technique. The results of the saturated and unsaturated consolidated drained triaxial tests were interpreted as curved surface shear strength envelope as shown in Figure 3.21 and the seven shear strength

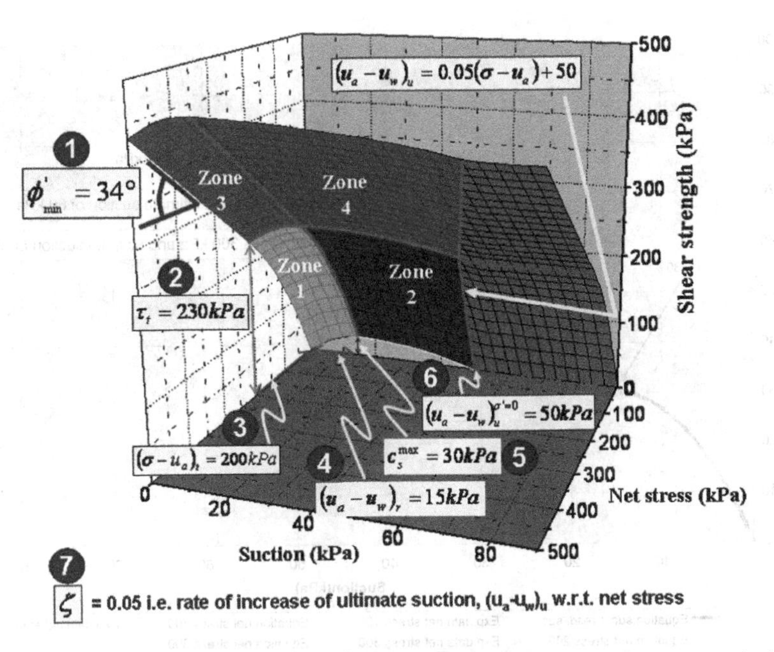

Figure 3.21 Perspective view from underneath the warped-surface extended Mohr–Coulomb envelope of the test material showing the shear strength parameters and the dimensions they represent (Md Noor, 2006).

Table 3.5 The seven shear strength parameters of the test material according to the warped-surface extended Mohr–Coulomb envelope

No.	*Parameter*	*Symbol*	*Value*
1	Maximum apparent shear strength	c_s^{max}	30 kPa
2	Residual suction	$(u_a - u_w)_r$	15 kPa
3	Transition effective stress	$(\sigma - u_w)_t$	200 kPa
4	Transition shear strength	τ_t	230 kPa
5	Effective minimum friction angle at failure	ϕ'_{minf}	34°
6	Rate of increase in ultimate suction with respect to net stress	ζ	0.05
7	Ultimate suction at zero net stress	$(u_a - u_w)_u^{\sigma'=0}$	50 kPa

parameters applied are presented in Table 3.5. Figure 3.22 shows the variation of the apparent shear strength with respect to suction at net stresses of 100, 200 and 300 kPa. Note that the ultimate suction increases with net stresses. In other words, the drop in the apparent shear strength is more gradual at higher net stresses. Figure 3.23 shows the linear variation of ultimate suction with respect to net stress as presented in three-dimensions in Figure 3.21. Figure 3.24 shows the contours of total shear strength variation with respect to suction at various net stresses. In addition, this is the interpretation

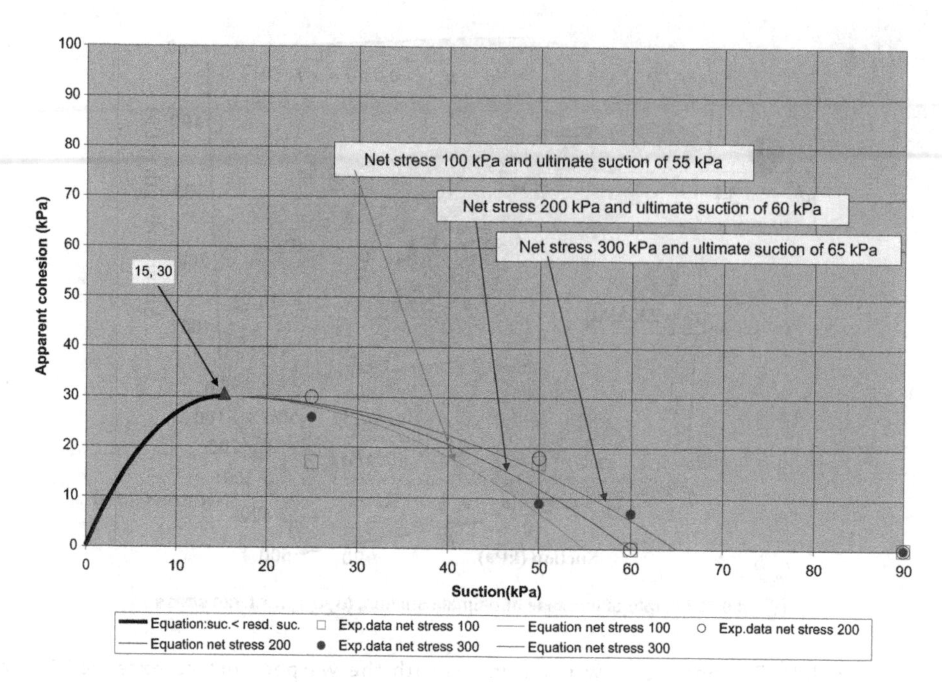

Figure 3.22 The variation of apparent shear strength with respect to suction using Equations 3.4 and 3.7 excluding the saturated shear strength (Md Noor, 2006).

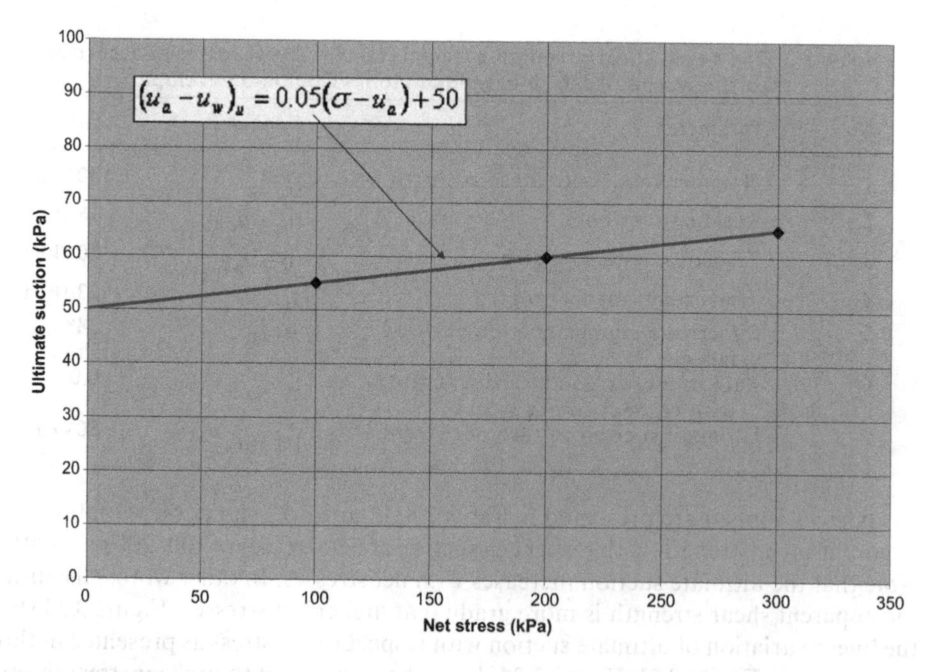

Figure 3.23 The linear variation of ultimate suction with respect to effective stress or net stress from the behaviour in Figure 3.22 (Md Noor, 2006).

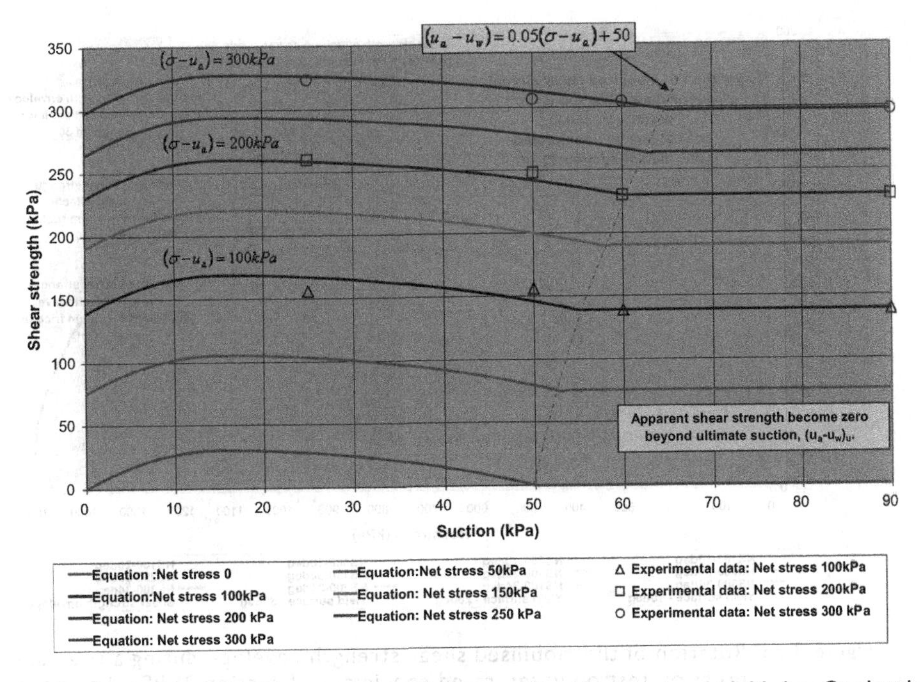

Figure 3.24 Analytical contours for the warped-surface extended Mohr–Coulomb envelope of the test material. Data points are the experimental data (Md Noor, 2006).

from the three-dimensional model in Figure 3.21. Figure 3.25 shows that the mobilised shear strength envelopes have been vertically lifted by 28 kN/m^2 due to the apparent shear strength at suction 25 kPa. The shape of the envelopes is identical to that of the envelopes at saturation as presented in Section 3.3.1, because the specimens are identical.

Figure 3.26 shows the stress–strain curves obtained at a suction of 25 kPa and net stresses of 100, 200 and 300 kPa using the multi-stage shearing technique. Figure 3.27 shows the plots of mobilised friction angle, ϕ'_{mob}, versus axial strain for triaxial tests on unsaturated specimens of suction 25 kPa and sheared at net stresses of 100, 200 and 300 kPa obtained using Technique B in Section 3.2.6. Through the same technique, the graphs of unique relationship $\phi'_{min_{mob}} - \varepsilon_a$ were obtained as shown in Figure 3.28. Then the deduced graph of unique relationship was obtained as in Figure 3.29. The mobilised surface envelope rotates homogeneously as a rigid surface; therefore, the unique relationship for the saturated and unsaturated conditions must be the same. This is substantiated by an identical graph of unique relationship for saturated conditions as shown in Figure 3.16 and for unsaturated conditions at a suction of 25 kPa as shown in Figure 3.29. If plotted to the same horizontal and vertical scale, those unique relationship graphs are in fact overlapping.

Figures 3.30–3.32 show the prediction of the deviator stresses during the plastic–elastic yielding in triaxial compression tests on an unsaturated specimen of suction 25 kPa at net stresses of 100, 200 and 300 kPa, respectively, using the elevated mobilised

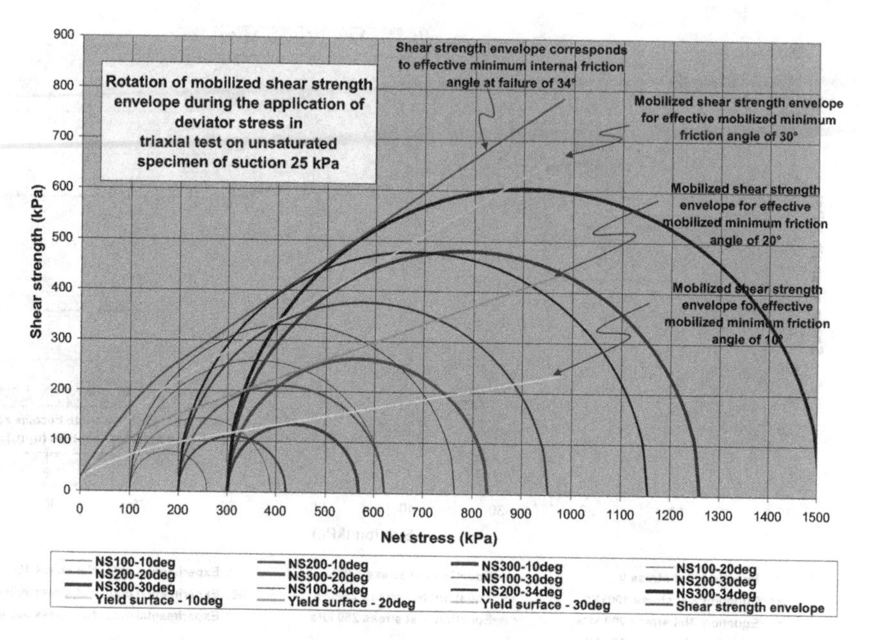

Figure 3.25 Rotation of the mobilised shear strength envelope during a triaxial compression test on unsaturated specimens of suction 25 kPa, i.e., note that all the envelopes are lifted vertically by a shear strength of 28 kN/m^2 that corresponds to the apparent shear strength at zero net stress and suction 25 kPa (Md Noor, 2006).

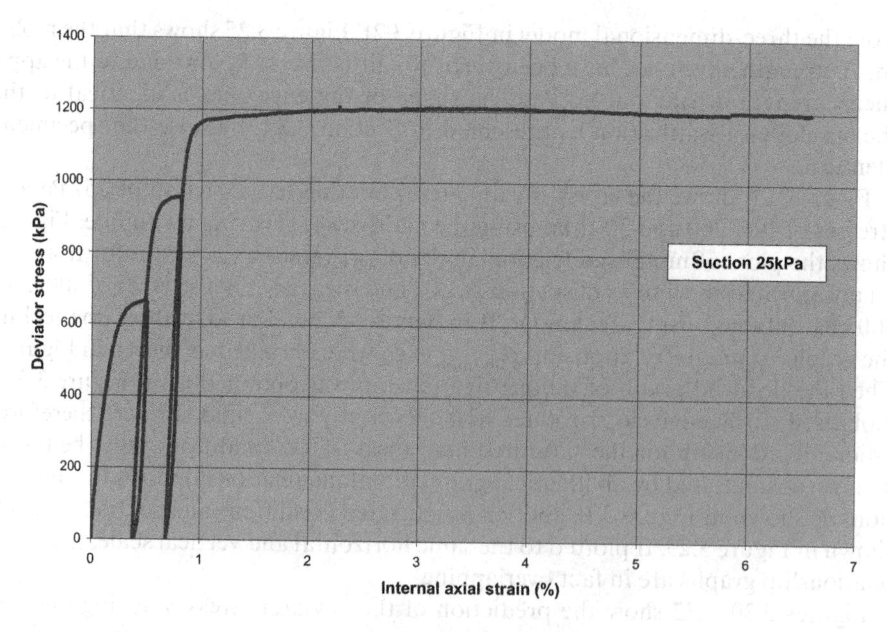

Figure 3.26 Graph of deviator stress versus internal axial strain obtained using the multi-stage shearing technique at a suction of 25 kPa (Test series C) (Md Noor, 2006).

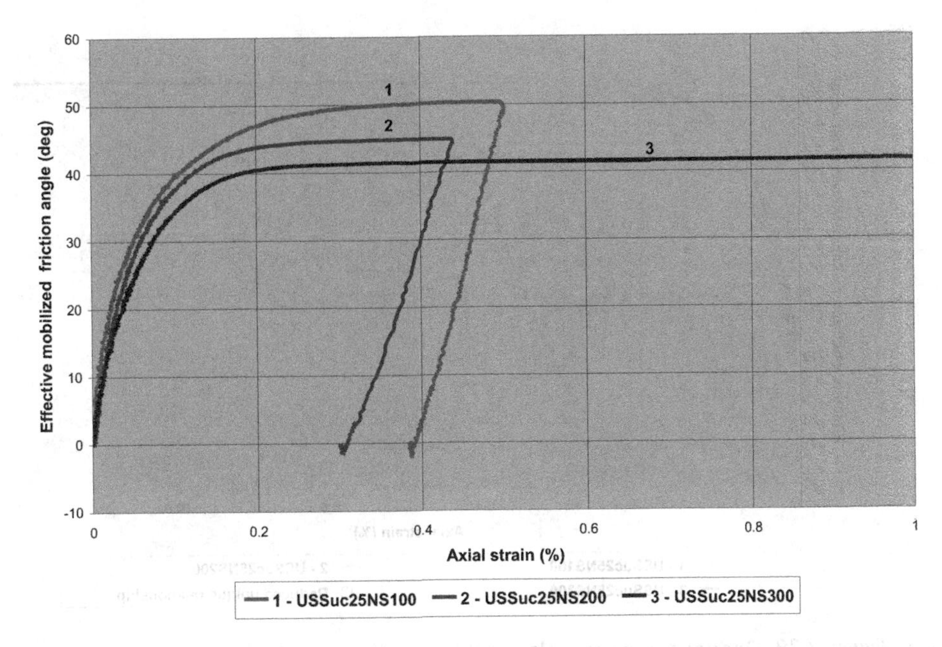

Figure 3.27 Plots of mobilised friction angle, ϕ'_{mob}, versus axial strain for triaxial tests on unsaturated specimens of suction 25 kPa and sheared at net stresses of 100, 200 and 300 kPa (Md Noor, 2006).

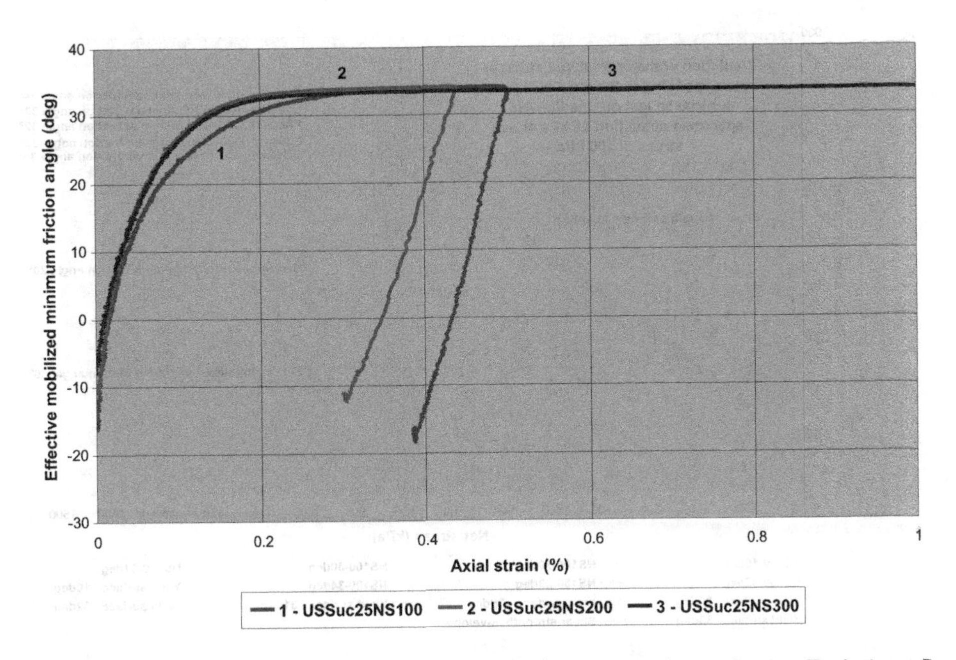

Figure 3.28 The graphs of unique relationship $\phi'_{\text{min}_{\text{mob}}} - \varepsilon_a$ derived using Technique B in Section 3.2.6 for triaxial compression tests on unsaturated specimens of suction 25 kPa and sheared at net stresses of 100, 200 and 300 kPa (Md Noor, 2006).

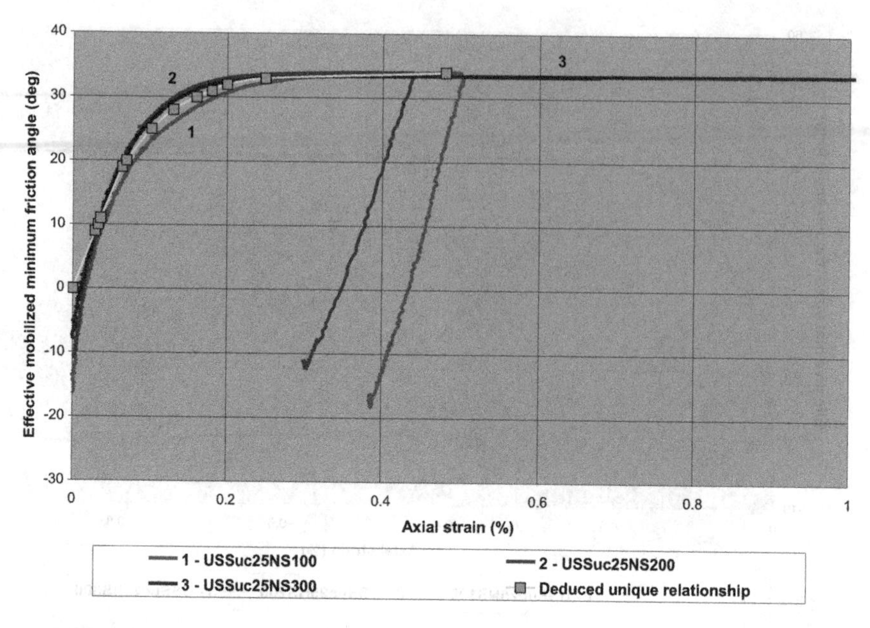

Figure 3.29 Deduced unique relationship of $\phi'_{\text{min}_{\text{mob}}} - \varepsilon_a$ for triaxial compression tests on unsaturated specimens at a suction of 25 kPa at net stresses of 100, 200 and 300 kPa (Md Noor, 2006).

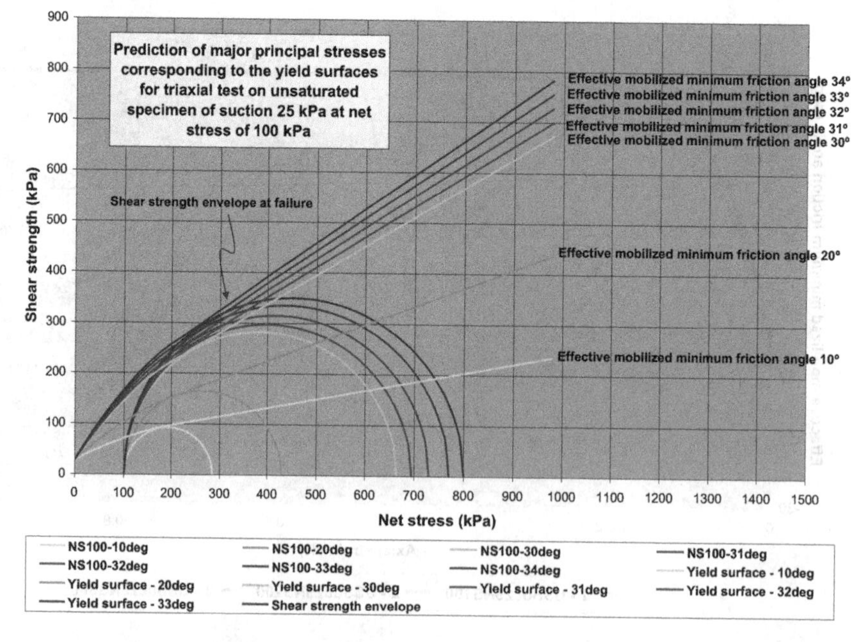

Figure 3.30 Prediction of major principal stresses during yielding in a triaxial compression test on an unsaturated specimen of suction 25 kPa at a net stress of 100 kPa (Md Noor, 2006).

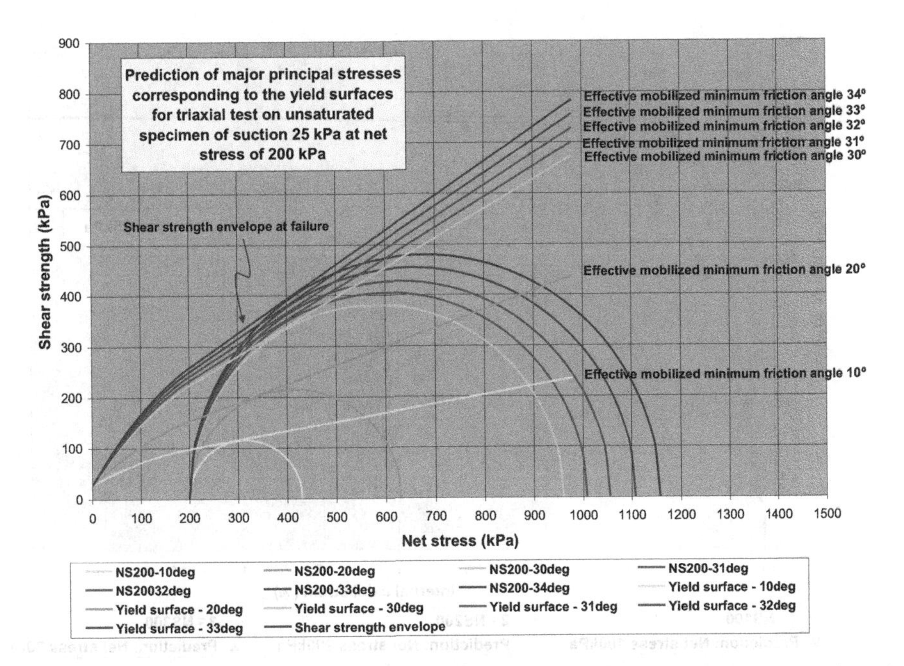

Figure 3.31 Prediction of major principal stresses during yielding in a triaxial compression test on an unsaturated specimen of suction 25 kPa at a net stress of 200 kPa (Md Noor, 2006).

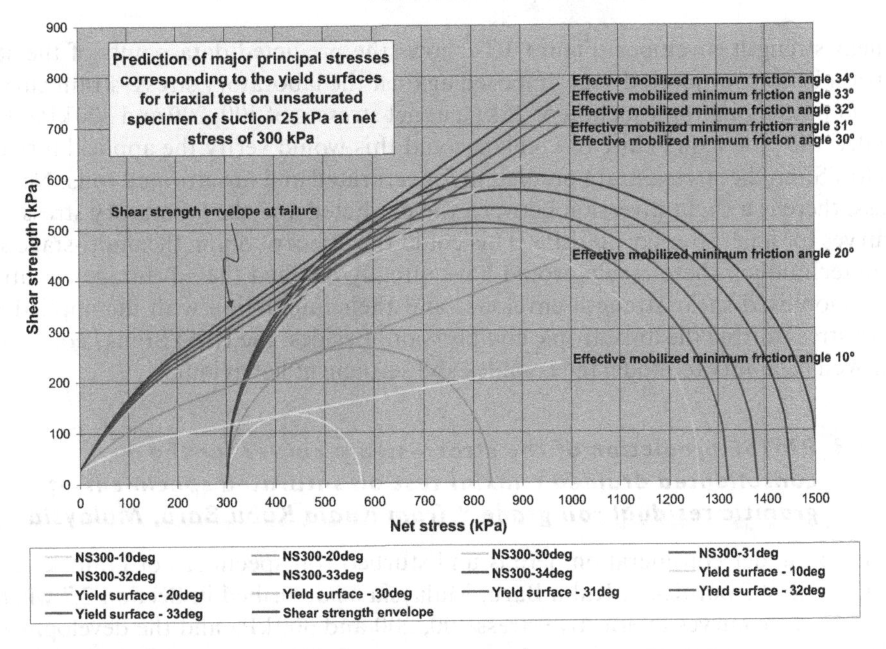

Figure 3.32 Prediction of major principal stresses during yielding in a triaxial compression test on an unsaturated specimen of suction 25 kPa at a net stress of 300 kPa (Md Noor, 2006).

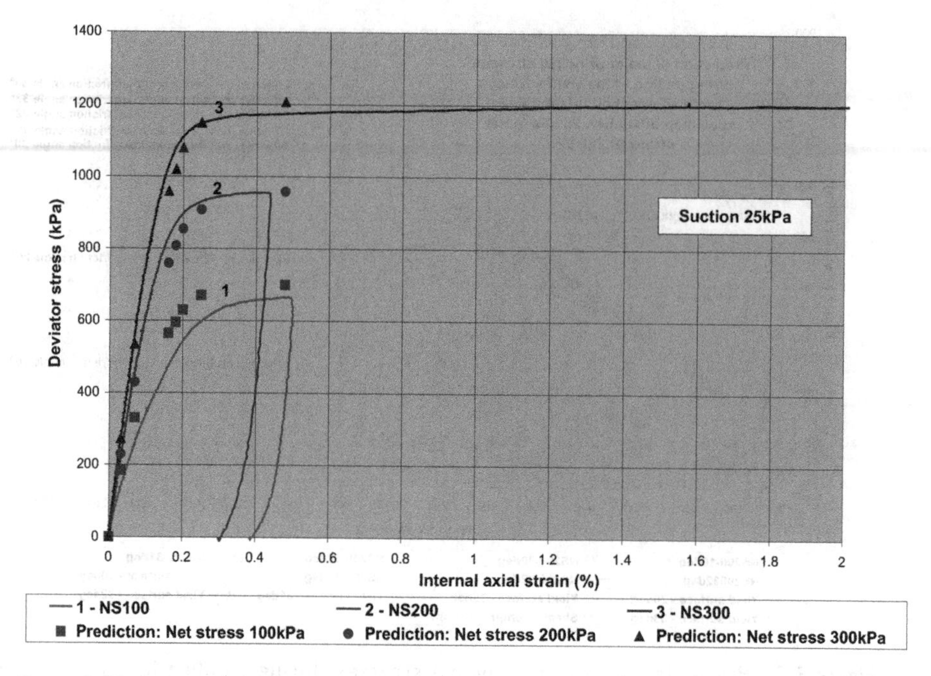

Figure 3.33 Prediction of the stress–strain behaviour up to failure point for a triaxial test on unsaturated specimens of suction 25 kPa at net stresses of 100, 200 and 300 kPa (Md Noor, 2006).

shear strength envelopes. Figure 3.33 shows the predicted data points of the stress–strain behaviour plotted superimposed against the laboratory stress–strain curves up to a failure point for a suction of 25 kPa at net stresses of 100, 200 and 300 kPa. Essentially, the plot shows close resemblance and this would verify the applicability of the RMYSF in the stress–strain prediction for saturated and unsaturated soils. Nevertheless, there is a slight diversion between the predicted and the laboratory stress–strain curves for a net stress of 100 kPa. This could the effect of doing the multi-stage shearing technique. These results would have already justified the advantages of utilising the mobilised shear strength envelopes and their interaction with the applied effective stress during the anisotropic compression. Besides, the RMYSF has conceptually proved its ability to model the complex soil settlement behaviour.

3.3.3 RMYSF prediction of the stress–strain curves for the consolidated drained triaxial test on saturated specimens of granitic residual soil grade V from Kuala Kubu Baru, Malaysia

The soil under consideration here is undisturbed soil specimens of granitic residual soil grade V from Kuala Kubu Baru, Malaysia as described in Section 2.7 where the stress–strain curves at effective stress 100, 200 and 300 kPa and the developments of the mobilised shear strength envelopes are presented (Rahman et al., 2017). The determined mobilised shear strength envelopes will be applied to predict the stress–strain curves at effective stresses of 100, 200 and 300 kPa as shown in Figures 3.34–3.36 and

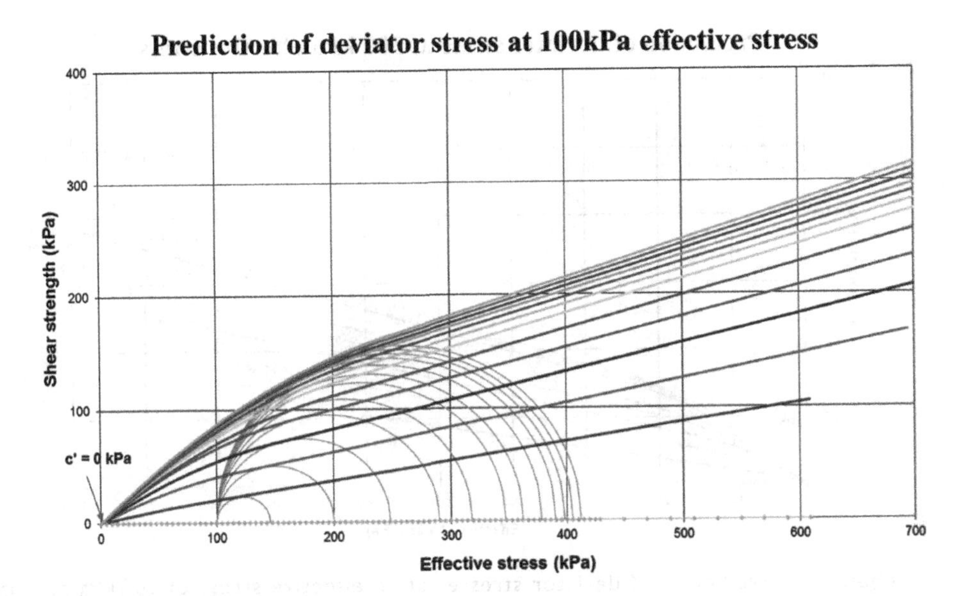

Figure 3.34 Prediction of deviator stresses at an effective stress of 100 kPa at axial strains of 1%–12% at increments of 1%.

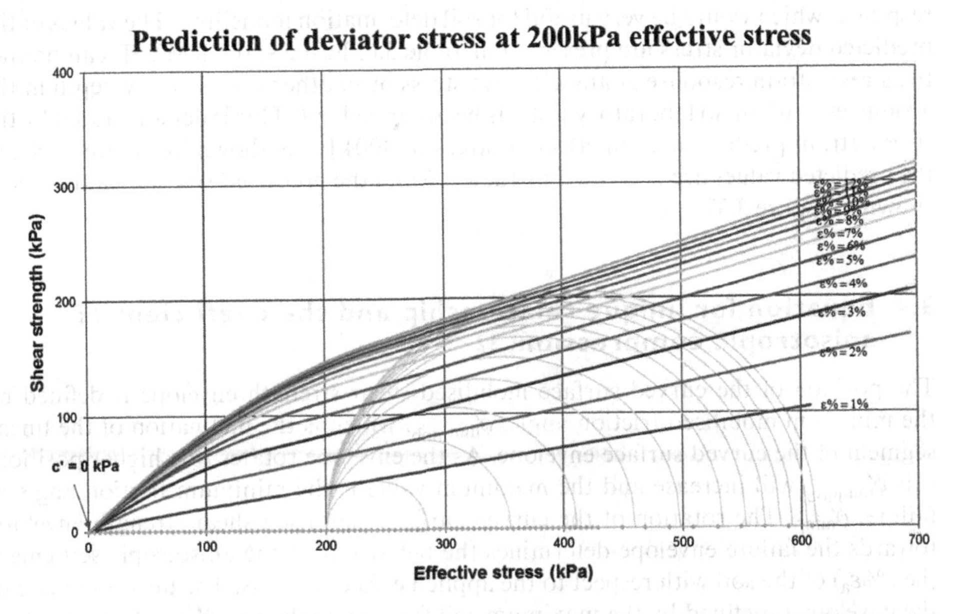

Figure 3.35 Prediction of deviator stresses at an effective stress of 200 kPa at axial strains of 1%–12% at increments of 1%.

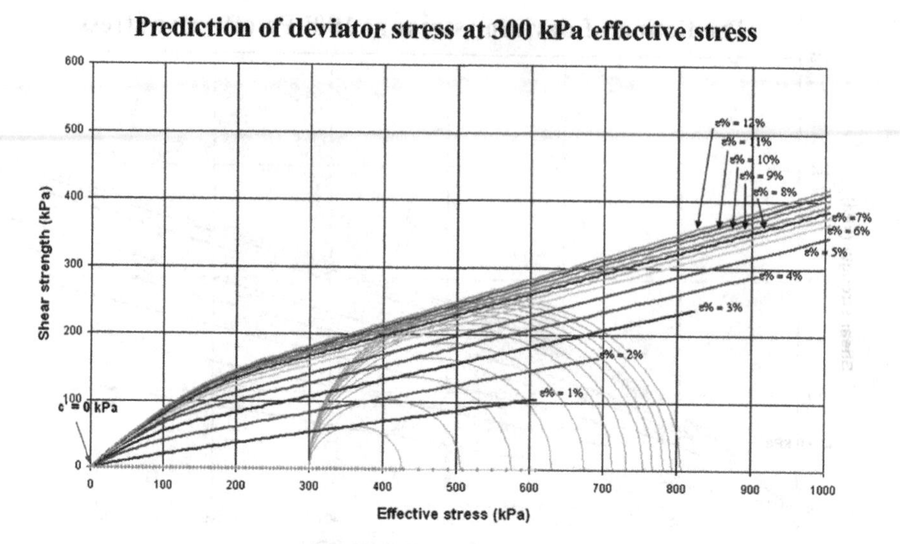

Figure 3.36 Prediction of deviator stresses at an effective stress of 300 kPa at axial strains of 1%–12% at increments of 1%.

then plotted superimposed against the stress–strain curves obtained from the consolidated drained triaxial tests as shown in Figure 3.37. Apparently, the prediction by RMYSF is very close with the laboratory curves.

This again substantiated the greater fidelity in the prediction of the stress–strain response, which would be very useful for soil deformation modelling. The values of the predicted deviator stress are presented in Table 3.6. Besides, the RMYSF can predict the stress–strain response at any effective stress or in other words at any depth in the ground even when no laboratory test has been carried out. This is demonstrated by the stress–strain prediction at an effective stress of 400 kPa as shown in Figure 3.38 and the predicted values are presented in Table 3.6 and the predicted stress–strain curve is shown in Figure 3.37.

3.4 Equation for unique relationship and the coefficient of anisotropic compression, Ω

The position of the curved surface mobilised shear strength envelope is defined by the minimum mobilised friction angle, $\phi'_{\min_{\mathrm{mob}}}$, which is the inclination of the linear segment of the curved surface envelope. As the envelope rotates to a higher position, this $\phi'_{\min_{\mathrm{mob}}}$ will increase and the maximum value is the minimum friction angle at failure, $\phi'_{\min_{\mathrm{f}}}$. The rotation of the curved surface mobilised shear strength envelope towards the failure envelope determines the behaviour of the anisotropic settlement (i.e., $\%\varepsilon_{\mathrm{a}}$) of the soil with respect to the applied effective stress. The maximum rise of the envelope is defined by the maximum soil friction angle, i.e., $\phi'_{\min_{\mathrm{f}}}$. The curvature of the unique relationship graph defines the rate of the soil anisotropic settlement with respect to the applied effective stress.

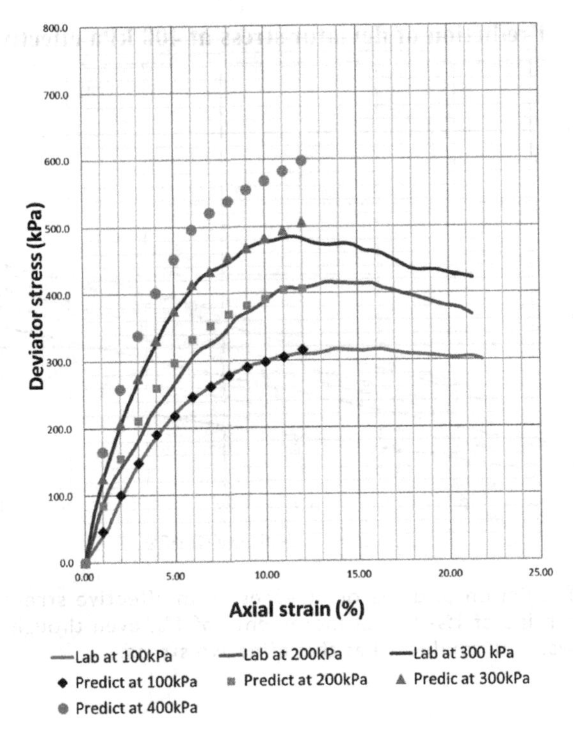

Figure 3.37 The predicted stress–strain response at effective stresses of 100, 200 and 300 kPa (i.e., data points) plotted superimposed against the stress–strain curves obtained from laboratory consolidated drained triaxial tests (i.e., full lines).

Table 3.6 Predicted values of deviator stresses at the corresponding axial strain as represented by the mobilised shear strength envelopes at effective stresses of 100, 200 and 300 kPa

Predicted stress–strain curve				
Axial strain (%)	Effective stress (kPa)			
	100	200	300	400
	Predicted deviator stress (kPa)			
0	0	0	0	0
1	46	85	126	165
2	100	155	206	259
3	148	211	275	338
4	190	260	331	402
5	218	298	375	452
6	247	333	414	497
7	262	353	433	521
8	278	369	455	538
9	290	382	468	555
10	298	392	482	568
11	305	405	494	583
12	315	407	506	598

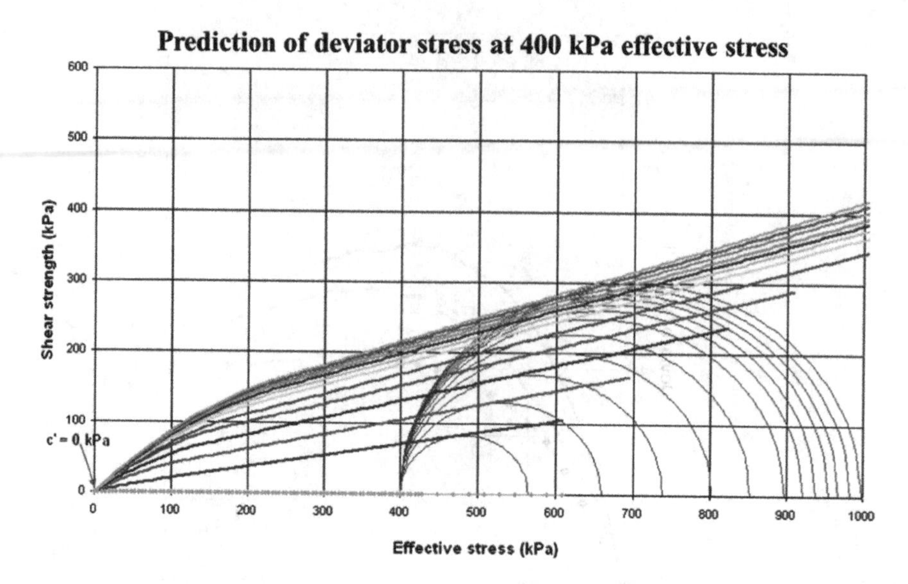

Figure 3.38 Prediction of deviator stresses at an effective stress of 400 kPa at axial strains of 1%–12% at increments of 1%, even though the laboratory test was not conducted at this effective stress.

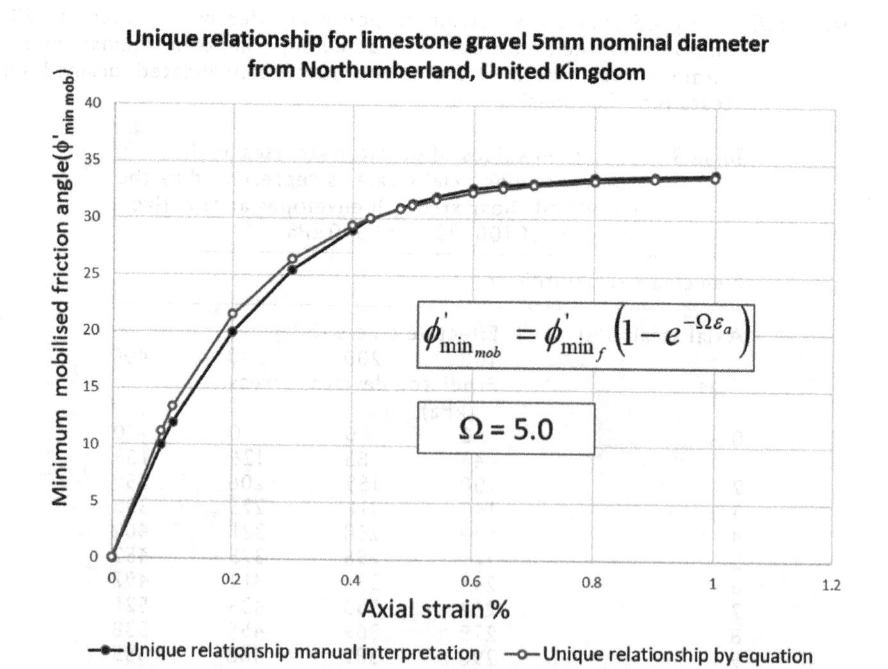

Figure 3.39 Unique relationship for a Blue Mountain limestone aggregate of 5 mm nominal diameter from Matfan, Northumberland, UK (Md Noor, 2006).

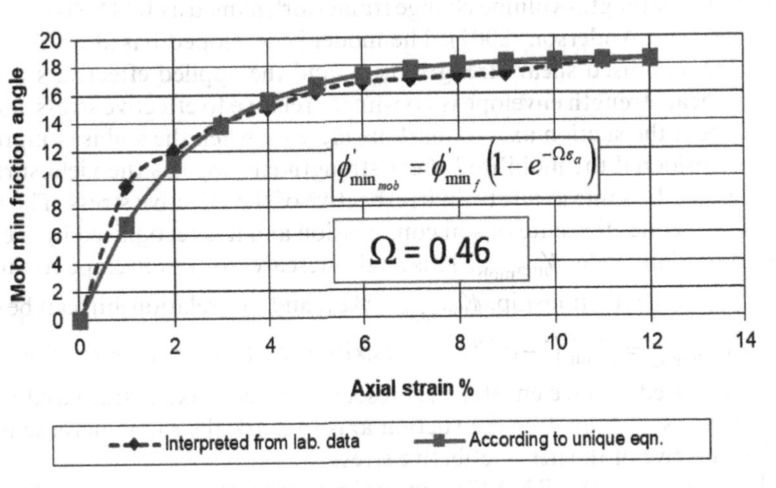

Figure 3.40 Unique relationship for granitic residual soil grade V from Kuala Kubu Baru, Malaysia (Rahman et al., 2017).

Equation 3.16 was introduced by Jais and Md Noor (2009) to represent the unique relationship $\phi'_{\min_{\text{mob}}} - \%\varepsilon_a$, where Ω is called the coefficient of anisotropic settlement of the soil. The higher the value of Ω, the stronger the soil. As examples, the coefficient of anisotropic settlement, Ω, for a Blue Mountain limestone aggregate of 5 mm nominal diameter, is 5.0 with a minimum friction angle at failure, $\phi'_{\min f}$, of 34° as shown in Figure 3.39 (Md Noor, 2006) and Ω is 0.46 for granitic residual soil grade V from Kuala Kubu Baru, Malaysia with a $\phi'_{\min f}$ of 18° (Figure 3.40; Rahman et al., 2017).

$$\phi'_{\min_{\text{mob}}} = \phi'_{\min f}\left(1-e^{-\Omega\varepsilon_a}\right) \tag{3.16}$$

3.5 Summary

This chapter started by reviewing soil volume change models that have been introduced by many researchers. This is the search to understand especially the occurrence of inundation collapse settlement, which occurs under effective stress decrease. This behaviour seems to contradict the effective stress concept of Terzaghi (1936). It is started by Bishop (1959), who tried to resolve this understanding by introducing the effective stress equation for unsaturated soils. Later, the behaviour was tried to be understood by applying two independent stress state variables, which are net stress and suction. The proposed soil volume change models are mainly semi-empirical like the model of Fredlund and Morgenstern (1976). This understanding cannot be resolved even when the critical state model of Alonso et al. (1990) was introduced, which later was denied by Wheeler et al. (2003). Nevertheless, none of those volume change models incorporates the role of shear strength in governing the behaviour. Basically, when the soil is compacted drier than the optimum moisture content, it still has the void spaces for the particles to undergo rearrangement and is very prone to exhibit inundation settlement. That is why even engineered backfill suffers wetting collapse when inundated.

Then a shear strength–volume change framework named as RMYSF was introduced by Md Noor and Anderson (2007). The model is developed based on the interaction between the mobilised shear strength, τ_{mob}, and the applied effective stress, σ'. This mobilised shear strength envelope is non-linear relative to effective stress and it rotates upwards about the suction axis to mark its increase when the soil is compressed. The RMYSF considered the mobilised shear strength envelope as the yield surface and it is representing the same axial strain irrespective of the effective stress. The position of the envelope defines the state of soil compression and it is recognised by the minimum mobilised friction angle, $\phi'_{min_{mob}}$. This angle increases as the envelope rotate upwards. There exist unique relationship, $\phi'_{min_{mob}} - \%\varepsilon_a$, and this relationship can be defined by equation $\phi'_{min_{mob}} = \phi'_{minf}\left(1 - e^{-\Omega\varepsilon_a}\right)$. The existence of this unique relationship proves that the mobilised surface envelope represents a specific axial strain and it rotates as a rigid surface spanning over the suction axis to mark the same increase in the axial strain irrespective of the net or effective stress.

By this technique, the RMYSF can make a good stress–strain prediction at any irrespective stress meaning at any depth in the ground. The RMYSF has proved to be able to make good prediction of soil stress–strain response for saturated and unsaturated conditions. This would be very useful for soil deformation modelling for a greater accuracy.

References

Alonso, E. E., Gens, A., and Hight, D. W. (1987). “General Report: Special problem soils.” *Proceedings of the 9th European Conference on Soil Mechanics and Foundation Engineering*, Dublin, Vol. 3, 1087–1146.

Alonso, E. E., Gens, A., and Josa, A. (1990). “A constitutive model for partially saturated soil.” *Geotechnique*, 40(3), 405–430.

Alonso, E. E., Lloret, A., Gens, A., and Yang, D. Q. (1995). “Experimental behaviour of highly expansive double-structure clay.” *Proceedings of the 1st International Conference on Unsaturated Soils*, Paris, Vol. 1, 11–16.

Barden, L., Madedor, A. O., and Sides, G. R. (1969). “Volume change characteristics of unsaturated clay.” *Journal Soil Mechanics Foundation Engineering, ASCE*, 95, 33–52.

Billam, J. (1972). “Some aspects of the behaviour of granular materials at high pressures.” *Stress-Strain Behaviour of Soils, Proceedings of the Roscoe Memorial Symposium, Cambridge University*, Henley on Thames, UK, 69–80.

Bishop, A. W. (1959). “The principle of effective stress.” *Teknisk Ukeblad*, 106(39), 859–863.

Bishop, A. W. (1966). “The strength of soils as engineering materials.” *Geotechnique*, 16(2), 91–130.

Bishop, A. W. (1972). “Shear strength parameters of undisturbed and remoulded soil specimens.” *Proceedings of the Roscoe Memorial Symposium, Cambridge University*, Henley on Thames, UK, 3–139.

Bishop, A. W., and Blight, G. E. (1963). “Some aspects of effective stress in saturated and unsaturated soils.” *Geotechnique*, 13(3), 177–197.

Escario, V., and Saez, J. (1973). “Measurements of the properties of swelling and collapsing soils under controlled suctions.” *Proceedings of the 3rd International Conference on Expansive Soils*, Haifa, Israel, Vol. 1, 195–200.

Fredlund, D. G. (2000). “The 1999 R.M. Hardy Lecture: The implementation of unsaturated soil mechanics into geotechnical engineering.” *Canadian Geotechnical Journal*, 37, 963–986.

Fredlund, D. G., and Morgenstern, N. R. (1976). "Constitutive relations for volume change in unsaturated soils." *Canadian Geotechnical Journal*, 13(3), 261–276.

Fredlund, D. G., and Rahardjo, H. (1993). *Soil mechanics for unsaturated soils*, Wiley, New York.

Fredlund, D. G., Morgenstern, N. R., and Widger, R. A. (1978). "Shear strength of unsaturated soils." *Canadian Geotechnical Journal*, 15(3), 313–321.

Gallipoli, D., Gens, A., Sharma, R. S., and Vaunat, J. (2003). "An elasto-plastic model for unsaturated soil incorporating the effects of matric suction and degree of saturation on mechanical behavior." *Geotechnique*, 53(1), 123–135.

Jennings, J. E., and Burland, J. B. (1962). "Limitations to the use of effective stresses partly saturated soils." *Geotechnique*, 12(2), 125–144.

Matyas, E. L., and Radhakrishna, H. S. (1968). "Volume change characteristics of partially saturated soils." *Geotechnique*, 18(4), 432–448.

Md Noor, M. J. (2006). "Shear strength and volume change behaviour of saturated and unsaturated soils." Ph.D. Thesis, University of Sheffield.

Md Noor, M. J., and Anderson, W. F. (2007). "A qualitative framework for loading and wetting collapses in saturated and unsaturated soils." *16th South East Asian Geotechnical Conference*, Kuala Lumpur, Malaysia.

Mohamed Jais, I. B., and Md Noor, M. J.. (2009). "Establishing a unique relationship between minimum mobilised friction angle and axial strain for anisotropic soil settlement model." *4th Asia Pacific Conference on Unsaturated Soils*, Newcastle, Australia, Theoretical and Numerical Advances in Unsaturated Soil Mechanics, 775–781.

Powrie, W. (1997). *Soil mechanics; concepts and applications*, Spon Press, Oxford, Great Britain.

Rahardjo, H., and Fredlund, D. G. (2003). "Ko - Volume change characteristics of an unsaturated soil with respect to various loading paths." *Geotechnical Testing Journal*, 26(1), 79–91.

Rahman, A. S. A., Md Noor, M. J., Jais, I. B. M., and Ibrahim, A. (2017) "Prediction of soil anisotropic stress-strain behaviour incorporating shear strength using improvise normalised stress-strain method." *4th International Conference on Civil and Enviromental Engineering for Sustainability*, Langkawi, Malaysia.

Sabbagh, A. (2000). "The lateral swelling pressure on the volumetric behaviour of natural expansive soils deposits." *Proceedings for the Unsaturated Soil for Asia*, Singapore, 709–714.

Sharma, R. S. (1998). "Mechanical behaviour of unsaturated highly expansive clays." Ph.D. Thesis, University of Oxford.

Tadepalli, R., Rahardjo, H., and Fredlund, D. G. (1992). "Measurement of matric suction and volume change during inundation of collapsible soil." *ASTM Geotechnical Testing Journal*, 15(2), 115–122.

Tang, G. X., and Graham, J. (2002). "A possible elastic-plastic framework for unsaturated soils with high-plasticity." *Canadian Geotechnical Journal*, 39, 894–907.

Wheeler, S. J., and Sivakumar, V. (1995). "An elasto-plasticity critical state framework for unsaturated silt soil." *Geotechnique*, 45(1), 35–53.

Wheeler, S. J., Sharma, R. S., and Buisson, M. S. R. (2003). "Coupling of hydraulic hysteresis and stress-strain behaviour in unsaturated soils." *Geotechnique*, 53(1), 41–54.

Chapter 4

Normalised Strain Rotational Multiple Yield Surface Framework

4.1 Limitations of RMYSF

A very important characteristic of the soil stress–strain behaviour is that the axial strain at failure increases with the increase of the effective confining pressure. In other words, the axial strain at failure is not common between effective confining pressures, i.e., it did not fail at the same axial strain. This effect is due to particle breakage, which increases the axial strain at failure at a high effective confining pressure. Particle breakage is prevailing at a higher effective confining pressure, which occurs along the inclined shear plane developed in the specimen during the shearing stage when close to failure condition of maximum deviator stress. This is like the reports of Bishop (1966), Indraratna et al. (1993) and Futai and Almeida (2005). This has been discussed in Section 1.1, Chapter 1. When the axial strain at failure is not the same, then it will limit the range for the stress–strain prediction according to the RMYSF. This is because the RMYSF can only predict up to the minimum failure axial strain among the axial strains at failure. Then at higher effective confining pressures, the prediction cannot be extended up to the maximum deviator stress. In the RMYSF, only a single axial strain at failure is applied to represent the various axial strains at failure. This will limit the prediction of stress–strain at a higher confining pressure. However, this limitation can be overcome by applying a normalised strain method. By applying the normalised strain method, the various strains at failure can be made common by normalising the axial strain for every stress–strain curve. This technique is called Normalised Strain Rotational Multiple Yield Surface Framework (NSRMYSF). This will be presented in the following section.

4.2 Introduction to the NSRMYSF

The NSRMYSF is an improvised method from the RMYSF to achieve a better accuracy in the deviator stress prediction. Besides, this method is to resolve the various strains at failure for the stress–strain curves at different effective stresses. When the stress–strain curves are normalised, then all the curves will have identical axial strain at failure. This would ease the plotting of the mobilised shear strength envelopes. The values of the axial strain for each stress–strain curve are multiplied by a specific conversion factor so that their failure axial strain is common. Finally, at the end of the process, when the predicted deviator stresses have been determined, then the axial strains are multiplied by the inverse factor to bring back the axial strains to their original values.

The procedures to determine the mobilised shear strength envelopes and to conduct the stress–strain prediction in the NSRMYSF are as follows:

1. Determine the stress–strain curves at various effective stresses, e.g., like effective stresses of 50, 100, 200 and 300 kPa using the consolidated drained or the consolidated undrained triaxial test.
2. Mark the maximum axial strain at failure, $\%\varepsilon_{\text{fmax}}^{\sigma'300}$, for the curve with the highest effective confining pressure, i.e., assume for an effective confining pressure of 300 kPa.
3. Mark the axial strain at failure for the rest of the stress–strain curves, e.g., $\%\varepsilon_{\text{f}}^{\sigma'50}$, $\%\varepsilon_{\text{f}}^{\sigma'100}$ and $\%\varepsilon_{\text{f}}^{\sigma'200}$ for effective stresses of 50, 100 and 200 kPa, respectively.
4. Calculate the normalise strain factor for each stress–strain curve; $\text{NSf}^{\sigma'50}, \text{NSf}^{\sigma'100}, \text{NSf}^{\sigma'200}$ and $\text{NSf}^{\sigma'300}$ as follows;

   a. For effective stress of 50 kPa, $$\text{NSf}^{\sigma'50} = \frac{\%\varepsilon_{\text{fmax}}^{\sigma'300}}{\%\varepsilon_{\text{f}}^{\sigma'50}} \tag{4.1a}$$

   b. For effective stress of 100 kPa, $$\text{NSf}^{\sigma'100} = \frac{\%\varepsilon_{\text{fmax}}^{\sigma'300}}{\%\varepsilon_{\text{f}}^{\sigma'100}} \tag{4.1b}$$

   c. For effective stress of 200 kPa, $$\text{NSf}^{\sigma'200} = \frac{\%\varepsilon_{\text{fmax}}^{\sigma'300}}{\%\varepsilon_{\text{f}}^{\sigma'200}} \tag{4.1c}$$

   d. For effective stress of 300 kPa, $$\text{NSf}^{\sigma'300} = \frac{\%\varepsilon_{\text{fmax}}^{\sigma'300}}{\%\varepsilon_{\text{f}}^{\sigma'300}} = 1.0 \tag{4.1d}$$

5. Calculate the normalised strain for every stress–strain curve by multiplying the whole range of the axial strains with the appropriate normalise strain factors; $\text{NSf}^{\sigma'50}, \text{NSf}^{\sigma'100}, \text{NSf}^{\sigma'200}$ and $\text{NSf}^{\sigma'300}$ as follows:
   e.g., for stress–strain curve with effective stress 50 kPa;

$$\%\varepsilon_{i\ \text{normalise}}^{\sigma'50} = \%\varepsilon_{i}^{\sigma'50} \times \frac{\%\varepsilon_{\text{fmax}}^{\sigma'300}}{\%\varepsilon_{\text{f}}^{\sigma'50}} \tag{4.2}$$

   where,
   $\%\varepsilon_{i}^{\sigma'50}$ is the axial strain at an arbitrary point, i
   $\%\varepsilon_{i\ \text{normalise}}^{\sigma'50}$ is the normalised strain at an arbitrary point, i.
6. Plot the normalised stress–strain curves, i.e., deviator stress versus normalised strain.
7. Select the axial strain values for the plotting of the yield surfaces or the mobilised shear strength envelopes, i.e., the mobilised shear strength envelopes will be representing a certain axial strain.
8. For each axial strain value, determine the value of deviator stress for the respective stress–strain curves.
9. Plot the Mohr circles for each axial strain values. If there are four stress–strain curves, then there will be four Mohr circles to guide the plot of each mobilised shear strength envelope.

10. Plot the mobilised shear strength envelope for each axial strain.
11. Then finally establish the set of mobilised shear strength envelopes or the yield surfaces for the specific soil.
12. To predict the stress–strain curve, then decide the effective confining pressure for the curve, i.e., σ'_3, which is the minor principal stress.
13. Draw the predicted Mohr circles to touch each mobilised envelope and note the diameter of each Mohr circle and note that the diameter is the magnitude of the predicted deviator stress.
14. To get the actual axial strain, multiply each normalise axial strain by its inverse factor, i.e., the reciprocal of the normalise strain factor, $NSf^{\sigma'??}$.
15. To determine the predicted stress–strain curve for a specific value of the effective confining pressure, plot the predicted deviator stress against the corresponding actual axial strain.
16. Compare the predicted stress–strain curve with the laboratory curve.

This procedure will be applied in the following sections.

4.3 Stress–strain behaviour of undisturbed Auckland residual clay soil

In this section, the anisotropic compression behaviour of Auckland residual clay will be assessed using both (1) the RMYSF and (2) the NSRMYSF as reported by Md Noor et al. (2017). The predicted stress–strain curves via both methods will be plotted against the stress–strain curves obtained from the laboratory tests. The accuracy between the methods will be compared and the advantages of the NSRMYSF will be emphasised.

4.3.1 *Application of conventional RMYSF to predict stress–strain behaviour of undisturbed Auckland residual clay soil*

Samples of undisturbed Auckland residual clay were taken from the Orewa site, located approximately 37 km north of Auckland City central. Undisturbed samples were obtained from a shallow sampling pit by pushing a 200-mm diameter by 200-mm length steel tube, fitted with a low angle cutting shoe (i.e., sharp cutting edge) into the ground at the desired level using a hydraulic jack (Md Noor et al., 2017). The tubes were then recovered and sample ends were levelled and sealed with a thin rubber disk between the soil and caps to preserve natural water content. The soil sample was extruded using a hydraulic jack and cut into four quarters using a band saw. Each quarter of material was trimmed to 75 mm in diameter and 150 mm length using a hand-operated soil lathe.

Consolidated drained small strain triaxial tests were conducted on fully saturated undisturbed Auckland residual clays at four different effective confining pressures. The small local axial strain measuring system consists of three numbers of the internal submersible LVDTs arranged at 120° apart assembled to the transducers holding system fixed at the specimen gauge height as shown in Figure 4.1a and b. Figure 4.2 shows the stress–strain curves at effective confining stresses of 60, 220, 500 and 800 kPa (Md Noor et al., 2017). The axial strains at failure are 1.4%, 2.0%, 2.3%

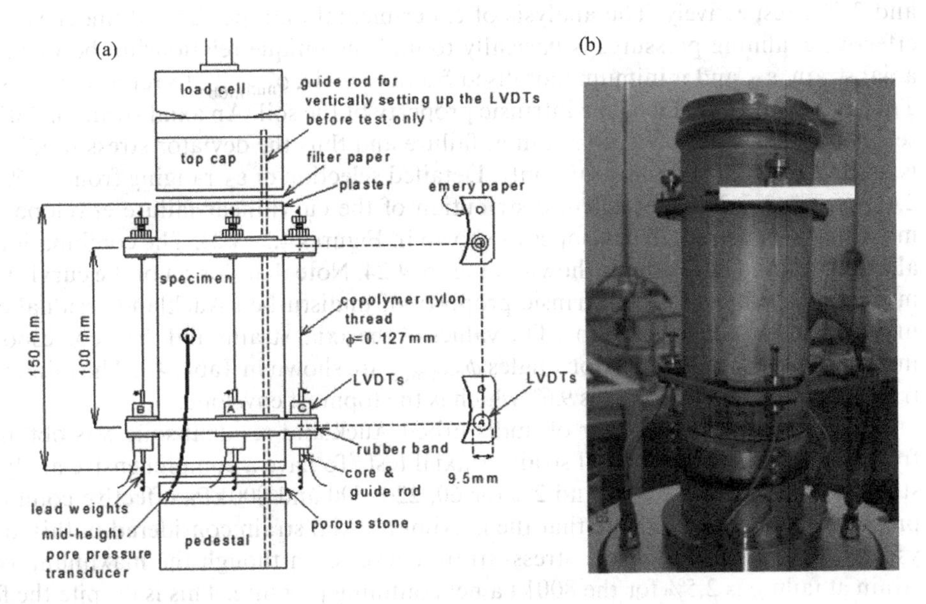

Figure 4.1 Triaxial test set-up using three submersible miniature LVDTs attached locally on the peripheral of 150-mm height and 75-mm diameter specimen (Md Noor et al., 2017). (a) Local transducers holding system. (b) Local transducers fitted on the specimen.

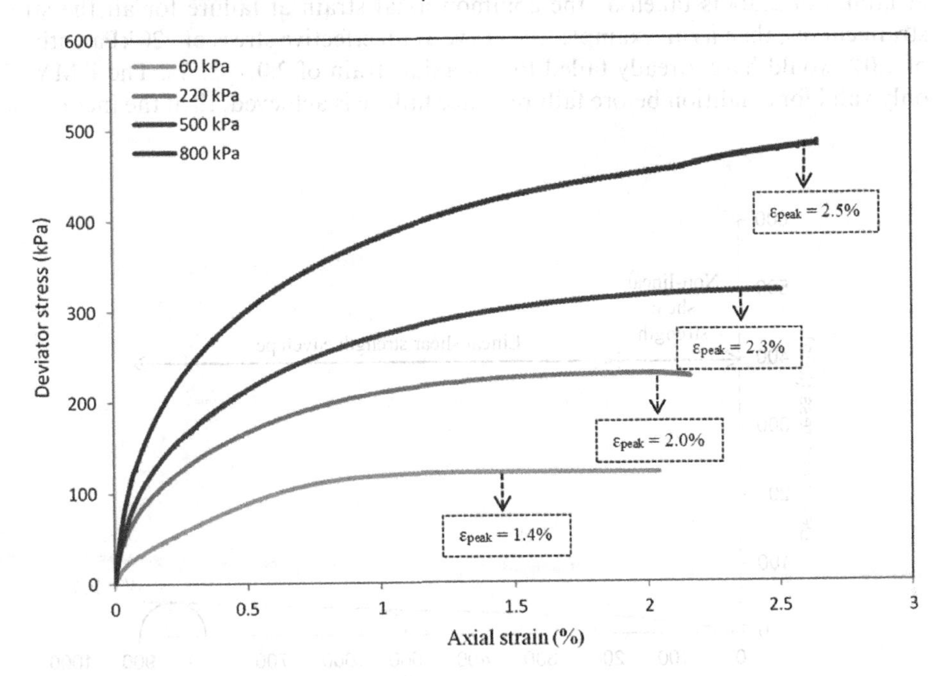

Figure 4.2 Stress–strain curves of laboratory small strain triaxial tests at various effective confining pressures, i.e., 60, 220, 500 and 800 kPa showing that axial strain at failure increases with effective confining pressure (Md Noor et al., 2017).

and 2.5%, respectively. The analysis of experimental data performed under various effective confining pressures is basically to find the unique relationship between the axial strain, ε_A, and minimum mobilised friction angle, $\phi'_{min_{mob}}$, to represent the soil. This unique relationship is the intrinsic property of the soil. An axial strain of 2.0% is selected as the common axial strain at failure and thus the deviator stress prediction is limited up to 2.0% axial strain only. Detailed selection of ε_A ranging from 0.05% to 2.0% strains produced excellent distribution of the curvilinear failure envelope and mobilised shear strength envelopes as shown in Figures 4.3–4.23. The combination of all the mobilised envelopes is shown in Figure 4.24. Note that this set of the curvilinear mobilised envelopes is the intrinsic property of undisturbed Auckland residual clay under various ε_A considerations. The values of the axial strains and their corresponding minimum mobilised friction angles $\phi_{min\ mob}$ are shown in Table 4.1. The minimum friction angle at failure ϕ'_{minf} is 9.6°, which is the topmost envelope.

The stress–strain behaviour of undisturbed Auckland residual soils was obtained from the local LVDTs of small strain triaxial test. To have a comprehensive analysis, strain values between 0.05% and 2% for 60, 220, 500 and 800 kPa effective confining pressures were selected. Note that the maximum axial strain considered in this analysis is 2% to represent all the stress–strain curves even though the maximum axial strain at failure is 2.5% for the 800 kPa net confining pressure. This is despite the fact that the lowest axial strain at failure is 1.4% for the curve with an effective stress of 60 kPa. This is insisted because the curve is almost horizontal from 1.4% to 2.0% axial strains (refer to the curve with an effective stress of 60 kPa). However, if an axial strain at failure of 2.5% is taken as the common axial strain at failure for all the stress–strain curves, then as an example, the curve at an effective stress of 220 kPa with ε_{peak} of 2.0% would have already failed for an axial strain of 2.0%–2.5%. The RMYSF is only valid for condition before failure. Once failure is achieved, then the incline shear

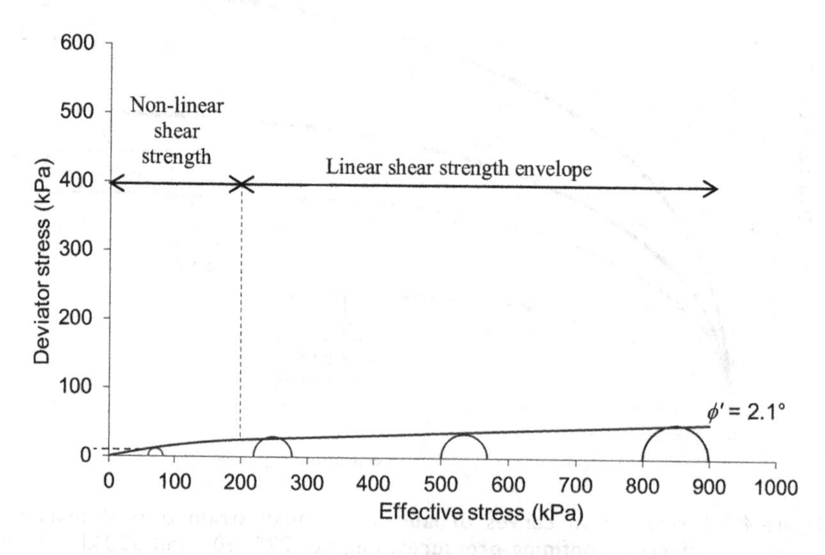

Figure 4.3 Curvilinear mobilised shear strength envelope at 0.05% strain of fully saturated condition undisturbed Orewa clay based on local LVDT.

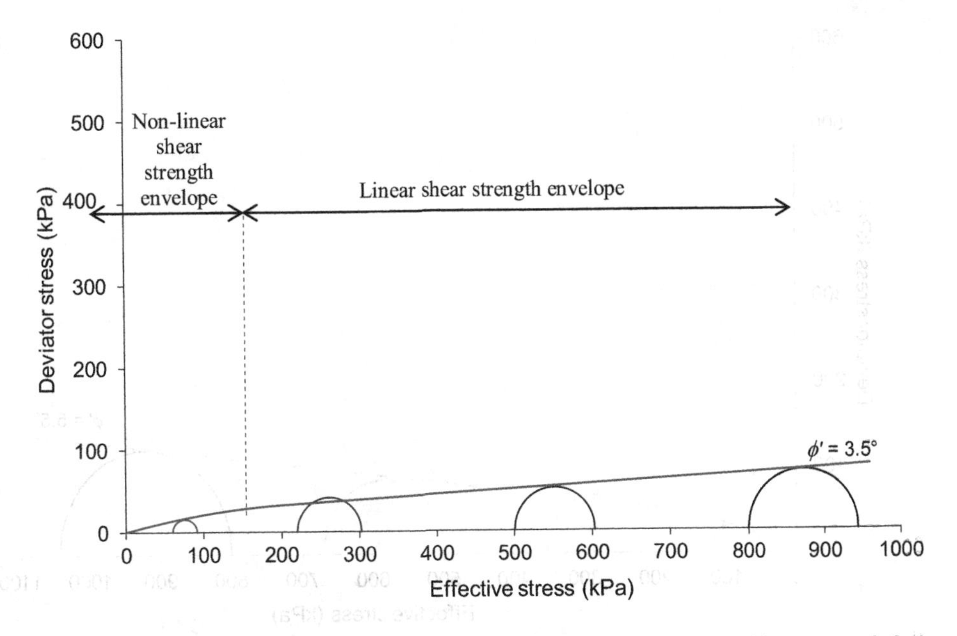

Figure 4.4 Curvilinear mobilised shear strength envelope at 0.1% strain of fully saturated condition undisturbed Orewa clay based on local LVDT.

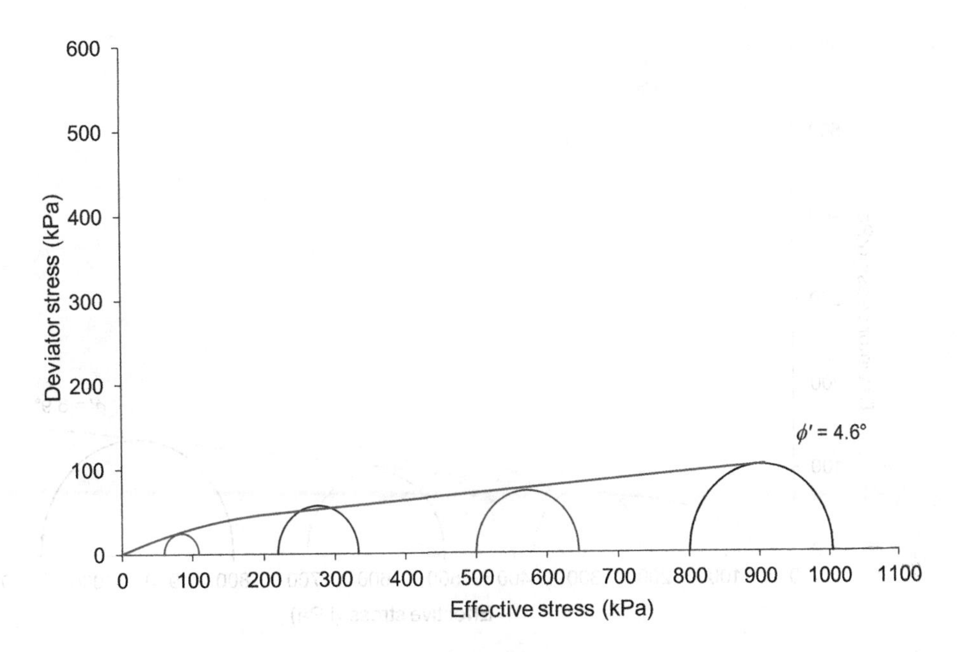

Figure 4.5 Curvilinear mobilised shear strength envelope at 0.2% strain of fully saturated condition undisturbed Orewa clay based on local LVDT.

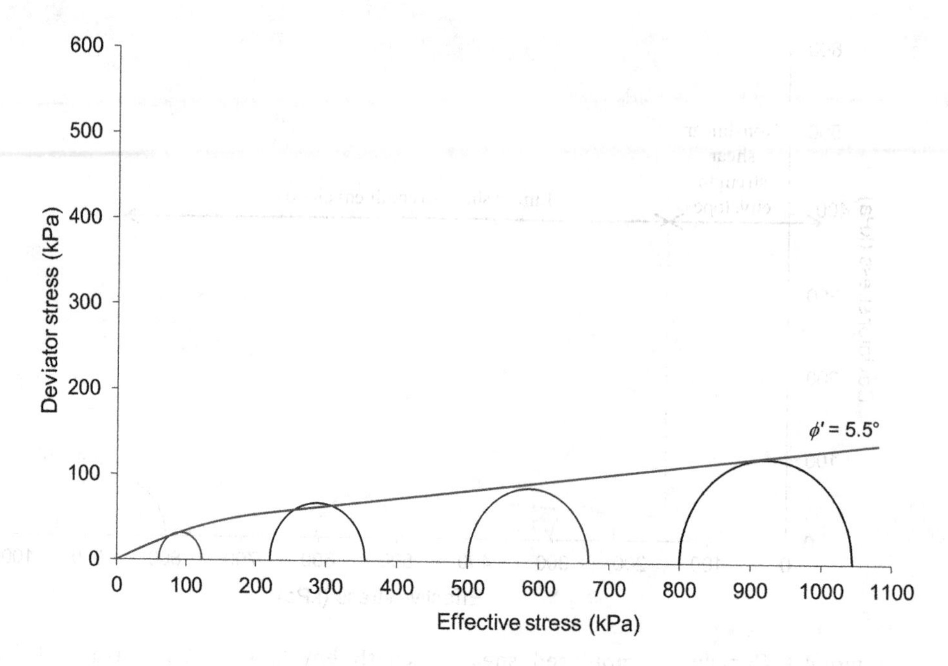

Figure 4.6 Curvilinear mobilised shear strength envelope at 0.3% strain of fully saturated condition undisturbed Orewa clay based on local LVDT.

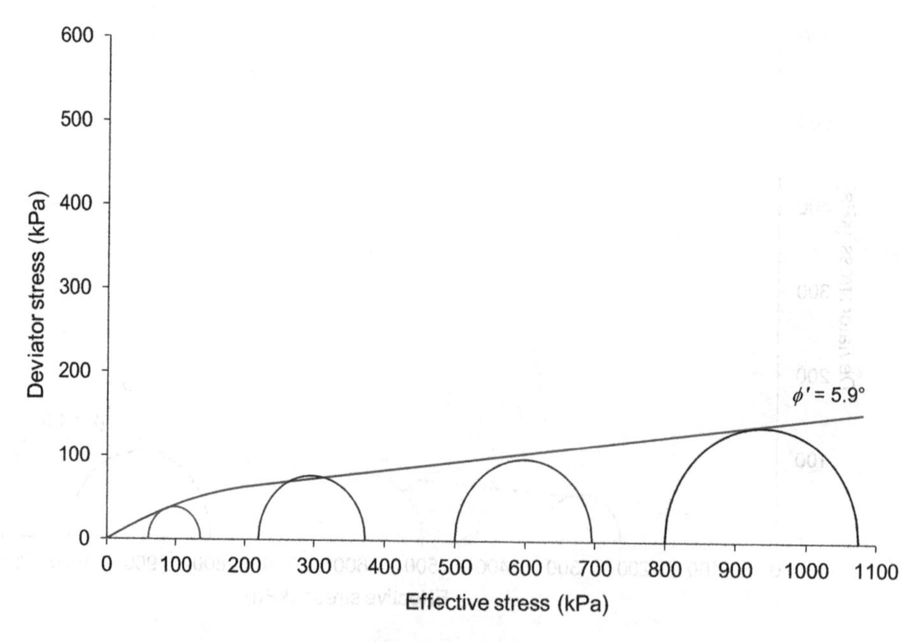

Figure 4.7 Curvilinear mobilised shear strength envelope at 0.4% strain of fully saturated condition undisturbed Orewa clay based on local LVDT.

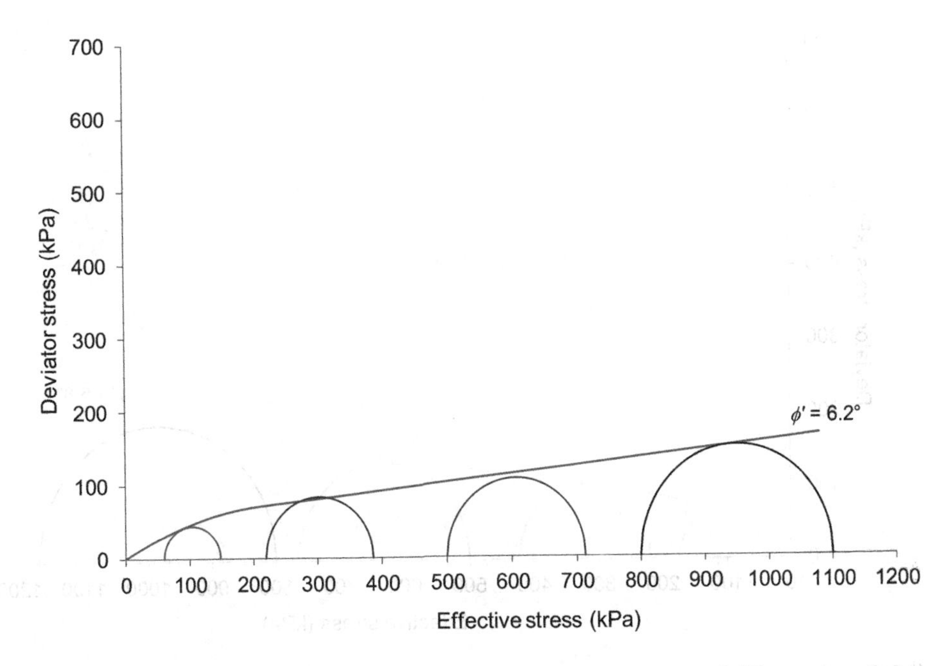

Figure 4.8 Curvilinear mobilised shear strength envelope at 0.5% strain of fully saturated condition undisturbed Orewa clay based on local LVDT.

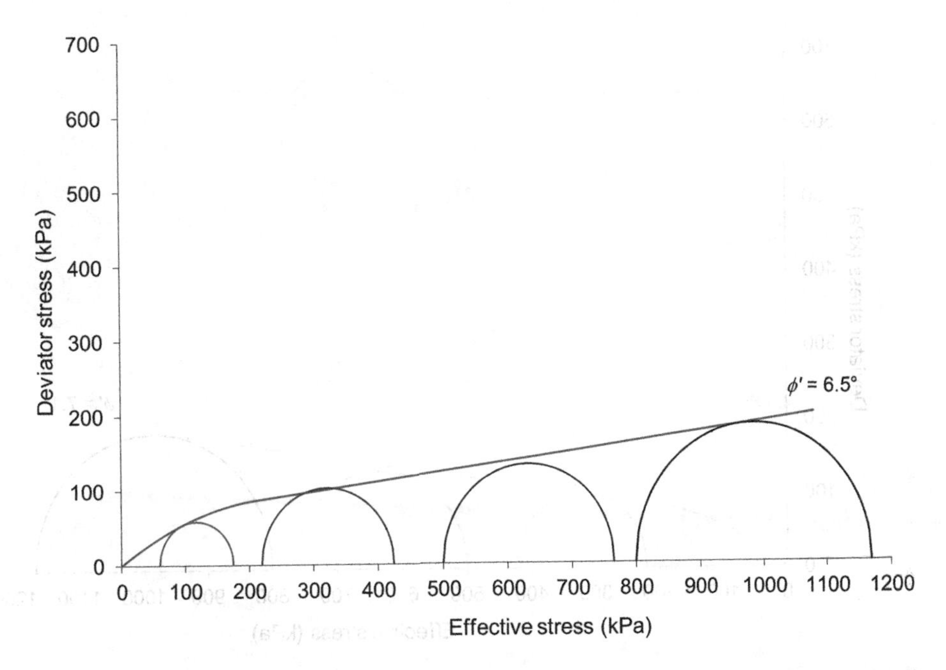

Figure 4.9 Curvilinear mobilised shear strength envelope at 0.6% strain of fully saturated condition undisturbed Orewa clay based on local LVDT.

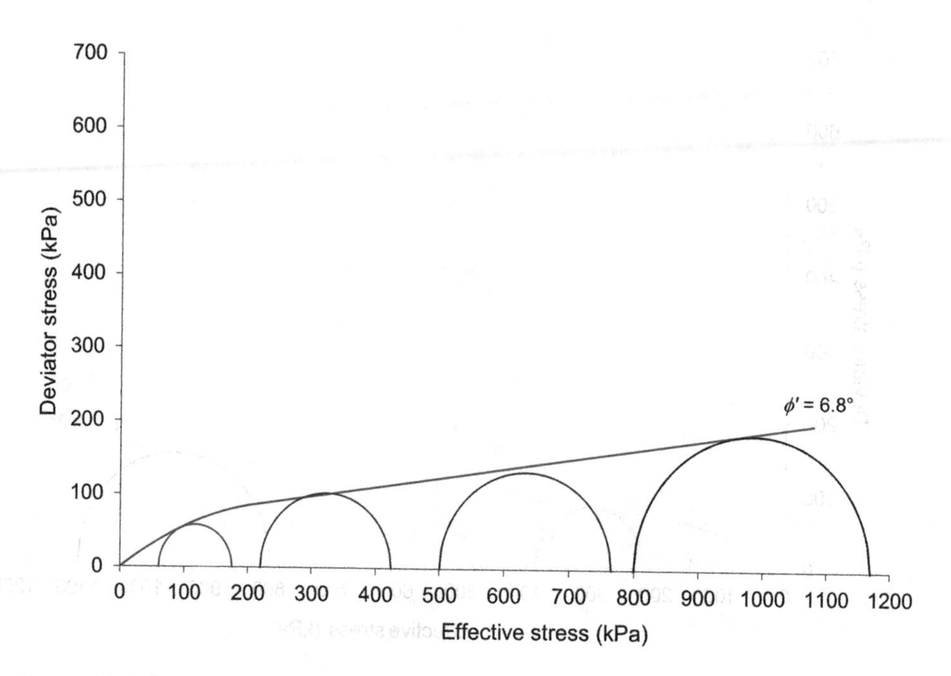

Figure 4.10 Curvilinear mobilised shear strength envelope at 0.7% strain of fully saturated condition undisturbed Orewa clay based on local LVDT.

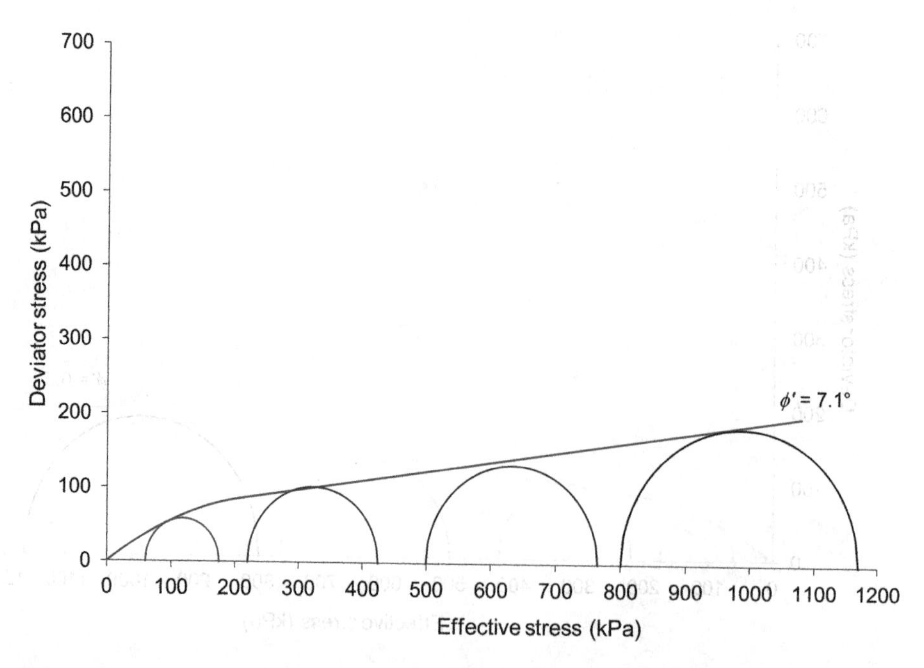

Figure 4.11 Curvilinear mobilised shear strength envelope at 0.8% strain of fully saturated condition undisturbed Orewa clay based on local LVDT.

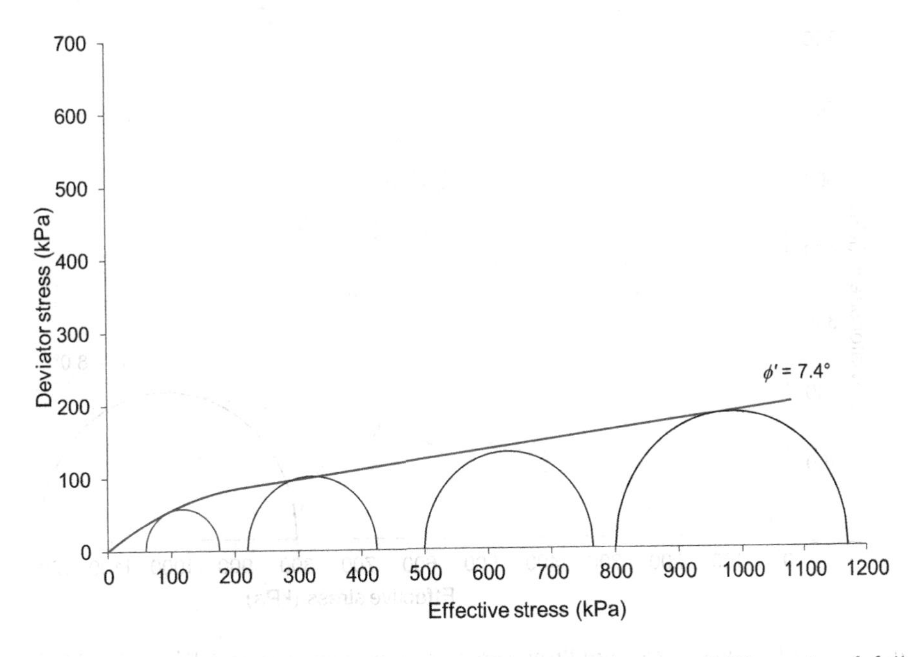

Figure 4.12 Curvilinear mobilised shear strength envelope at 0.9% strain of fully saturated condition undisturbed Orewa clay based on local LVDT.

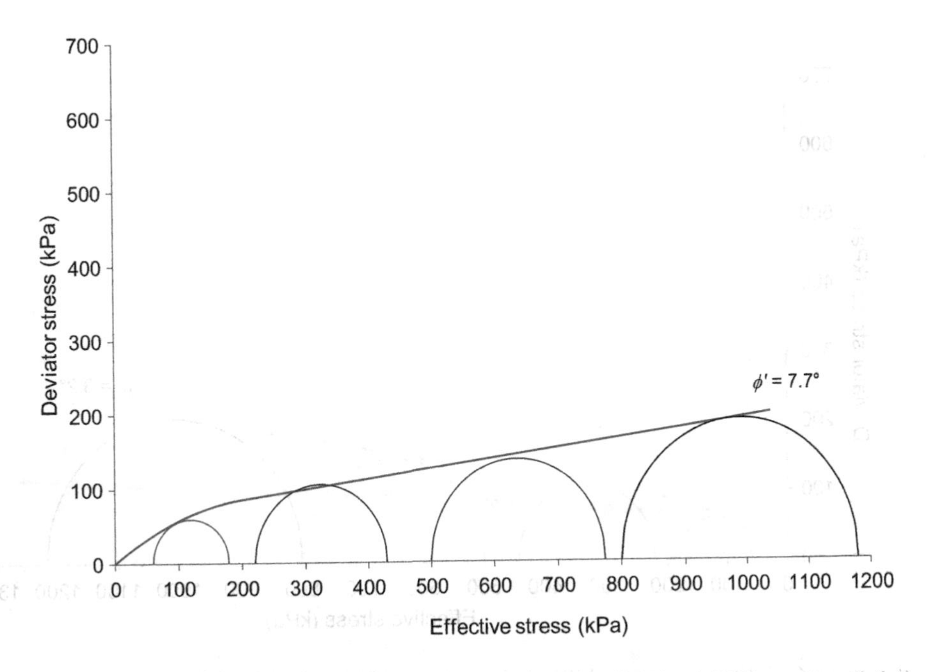

Figure 4.13 Curvilinear mobilised shear strength envelope at 1.0% strain of fully saturated condition undisturbed Orewa clay based on local LVDT.

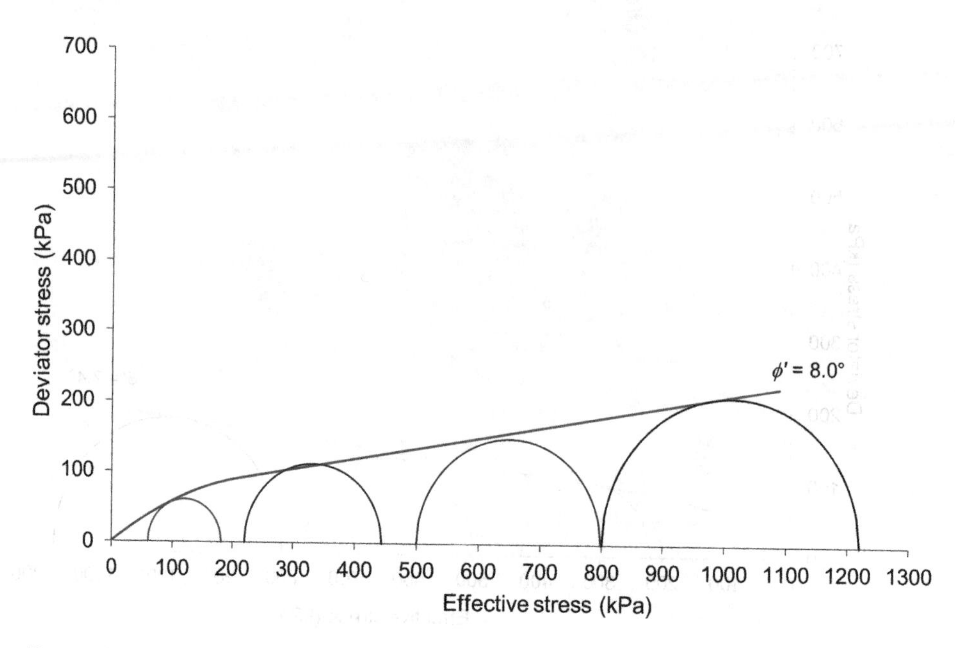

Figure 4.14 Curvilinear mobilised shear strength envelope at 1.1% strain of fully saturated condition undisturbed Orewa clay based on local LVDT.

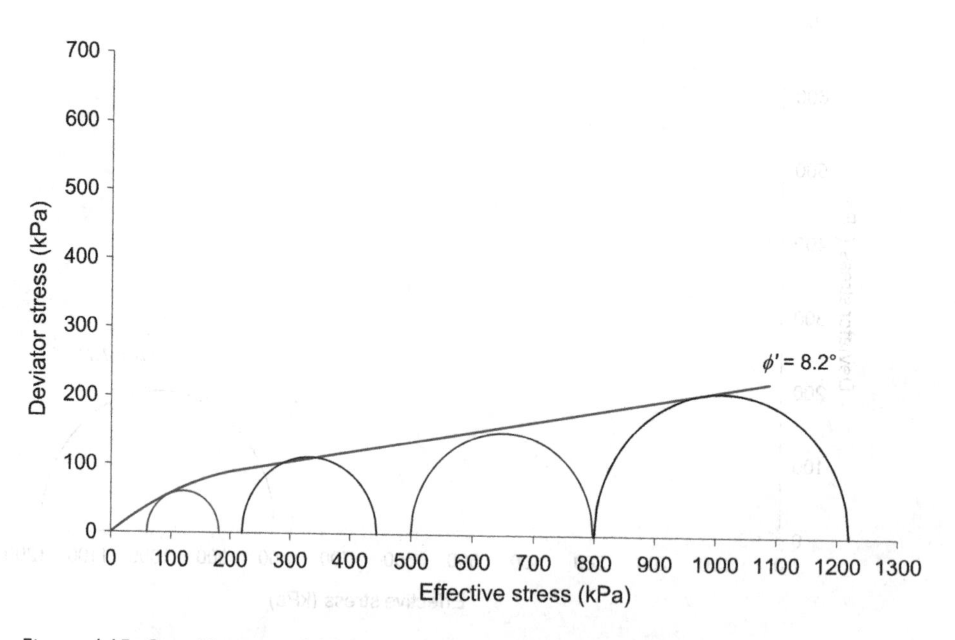

Figure 4.15 Curvilinear mobilised shear strength envelope at 1.2% strain of fully saturated condition undisturbed Orewa clay based on local LVDT.

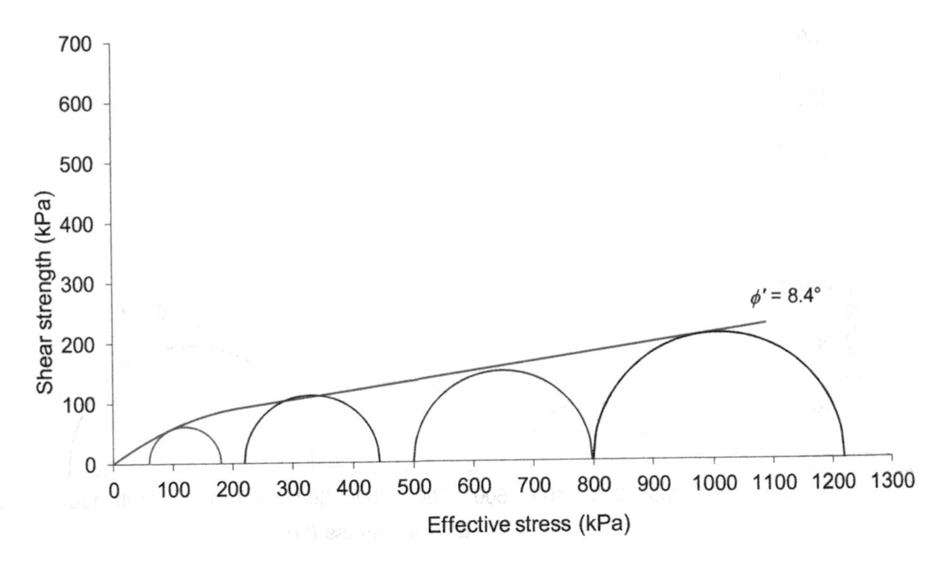

Figure 4.16 Curvilinear mobilised shear strength envelope at 1.3% strain of fully saturated condition undisturbed Orewa clay based on local LVDT.

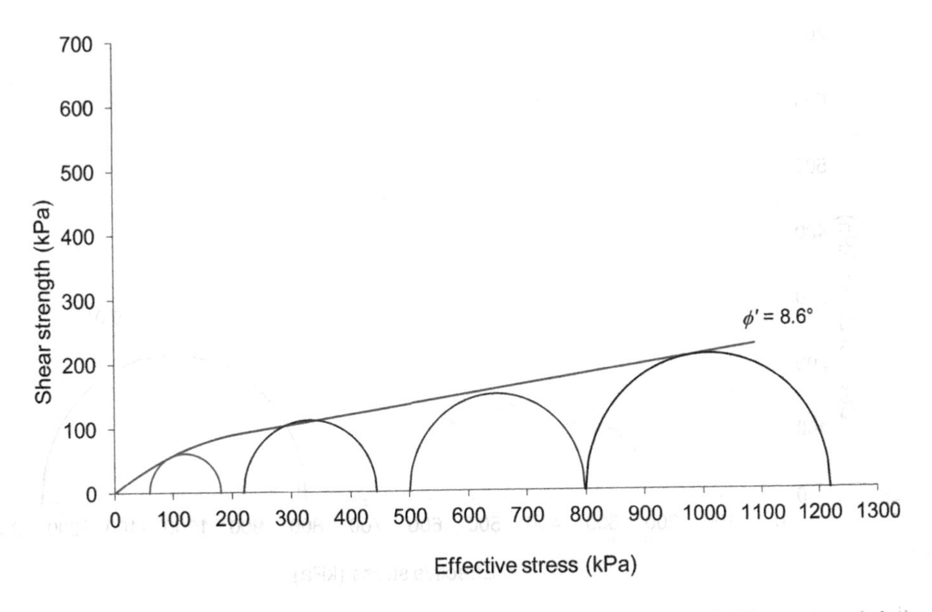

Figure 4.17 Curvilinear mobilised shear strength envelope at 1.4% strain of fully saturated condition undisturbed Orewa clay based on local LVDT.

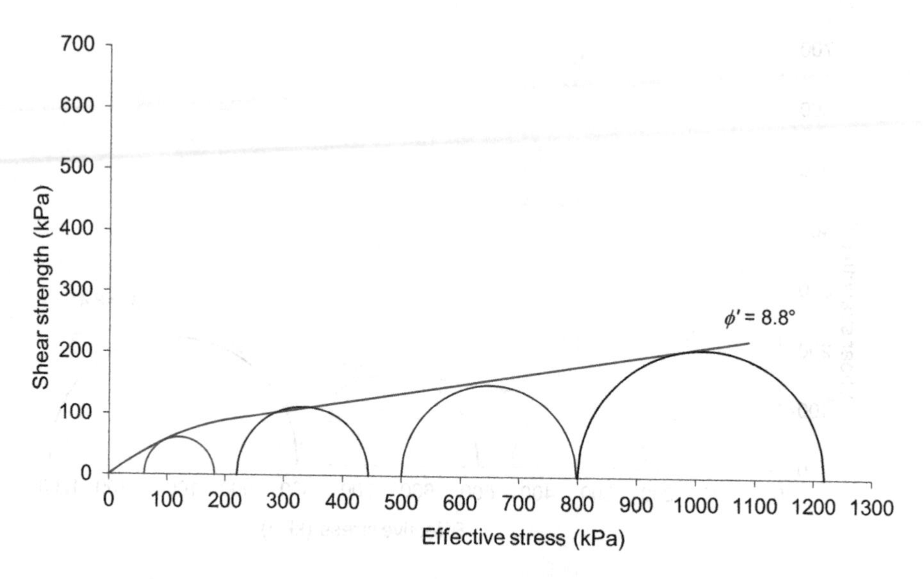

Figure 4.18 Curvilinear mobilised shear strength envelope at 1.5% strain of fully saturated condition undisturbed Orewa clay based on local LVDT.

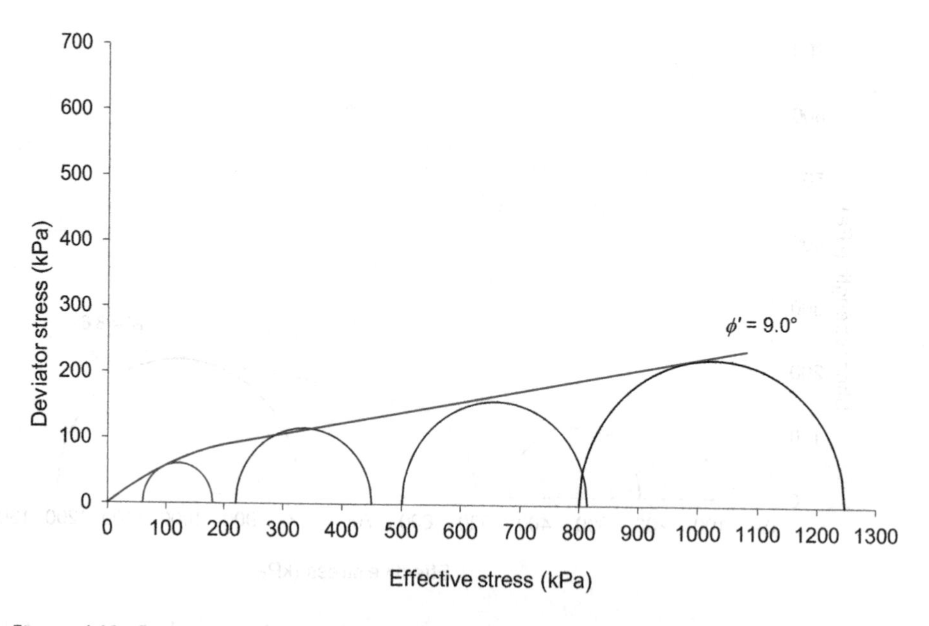

Figure 4.19 Curvilinear mobilised shear strength envelope at 1.6% strain of fully saturated condition undisturbed Orewa clay based on local LVDT.

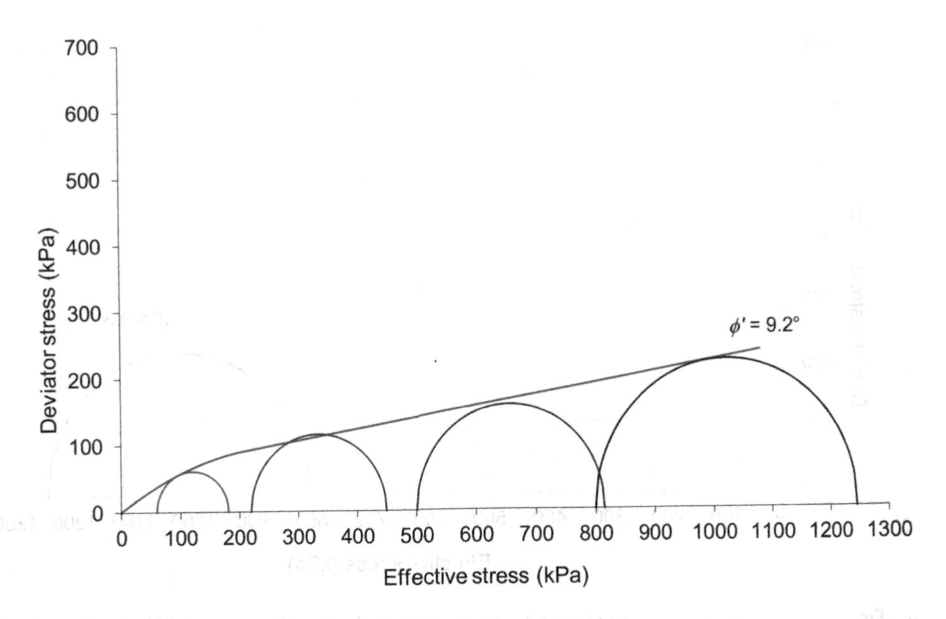

Figure 4.20 Curvilinear mobilised shear strength envelope at 1.7% strain of fully saturated condition undisturbed Orewa clay based on local LVDT.

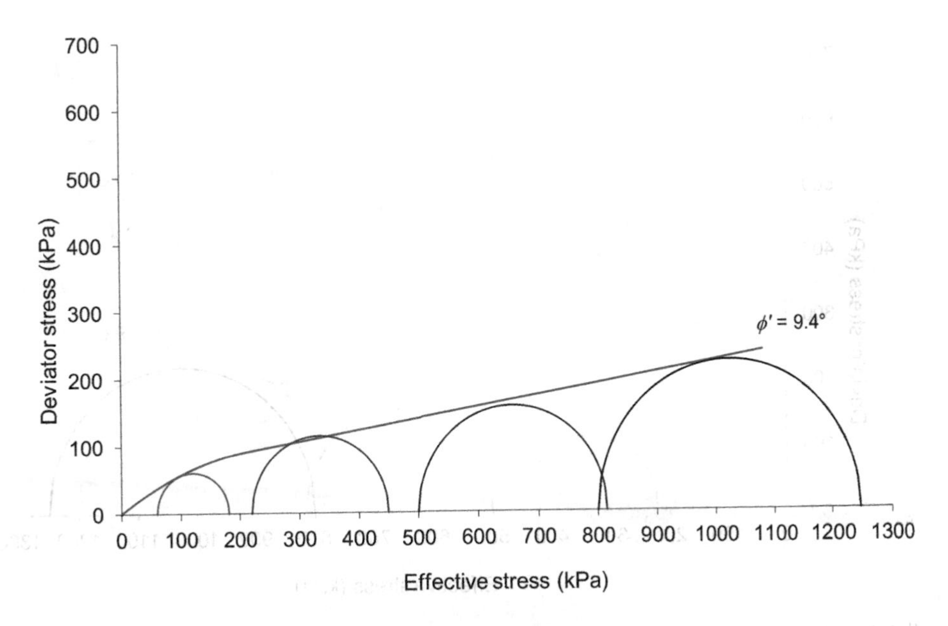

Figure 4.21 Curvilinear mobilised shear strength envelope at 1.8% strain of fully saturated condition undisturbed Orewa clay based on local LVDT.

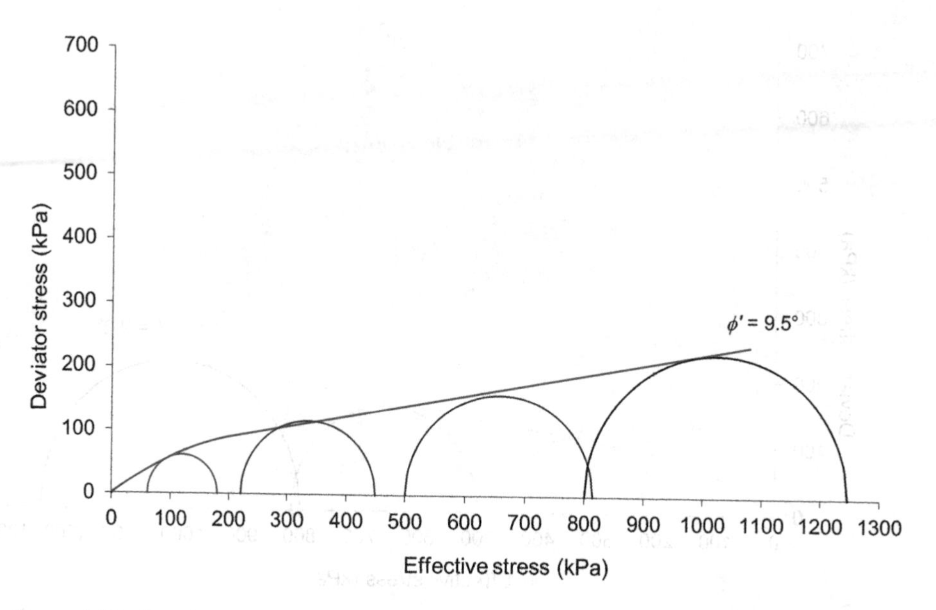

Figure 4.22 Curvilinear mobilised shear strength envelope at 1.9% strain of fully saturated condition undisturbed Orewa clay based on local LVDT.

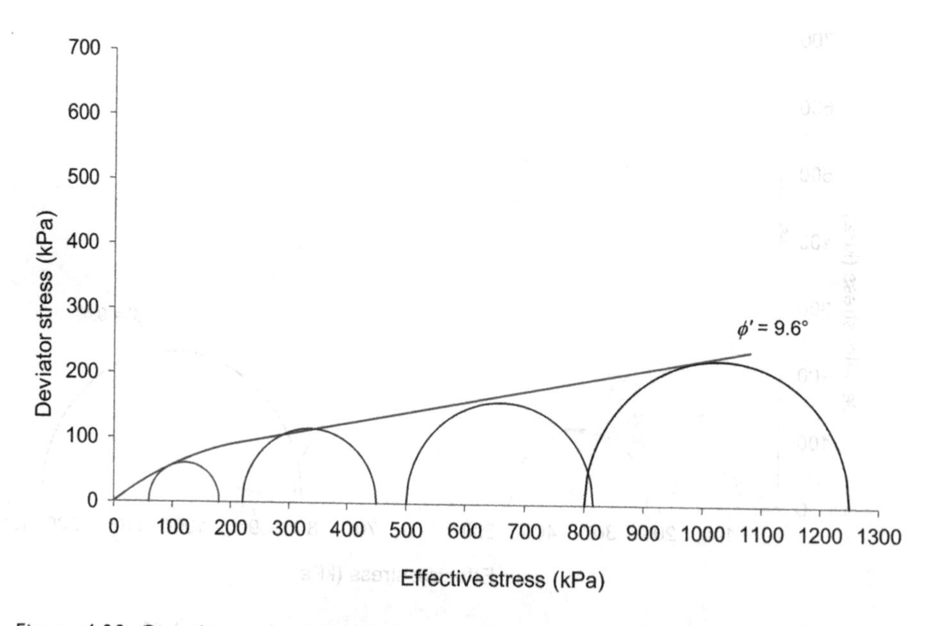

Figure 4.23 Curvilinear mobilised shear strength envelope at 2.0% strain of fully saturated condition undisturbed Orewa clay based on local LVDT.

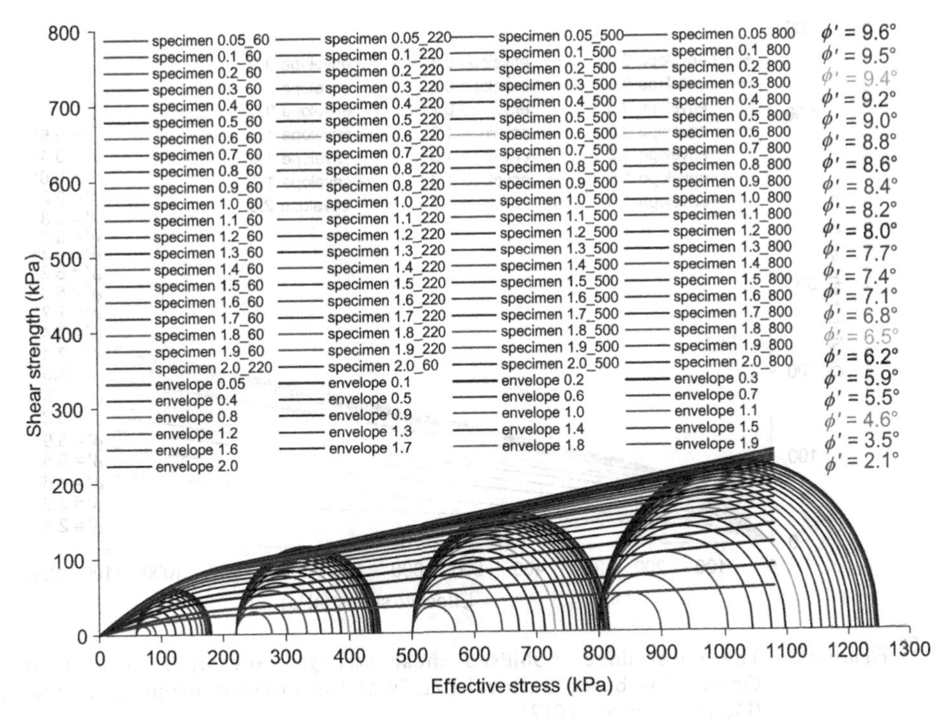

Figure 4.24 The superimposed mobilised shear strength envelopes for the selected axial strains, ε_A, ranging from 0.05% to 2.0% for undisturbed Orewa clay based on local LVDTs (Md Noor et al., 2017).

plane has already developed and the soil compression is not because of the particle rearrangement but the shearing along the shear plane. Thence, this will nullify the stress–strain prediction.

In the RMYSF, the best representative maximum axial strain needs to be selected and the prediction of the stress–strain curves is limit to this selected maximum axial strain. Mobilised curvilinear shear strength envelopes are drawn for various axial strains up to 2.0% as shown in Figures 4.3–4.23. The minimum mobilised friction angles $\phi'_{\text{min}_{\text{mob}}}$ corresponding to the specific axial strains, ε_A, are determined. The whole set of the mobilised shear strength envelopes are plotted together in Figure 4.24, and in Figure 4.25 are the mobilised envelopes without the Mohr circles. Note that the topmost envelope is the envelope at failure, which represents the axial strain of 2.0%, and the second highest is the mobilised envelope representing the axial strain of 1.9%. The minimum friction angle at failure ϕ'_{minf} is 9.6°.

Figures 4.26–4.29 show the predicted Mohr circles for axial strains of 0.05%–2.0% for effective stresses of 60, 220, 500 and 800 kPa, respectively. The predicted deviator stresses, which are the diameter of the Mohr circles, are presented in Table 4.1 together with the values of the deviator stresses obtained from the laboratory tests. The values of the deviator stresses obtained from the laboratory tests and the predicted values

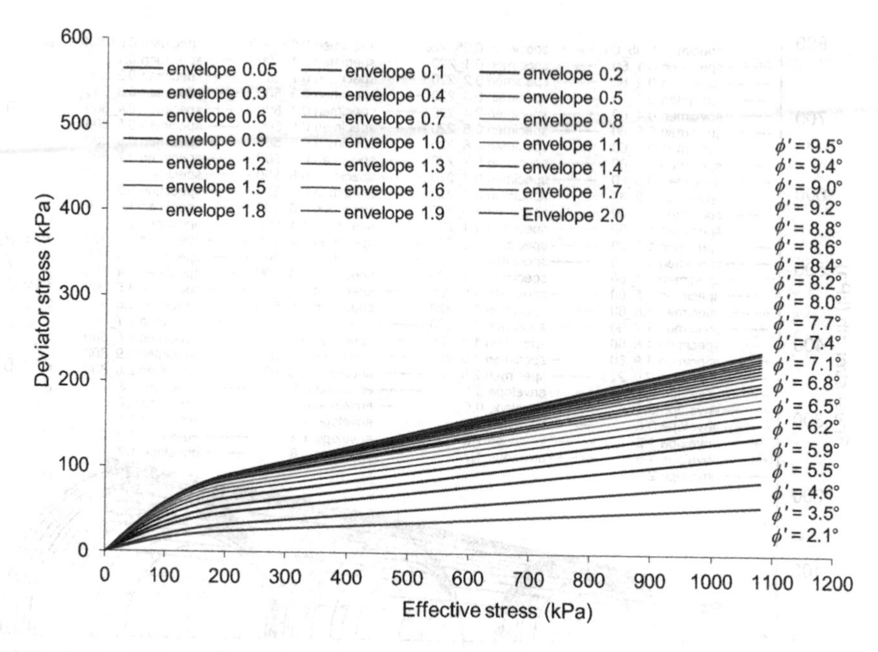

Figure 4.25 The determined mobilised shear strength envelopes of undisturbed Orewa clay based on local LVDTs as the intrinsic property of the soil (Md Noor et al., 2017).

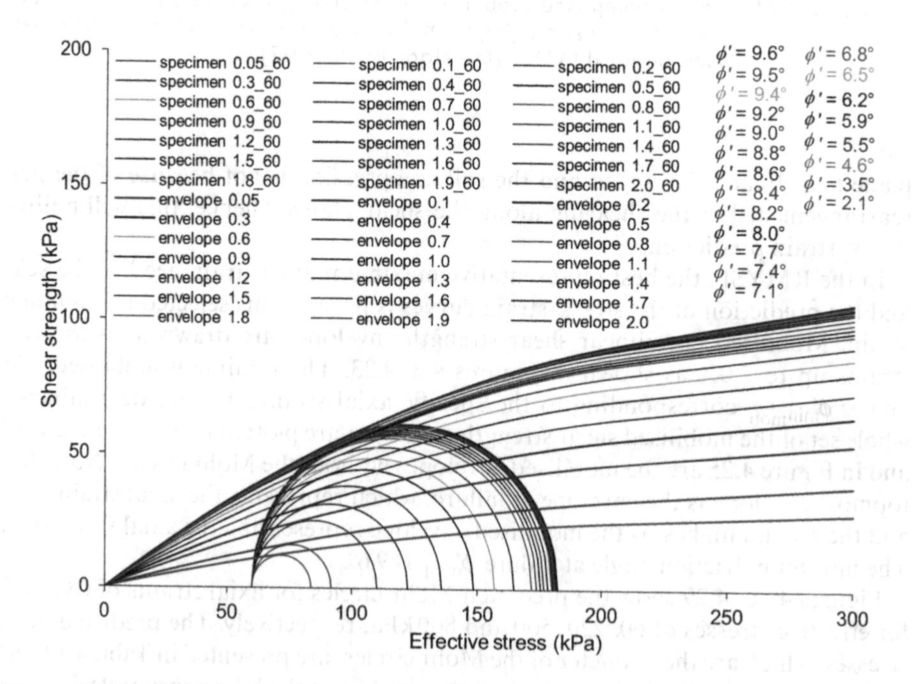

Figure 4.26 Prediction of deviator stresses at an effective confining pressure of 60 kPa (Md Noor et al., 2017).

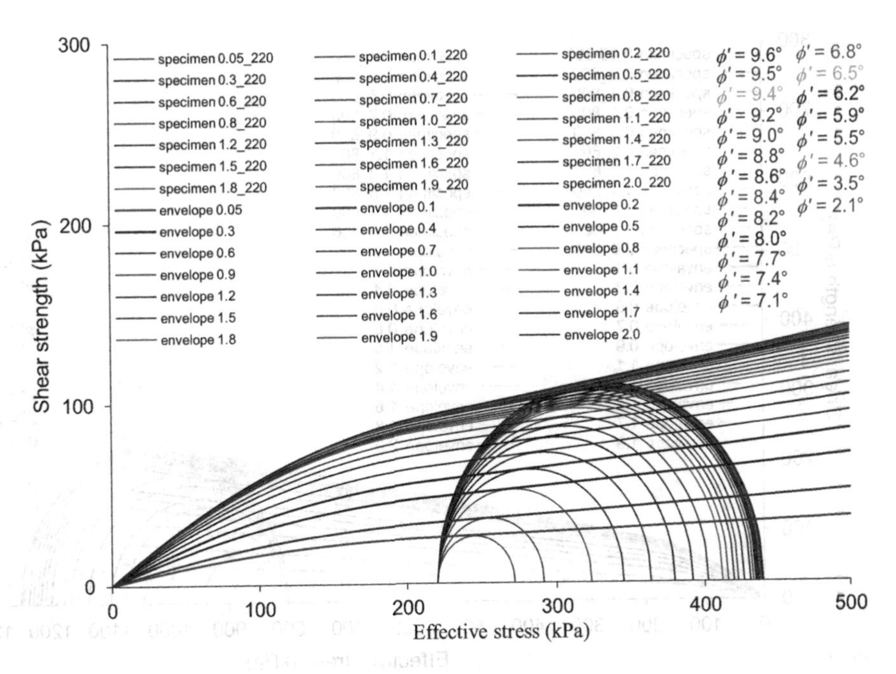

Figure 4.27 Prediction of deviator stresses at an effective confining pressure of 220 kPa (Md Noor et al., 2017).

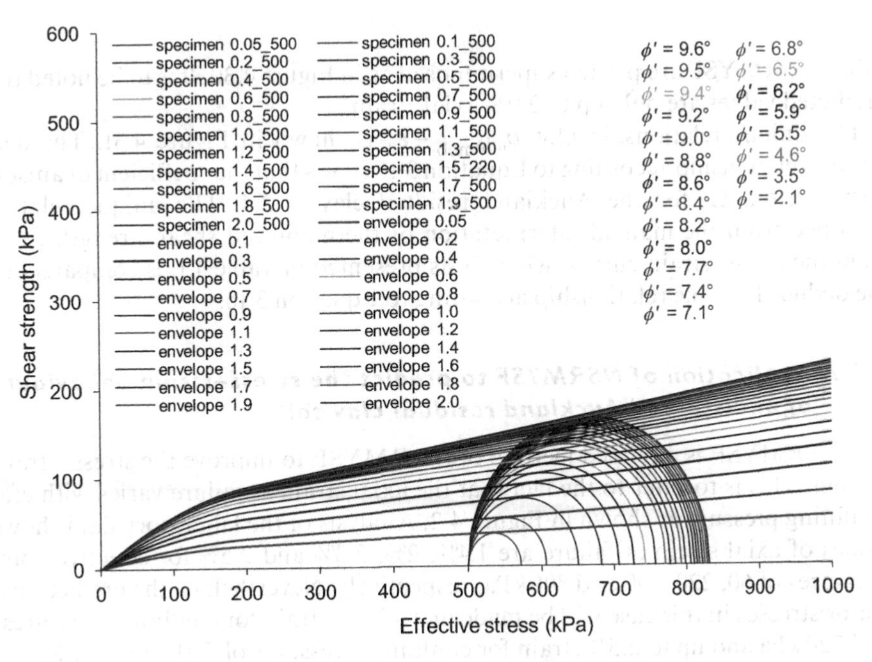

Figure 4.28 Prediction of deviator stresses at an effective confining pressure of 500 kPa (Md Noor et al., 2017).

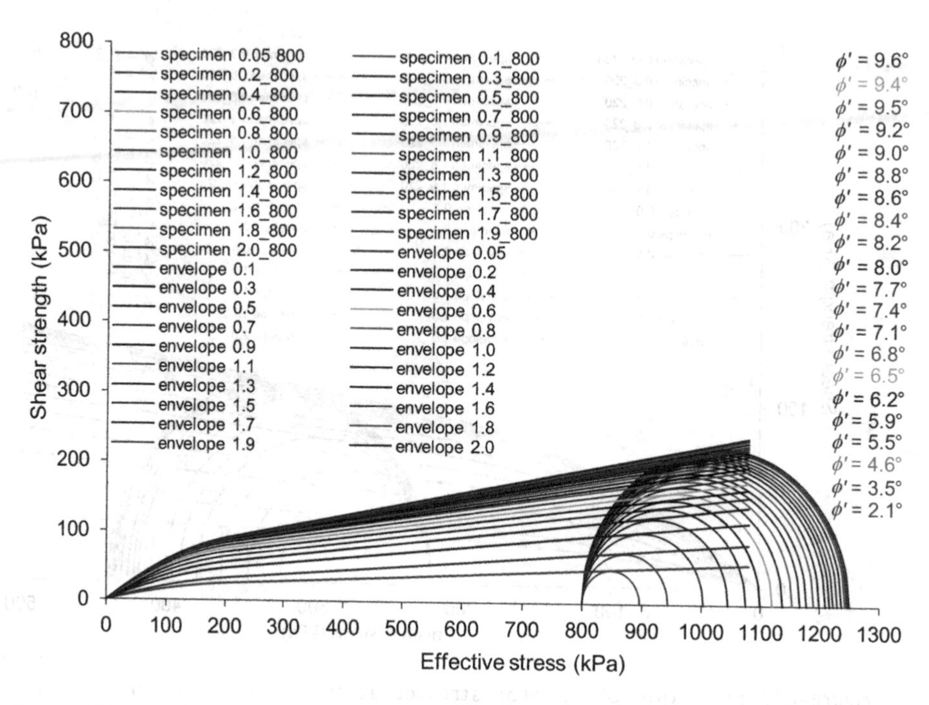

Figure 4.29 Prediction of deviator stresses at an effective confining pressure of 800 kPa (Md Noor et al., 2017).

using the RMYSF are plotted superimposed as in Figure 4.30. It is to be noted that the predicted values are only up to 2.0% axial strain.

The unique relationship plot $\phi'_{\text{min}_{\text{mob}}} - \varepsilon_a$ is shown in Figure 4.31. The deduced unique relationship according to Equation 3.16 shows that the coefficient of anisotropic compression, Ω, for the Auckland residual clay is 2.4. The unique relationship obtained from the manual interpretation of the mobilised shear strength envelopes from the stress–strain curves, which is as presented in Table 4.1, is compared against the deduced unique relationship according to Equation 3.16.

4.3.2 *Application of NSRMYSF to predict the stress–strain behaviour of undisturbed Auckland residual clay soil*

The NSRMYSF is a refined method of the RMYSF to improve the stress–strain prediction. This is to cater to the fact that the axial strain at failure varies with effective confining pressure as shown in Figure 4.2. Analysis of the laboratory data shows that values of axial strain at failure are 1.4%, 2%, 2.3% and 2.5% for effective confining pressures of 60, 220, 500 and 800 kPa, respectively. Nevertheless, the prediction of deviator stresses in this case will be made up to 2.0% strain for confining pressures of 60 and 220 kPa and up to 2.5% strain for confining pressures of 500 and 800 kPa.

The normalised strain for a specific stress–strain curve is calculated based on the ratio of maximum strain at failure over failure strain of each curve multiplied by the

Table 4.1 Comparison between predicted deviator stresses using the RMYSF and laboratory small strain triaxial test results (Md Noor, 2017)

Effective confining pressure (kPa)		*60*		*220*		*500*		*800*	
		Deviator stress (kPa)		*Deviator stress (kPa)*		*Deviator stress (kPa)*		*Deviator stress (kPa)*	
Axial strain (%)	$\phi'_{min\,mob}$	*Lab. test*	*Predict RMYSF*	*Lab. test*	*Predict RMYSF*	*Lab. test*	*Predict RMYSF*	*Lab. test*	*Predict RMYSF*
0.00	0.00	0.00	0.00	0.00	0.00	0.00	0.00	0.00	0.00
0.05	2.10	21.91	24.0	57.70	51.5	68.46	72.0	99.08	96.0
0.10	3.50	31.36	32.0	82.52	71.0	103.58	106.0	144.42	144.0
0.20	4.60	48.78	50.0	113.20	103.0	144.72	150.5	204.60	203.0
0.30	5.50	63.26	62.5	134.80	123.0	171.70	182.5	244.40	245.0
0.40	5.90	76.50	76.0	152.44	143.5	195.26	206.5	275.20	275.0
0.50	6.20	88.64	88.5	166.80	160.0	213.60	226.5	300.20	300.0
0.60	6.50	98.57	97.0	178.58	172.0	229.83	242.0	321.64	320.0
0.70	6.80	106.58	105.0	188.74	182.0	243.49	256.0	339.89	337.0
0.80	7.10	112.17	112.0	197.10	192.0	254.46	270.0	355.85	354.0
0.90	7.40	115.63	114.0	204.31	196.0	264.97	277.0	369.68	366.5
1.00	7.70	118.34	115.5	209.80	200.0	273.76	284.0	379.60	378.0
1.10	8.00	119.61	117.0	214.78	203	280.97	292	391.98	390.0
1.20	8.20	120.41	118.5	218.38	207	287.09	298.5	402.56	400.0
1.30	8.40	121.02	120.5	220.58	211	292.76	305.5	412.02	410.0
1.40	8.60	121.35	121.5	223.67	213.5	298.27	310.5	419.11	417.5
1.50	8.80	121.08	122.0	225.55	215	301.60	314.5	424.40	423.0
1.60	9.00	121.00	121.5	226.64	216	306.48	318.0	432.23	430.0
1.70	9.20	120.88	121.0	227.88	217	309.20	322.0	437.98	437.0
1.80	9.40	121.04	120.0	229.01	218	311.20	325.5	443.82	443.0
1.90	9.50	120.73	121.0	229.45	219	314.54	325.5	446.38	446.0
2.00	9.60	120.50	122.5	229.83	220	315.36	331.5	450.59	450.0

actual strain. The actual strains at failure of 1.4%, 2%, 2.3% and 2.5% are used in normalising the stress–strain curves. The calculation of the normalised strain is as discussed in Section 4.2, which is as follows:

$$\text{Normalised strain}_i = \frac{\text{Maximum strain at failure}}{\text{Failure strain}_i} \times \text{strain}_i$$

Figure 4.32 shows the actual stress–strain curves (full lines) against the normalised stress–strain curves (dotted lines). In the NSRMYSF, the mobilised shear strength envelopes are derived from the normalised stress–strain curves. In other words, there is a slight difference between the mobilised shear strength envelopes derived from the actual stress–strain curves compared to the normalised stress–strain curves when the dotted lines are slightly lowered from the full line graphs.

Figure 4.33 shows the excellent agreement between the predicted stress–strain curves (data points) using the NSRMYSF against the laboratory stress–strain curves (continuous line). The values of the predicted data are presented in Table 4.2.

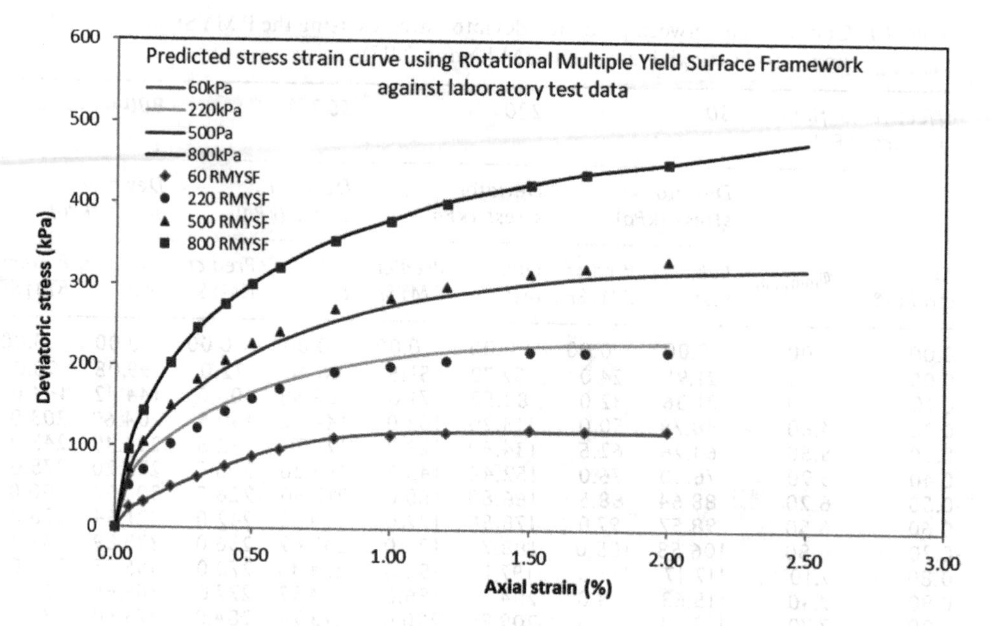

Figure 4.30 Comparison between predicted deviator stress values using Rotational Multiple Yield Surface Framework using the determined mobilised shear strength envelopes and laboratory data for Auckland residual soil, i.e., Orewa clay (Md Noor et al., 2017).

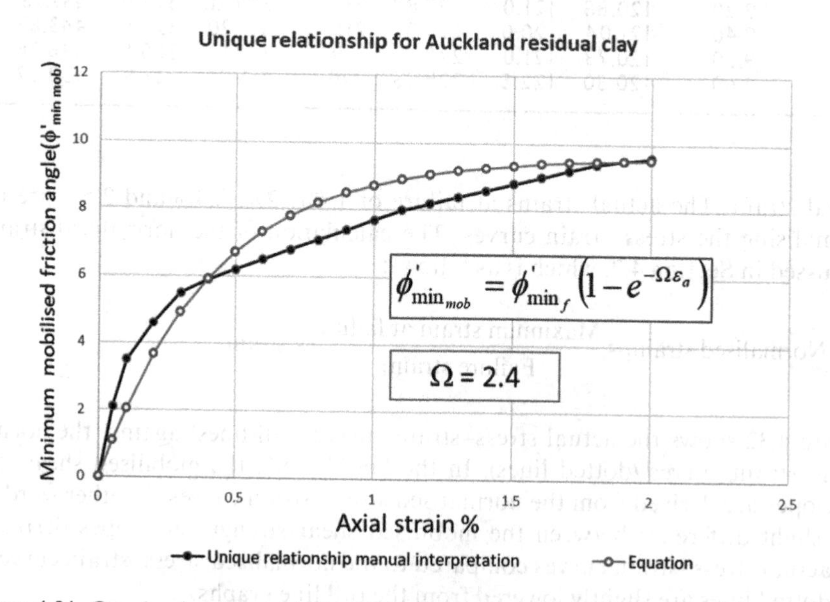

Figure 4.31 Graph of unique relationship plotted manually against the graph plotted using the Equation of Jais and Md Noor (2009) for undisturbed Orewa residual soil with a coefficient of anisotropic compression, Ω, of 2.4 (Md Noor et al., 2017).

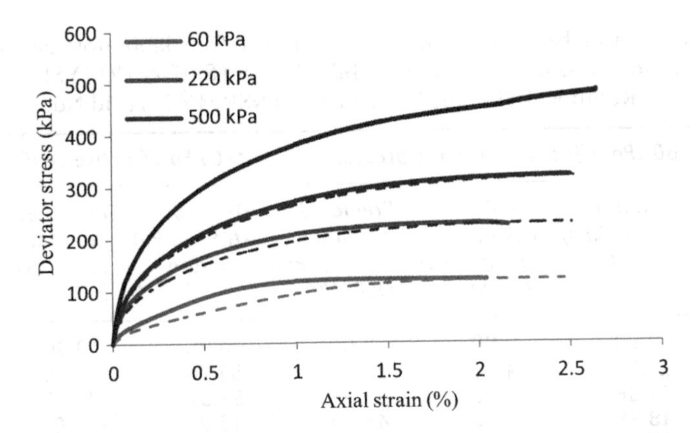

Figure 4.32 Stress–strain curves based on small strain measurement (continuous line) against normalised stress–strain (dotted lines) curves (Md Noor et al., 2017).

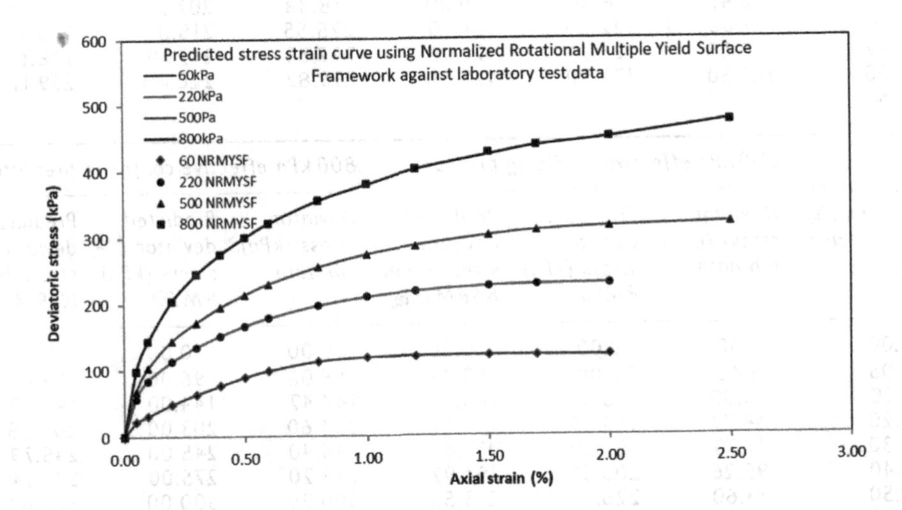

Figure 4.33 Predicted stress–strain curves (data points) using NSRMYSF plotted against the laboratory stress–strain curves (continuous line) show excellent agreement (Md Noor et al., 2017).

Good agreement between predicted deviator stresses (dotted points) and laboratory data (continuous line) at 60, 220, 500 and 800 kPa as shown in Figure 4.33 proved the effectiveness and better accuracy by the NSRMYSF in predicting soil stress–strain behaviour at any arbitrary effective stress value. Besides, the good prediction applies to a wide range of stress from low to high effective stresses when the curvilinear mobilised shear strength envelope is deployed. The existence of the unique relationship $\phi'_{\text{min}_{\text{mob}}} - \varepsilon_a$ irrespective of the values of the effective confining pressure applied during

Table 4.2 Comparison between deviator stress obtained from laboratory data and predicted deviator stress using Rotational Multiple Yield Surface (RMYSF) and Normalised Strain Rotational Multiple Yield Surface (NSRMYSF) (Md Noor, 2017)

Inversed normalised strains (%) (i.e., actual strain)	*60 kPa effective confining pressure*			*220 kPa effective confining pressure*		
	Deviator stress (kPa) lab data	*Predicted deviator stress (kPa) RMYSF*	*Predicted deviator stress (kPa) NSRMYSF*	*Deviator stress (kPa) lab data*	*Predicted deviator stress (kPa) RMYSF*	*Predicted deviator stress (kPa) NSRMYSF*
0.00	0.00	0.00	0.00	0.00	0.00	0.00
0.05	21.91	24.00	22.10	57.70	51.50	57.60
0.10	31.36	32.00	31.86	85.52	71.00	84.72
0.20	48.78	50.00	48.79	113.20	103.00	113.31
0.30	63.26	62.50	63.51	134.80	123.00	134.59
0.40	76.50	76.00	76.72	152.44	143.50	152.42
0.50	88.64	88.50	88.86	166.80	160.00	166.64
0.60	98.57	97.00	98.60	178.58	172.00	178.69
0.80	112.17	112.00	112.00	197.10	192.00	196.93
1.00	118.34	115.50	118.49	209.80	200.00	210.62
1.20	120.41	118.50	120.00	218.38	207.00	218.33
1.50	121.08	122.00	121.00	225.55	218.00	225.46
1.70	120.88	121.00	121.00	227.88	219.00	228.13
2.00	120.50	122.50	122.50	229.83	220.00	229.11
2.50						

Inversed normalised strains (%) (i.e., actual strain)	*500 kPa effective confining pressure*			*800 kPa effective confining pressure*		
	Deviator stress (kPa) lab data	*Predicted deviator stress (kPa) RMYSF*	*Predicted deviator stress (kPa) NSRMYSF*	*Deviator stress (kPa) lab data*	*Predicted deviator stress (kPa) RMYSF*	*Predicted deviator stress (kPa) NSRMYSF*
0.00	0.00	0.00	0.00	0.00	0.00	0.00
0.05	68.46	72.00	68.02	99.08	96.00	99.15
0.10	103.58	106.00	103.44	144.42	144.00	144.23
0.20	144.72	150.50	144.16	204.60	203.00	204.88
0.30	171.70	182.50	171.45	244.40	245.00	245.79
0.40	195.26	206.50	194.92	275.20	275.00	275.14
0.50	213.60	226.50	213.55	300.20	300.00	300.67
0.60	229.83	242.00	229.91	321.64	320.00	321.44
0.80	254.46	270.00	254.80	355.85	354.00	355.85
1.00	273.76	284.00	273.45	379.60	378.00	381.18
1.20	287.09	298.50	286.43	402.56	400.00	402.60
1.50	301.60	314.50	301.52	424.40	423.00	427.15
1.70	309.20	322.00	308.80	437.98	437.00	437.98
2.00	315.36	331.50	315.37	450.59	450.00	450.59
2.50	320.50		320.67	476.45		474.00

the tests proves that there is a simultaneous rotation of the surface envelope about the suction axis during the compression of the specimen in the shearing stage. The existence of this unique relationship is the keystone of the RMYSF introduced by Md Noor and Anderson (2006) and the NSRMYSF introduced by Md Noor et al. (2017).

Therefore, the prediction of stress–strain behaviour at any effective stress is possible by utilizing this unique relationship. Moreover, the good prediction obtained from this unique relationship can lead to a greater accuracy of soil deformation modelling.

4.4 Application of the NSRMYSF to predict the stress–strain behaviour of remoulded granitic residual soils grade V from Kuala Kubu Baharu, Malaysia

Consolidated drained triaxial tests were conducted at effective stresses of 50, 100, 200 and 300 kPa on four remoulded identical specimens of saturated granitic residual soil grade V taken from Kuala Kubu Baharu, Malaysia and the soil is classified as silty SAND (Rahman et al., 2017). The samples were taken at Kuala Kubu Baharu, Selangor at coordinates of 3°34′06.17″N; 101°41′51.50″E at depth of 1.5 m below the ground surface. The dimensions of specimens are 50 mm diameter and 100 mm height. The remoulded specimens were prepared using the same initial moisture content and weight, and then compacted in three layers using a rod of size 25 mm diameter and 350 mm long weighing 200 g. Twenty-five numbers of blows were applied at each layer.

The stress–strain curves obtained from the four triaxial tests are shown in Figure 4.34. Table 4.3 shows actual axial strain and the corresponding values of deviator stress at

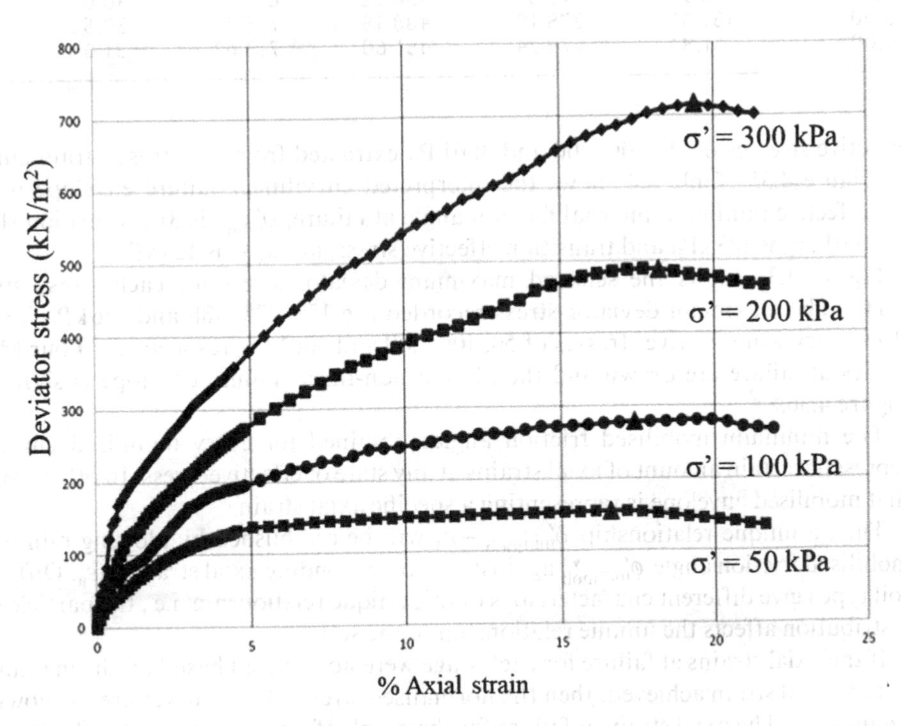

Figure 4.34 Stress–strain curves at effective stresses of 50, 100, 200 and 300 kPa obtained from consolidated drained triaxial tests on remoulded granitic residual soil grade V from Kuala Kubu Baharu, Malaysia showing the marked axial strains at failure (Rahman et al., 2017).

Table 4.3 Actual axial strain and the corresponding values of deviator stress at effective stresses of 50, 100, 200 and 300kPa extracted from the stress–train curves in Figure 4.35 (Rahman et al., 2017)

Actual axial strain, %	*Deviator stress for 50kPa*	*Deviator stress for 100kPa*	*Deviator stress for 200kPa*	*Deviator stress for 300kPa*	*Minimum mobilised friction angle $\phi'_{min_{mob}}$*
1.00	53.85	82.44	123.95	189.63	12.0
2.00	86.98	118.56	169.36	244.85	14.0
3.00	110.17	165.43	210.15	297.57	16.0
4.00	124.68	185.13	246.52	338.15	17.0
5.00	136.83	198.03	278.79	380.63	18.0
6.00	139.13	212.77	305.87	423.73	20.0
7.00	143.33	222.44	329.05	457.64	22.0
8.00	144.94	230.43	351.52	490.06	23.0
9.00	147.62	242.71	370.11	512.97	24.0
10.00	148.69	246.37	387.62	541.10	25.0
11.00	148.80	254.14	402.43	563.63	26.0
12.00	150.42	259.49	416.35	586.87	27.0
13.00	151.32	263.67	434.79	607.04	28.0
14.00	152.12	268.93	453.71	628.54	28.5
15.00	153.04	272.56	469.80	654.98	29.0
16.00	154.42	274.30	480.97	674.60	29.5
17.00	154.79	277.54	486.56	688.89	30.0
18.00	151.81	278.19	488.18	705.14	30.5
19.00	151.84	279.74	481.60	713.67	31.0

effective stresses of 50, 100, 200 and 300kPa extracted from the stress–strain curves in Figure 4.35. Table 4.4 shows the interpreted curvilinear failure envelope where the effective minimum internal friction angle at failure, ϕ'_{minf}, is 31°, transition shear strength, τ_t, is 183kPa and transition effective stress, $(\sigma - u_w)_t$, is 124kPa.

Figure 4.35 shows the selected maximum deviator stress for each stress–strain curve. The maximum deviator stress recorded are 155, 278, 488 and 716kPa for the stress curves at effective stresses of 50, 100, 200 and 300kPa, respectively. Four Mohr circles at failure are drawn and the plotted non-linear failure envelope is shown in Figure 4.36.

The minimum mobilised friction angles obtained for every mobilised envelope represent certain amount of axial strains at any state of effective stress. In other words, that mobilised envelope is representing a specific axial strain.

Then a unique relationship $\phi'_{min_{mob}} - \varepsilon_a$ will be established by plotting minimum mobilised friction angle $\phi'_{min_{mob}}$ against the corresponding axial strain, $\%\varepsilon_a$. Different soil types give different characteristics of this unique relationship, i.e., the particle size distribution affects the unique relationship of the soil.

If the axial strains at failure for each stage were normalised based on the maximum failure axial strain achieved, then the normalised stress–strain curves are as shown in Figure 4.37. The axial strain at failure for the graph of effective stress 300kPa is taken as the targeted common axial strain for all stress–strain curves. Then the axial strain values for the rest of the stress–strain curves (i.e., stress–strain at effective stresses 50, 100 and 200kPa) are multiplied by their specific normalising factor so that the

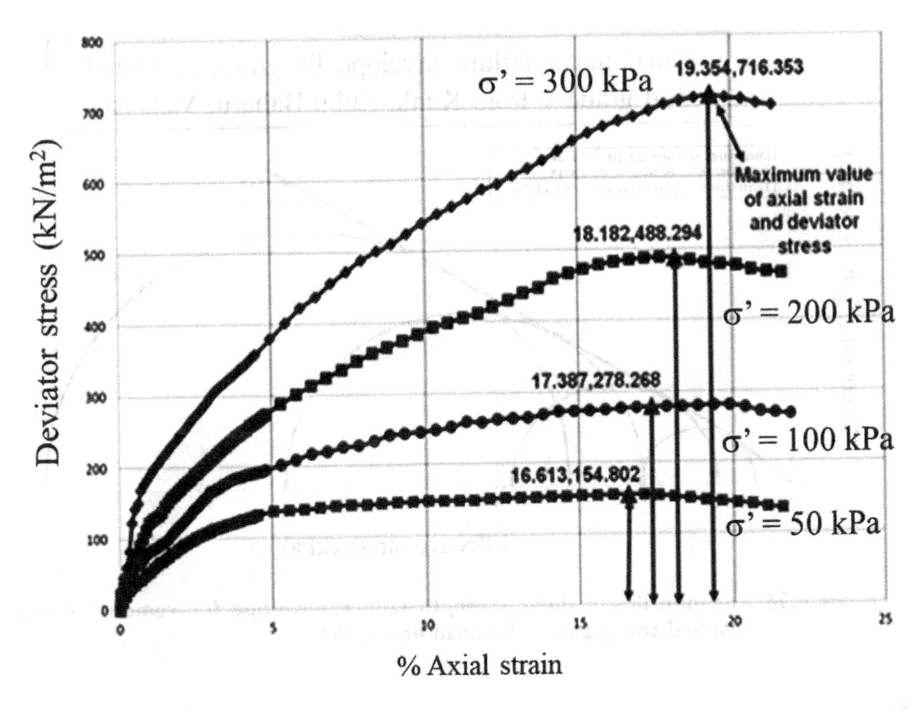

Figure 4.35 Stress–strain curves at effective stresses of 50, 100, 200 and 300 kPa showing the values of axial strains and deviator stresses at failure (Rahman et al., 2017).

Table 4.4 Effective shear strength parameters for non-linear failure envelope at saturation

Targeted effective stress (kPa)	*Condition of failure*			*Shear strength parameters for non-linear envelope*		
	Deviator stress (kPa)	*Pore water pressure (kPa)*	*Cell pressure (kPa)*	ϕ'_{minf}	τ_t *(kPa)*	$(\sigma - u_w)_t$ *(kPa)*
50	155	439	500	31°	183	124
100	278	443	550			
200	488	450	650			
300	716	451	750			

normalised curve will be stretched to the right to have the same maximum strain at failure as the stress–strain curve of effective stress 300 kPa. The normalised stress–strain curves are shown in Figure 4.37. The figure also shows the peak deviator stresses for 50, 100, 200 and 300 kPa effective stress curves that have been normalised. The common normalised axial strain at failure for all stress–strain curves is 19.354%.

Table 4.5 shows the normalised axial strains and their conversion factor and the corresponding deviator stresses for the stress–strain curves at effective stresses of 50, 100, 200 and 300 kPa. The normalised axial strain is calculated by multiplying the

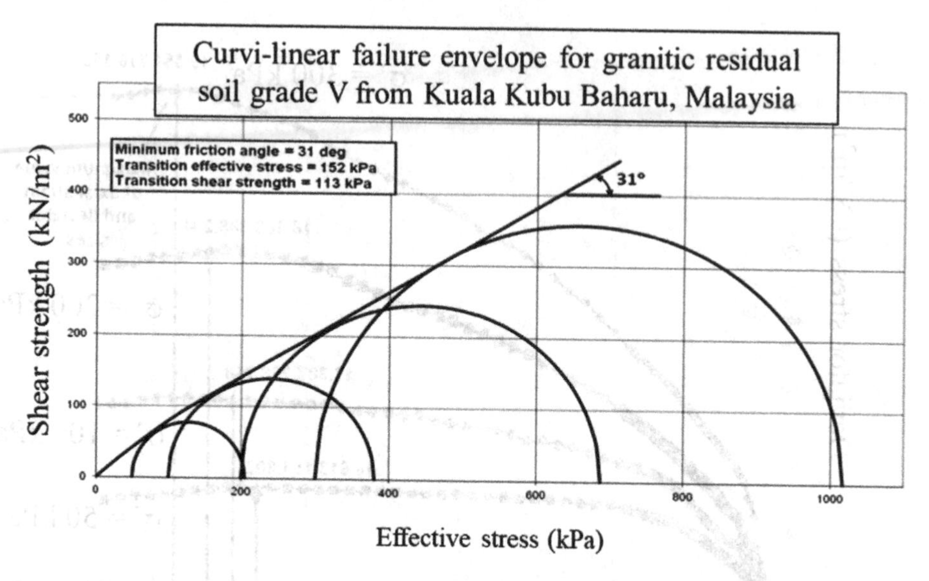

Figure 4.36 The non-linear shear strength failure envelope for remoulded granitic residual soil grade V (Rahman et al., 2017).

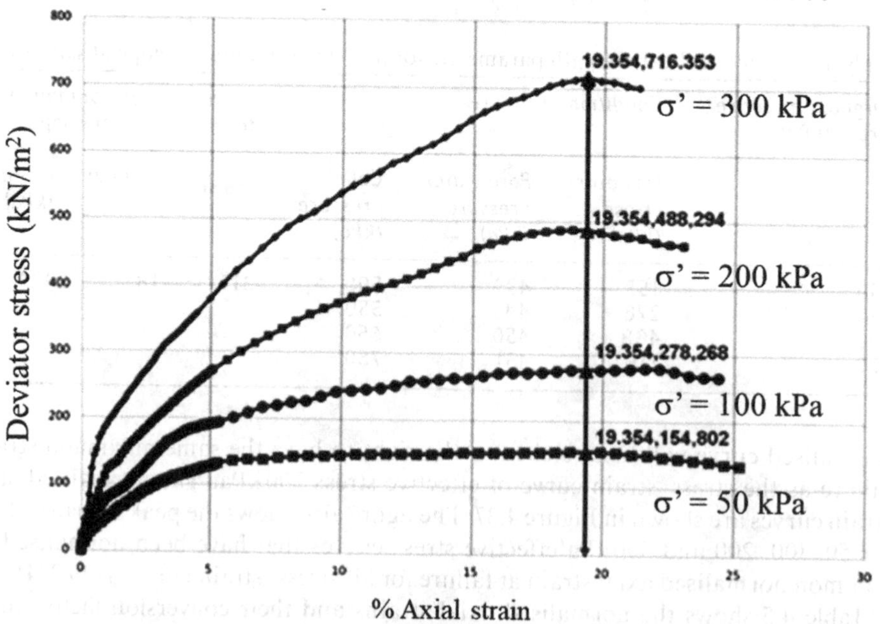

Figure 4.37 Normalised stress–strain curves at effective stresses of 50, 100, 200 and 300 kPa (Rahman et al., 2017).

Table 4.5 Normalised axial strain and their conversion factor and the corresponding deviator stress for stress – strain graphs of effective stress of 50, 100, 200 and 300 kPa (Rahman et al., 2017)

Normalised axial strain, %	*Deviator stress for a curve of 50 kPa*	*Deviator stress for a curve of 100 kPa*	*Deviator stress for a curve of 200 kPa*	*Deviator stress for a curve of 300 kPa*	*Minimum mobilised friction angle*
Normalised conversion factor	1.165	1.113	1.064	1.000	
1.00	49.41	79.17	119.45	189.63	12.0
2.00	79.49	109.44	164.01	244.87	14.0
3.00	101.78	150.60	202.81	297.57	16.0
4.00	116.88	179.61	238.41	338.15	17.0
5.00	128.70	191.81	269.87	380.63	18.0
6.00	137.54	203.40	296.53	423.73	20.0
7.00	139.13	217.50	319.61	457.64	22.0
8.00	142.79	224.56	340.70	490.06	23.0
9.00	144.95	231.47	360.58	512.97	24.0
10.00	146.23	242.65	376.90	541.10	25.0
11.00	147.90	245.73	393.43	563.63	26.0
12.00	148.68	251.37	406.20	586.87	27.0
13.00	148.90	258.61	419.86	607.04	28.0
14.00	150.48	262.72	437.94	628.54	28.5
15.00	151.40	264.75	456.07	654.98	29.0
16.00	151.66	271.81	470.21	672.60	29.5
17.00	152.56	272.70	480.70	688.89	30.0
18.00	153.66	275.08	486.16	705.14	30.5
19.00	154.66	277.67	488.08	713.67	31.0
19.35	154.80	278.27	488.29	716.35	31.0

actual axial strain with a conversion factor for each of the stress–strain curves. As an example, to normalise the stress–strain curve for an effective stress of 50 kPa, each of the axial strains must be multiplied by the normalising factor of 1.165 and thus the axial strain will change to a new value but the value of deviator stress remains the same. Thence, the curve will be stretched to the right. Similarly, the normalising factors are 1.113, 1.064 and 1.00 for curves of effective stresses 100, 200 and 300 kPa, respectively. For the curve at an effective stress of 300 kPa, the normalised axial strain will remain the same, because the normalising factor is unity. Figure 4.38 presents the mobilised shear strength envelopes plotted for axial strains 1%–19% with an increment of 1% and their corresponding minimum mobilised friction angle, $\phi'_{min_{mob}}$.

Figure 4.39a–d shows the Mohr circles for the prediction of the deviator stresses at the respective axial strains for effective stresses of 50, 100, 200 and 300 kPa, respectively. Data in Table 4.6 present the predicted deviator stresses at effective stresses of 50, 100, 200 and 300 kPa and the corresponding normalised axial strain with the inverse factor. Then finally, the plotting of the actual predicted stress–strain curves is by applying the actual strain versus the predicted deviator stresses.

Figure 4.40 shows the predicted deviator stresses plotted against normalised axial strains and superimposed against the laboratory normalised stress–strain curves

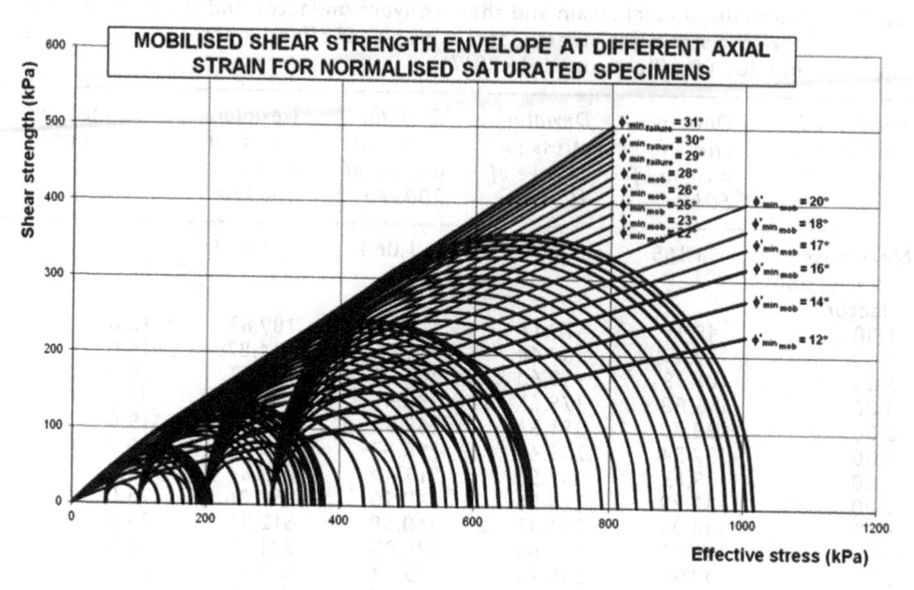

Figure 4.38 Failure envelopes and the mobilised shear strength envelopes for axial strains 1%–19% with an increment of 1% and their corresponding minimum mobilised friction angle $\phi'_{\text{min}_{\text{mob}}}$ for saturated specimens determined from consolidated drained triaxial tests at effective stresses of 50, 100, 200 and 300 kPa (Rahman et al., 2017).

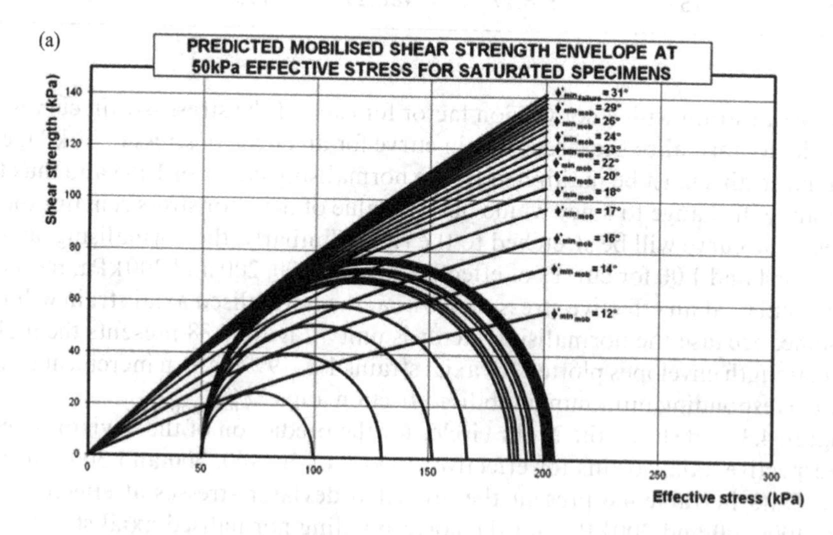

Figure 4.39 Predicted deviator stresses at 50, 100, 200 and 300 kPa effective stresses (Rahman et al., 2017). (a) Predicted deviator stress at an effective stress of 50 kPa. (b) Predicted deviator stress at an effective stress of 100 kPa. (c) Predicted deviator stress at an effective stress of 200 kPa. (d) Predicted deviator stress at an effective stress of 300 kPa.

(Continued)

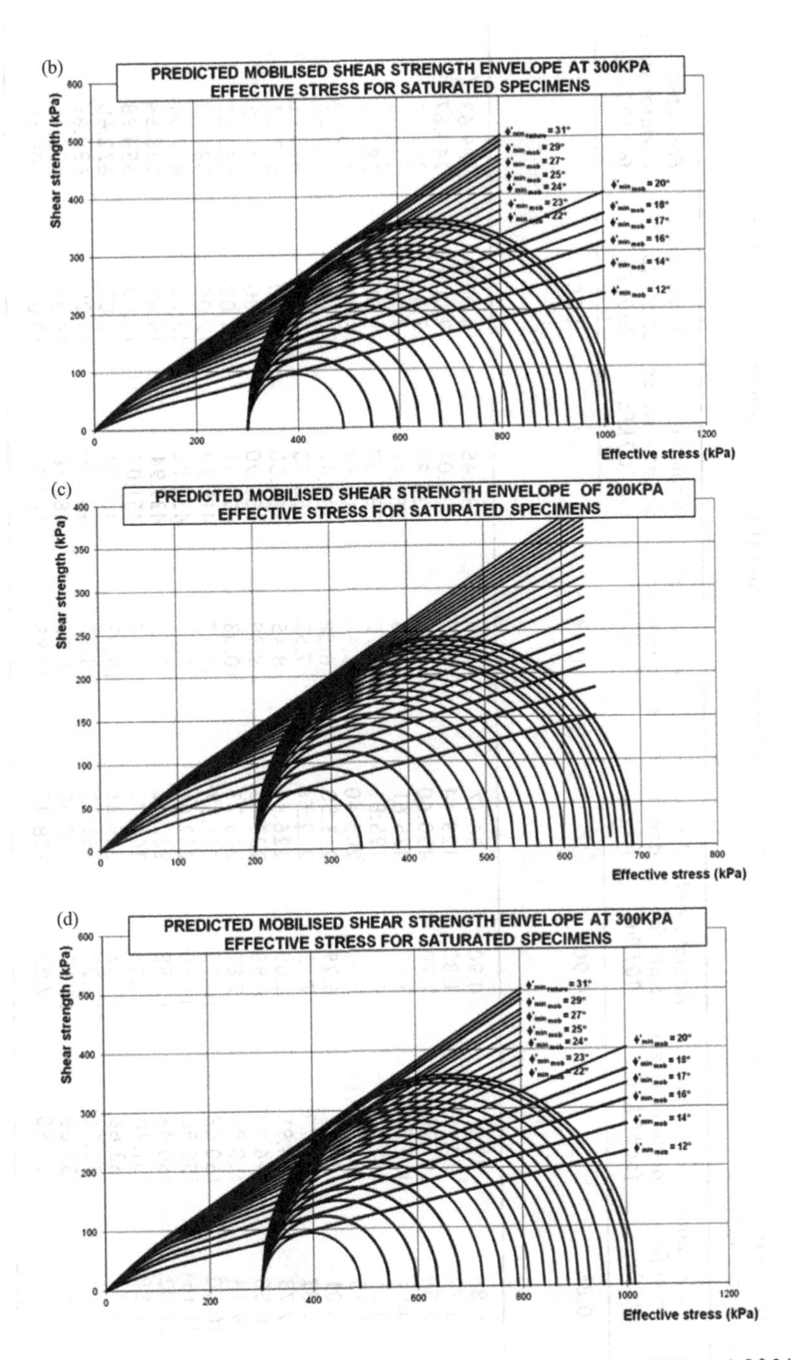

Figure 4.39 (Continued) Predicted deviator stresses at 50, 100, 200 and 300 kPa effective stresses (Rahman et al., 2017). (a) Predicted deviator stress at an effective stress of 50 kPa. (b) Predicted deviator stress at an effective stress of 100 kPa. (c) Predicted deviator stress at an effective stress of 200 kPa. (d) Predicted deviator stress at an effective stress of 300 kPa.

Table 4.6 Predicted deviator stress at effective stress of 50, 100, 200 and 300kPa and the corresponding normalised axial strains and their inverse factors

Normalised axial strain, %	*Actual axial strain for 50 kPa*	*Predicted deviator stress for 50 kPa*	*Actual axial strain for 100 kPa*	*Predicted deviator stress for 100 kPa*	*Actual axial strain for 200 kPa*	*Predicted deviator stress for 200 kPa*	*Actual axial strain for 300 kPa*	*Predicted deviator stress for 300 kPa*
Normalised inverse factor	0.86		0.90		0.94		1.00	
1.00	0.86	49.41	0.90	86.17	0.94	138.45	1.00	189.63
2.00	1.72	79.49	1.80	123.44	1.88	186.01	2.00	244.87
3.00	2.58	101.78	2.70	150.60	2.82	225.81	3.00	297.57
4.00	3.43	116.88	3.59	175.61	3.76	258.41	4.00	338.15
5.00	4.29	135.70	4.49	195.81	4.70	285.87	5.00	380.63
6.00	5.15	137.54	5.39	205.40	5.64	310.53	6.00	423.73
7.00	6.01	139.13	6.29	213.50	6.58	334.61	7.00	457.64
8.00	6.87	142.79	7.19	222.56	7.52	350.70	8.00	490.06
9.00	7.73	144.95	8.09	226.47	8.45	365.58	9.00	512.97
10.00	8.58	146.23	8.98	236.65	9.39	382.90	10.00	541.10
11.00	9.44	150.90	9.88	245.73	10.33	399.43	11.00	563.63
12.00	10.30	150.68	10.78	251.37	11.27	415.20	12.00	586.87
13.00	11.16	150.90	11.68	252.61	12.21	430.86	13.00	607.04
14.00	12.02	150.48	12.58	262.72	13.15	437.94	14.00	628.54
15.00	12.88	151.40	13.48	264.75	14.09	451.07	15.00	654.98
16.00	13.73	153.66	14.37	265.81	15.03	460.21	16.00	672.60
17.00	14.59	152.56	15.27	269.70	15.97	468.70	17.00	688.89
18.00	15.45	153.66	16.17	270.08	16.91	480.16	18.00	705.14
19.00	16.31	154.80	17.07	278.27	17.85	488.29	19.00	705.14

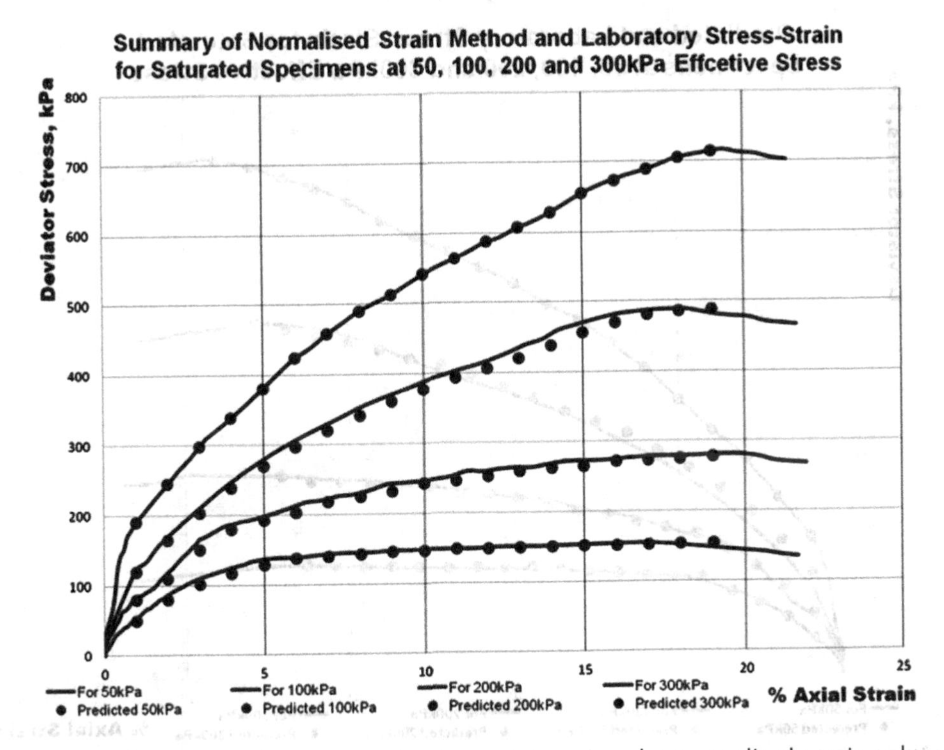

Figure 4.40 Predicted stress–strain data points using the normalised strain values plotted against the laboratory stress–strain curves at effective stresses of 50, 100, 200 and 300 kPa (Rahman et al., 2017).

at effective stresses of 50, 100, 200 and 300 kPa. The predicted deviator stresses are slightly underestimated for the graphs of effective stresses of 100 and 200 kPa.

However, the true predicted stress–strain curves are achieved when the normalised axial strains are reverted to the actual axial strain by multiplying the normalised axial strains with the inverted conversion factor. The result of the true predicted stress–strain curves obtained by applying the NSRMYSF is as shown in Figure 4.41. Essentially, the predicted and the laboratory stress–strain curves showed an excellent fit. Figure 4.42 shows the unique relationship graph for the remoulded granitic residual soil grade V from Kuala Kubu Baharu with the coefficient of anisotropic compression, Ω, of 0.21.

4.5 Application of normalised strain RMYSF to predict the stress–strain behaviour of compacted tropical sedimentary residual sandy soil

This is about characterising the anisotropic compression of a tropical residual sandy soil by the interaction between the mobilised shear strength, which is developed within the soil mass when it is being compressed, and the applied effective stress, which is the

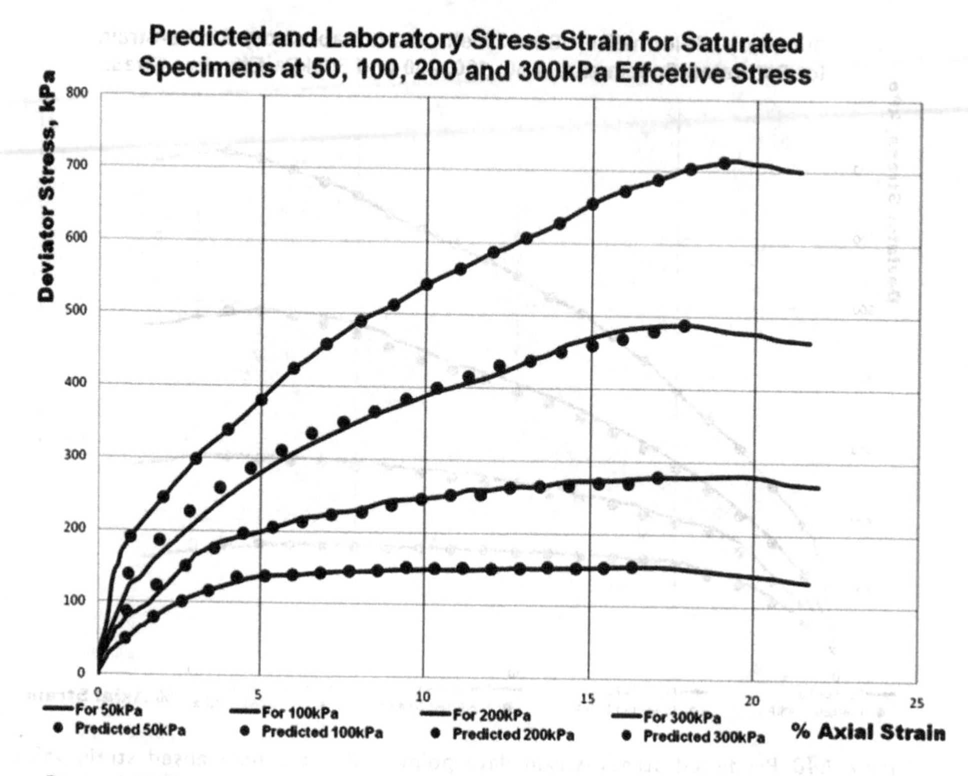

Figure 4.41 Predicted stress–strain data points using the NSRMYSF against the laboratory stress–strain curves for effective stresses of 50, 100, 200 and 300 kPa (Rahman et al., 2017).

concept adopted in the NSRMYSF as reported by Alias et al. (2019). The anisotropic compression is observed in consolidated drained triaxial tests conducted on four remoulded fully saturated specimens with 50 mm diameter and 100 mm height. The soil samples were taken from Semenyih, Hulu Langat district in Selangor with coordinates of 3°0
35.975″N; 101°51′51.145″E from a depth of 1.0 m below the ground surface, to represent the original residual soil.

The soil is reddish yellowish-brown well-graded SAND. The soil is derived from weathered sandstone of weathering grade VI and can be described as well-graded SAND according to the BSCS and SW according to USCS. The soil consists of less than 5% of fines and 72.4% sand particles. The specific gravity of the soil is equal to 2.57. The initial moisture content values range from 20% to 25%.

The soil specimens were subjected to saturation stage, consolidation and compression at the rate of 0.01 mm/min. The targeted effective stresses of 50, 100, 200 and 300 kPa were applied to obtain Mohr circles, which represent driving variables of the anisotropic compression. Shear strength parameters of the tested soil were defined according to the curved-surface envelope shear strength model (CSESSSM) of Md Noor and Anderson (2006). The analysis gave the values of effective internal minimum

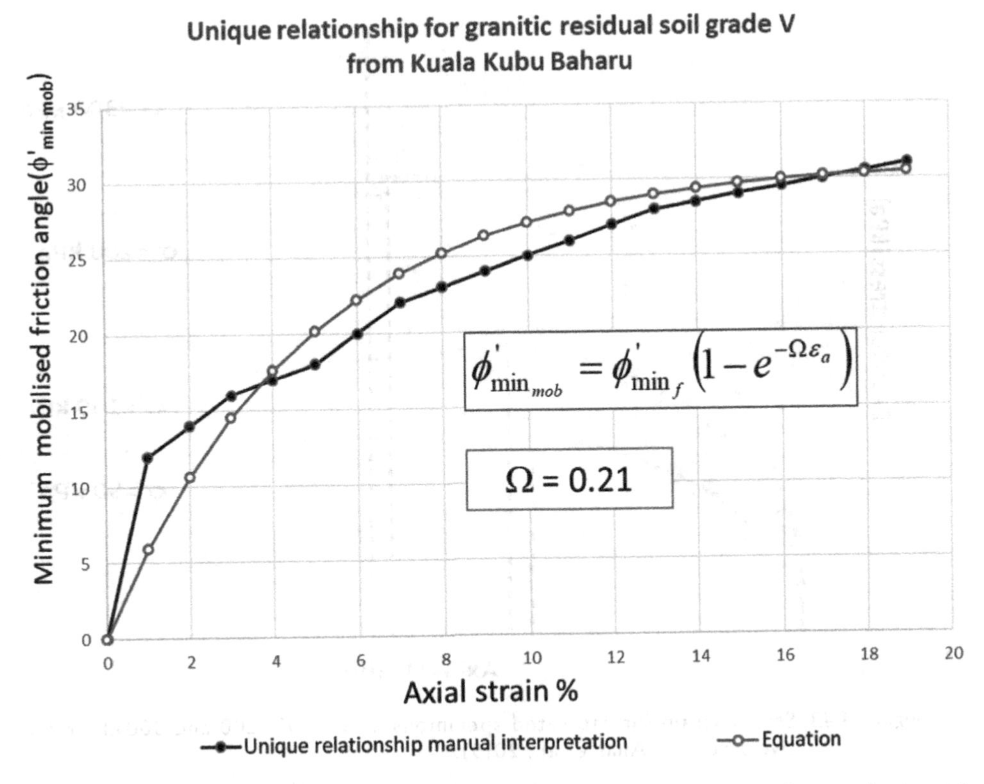

Figure 4.42 Graph of unique relationship plotted manually against the graph plotted using the Equation of Jais and Md Noor (2009) for granitic residual soil grade V with a coefficient of anisotropic compression Ω of 0.21 (Rahman et al., 2017).

friction angle at failure, ϕ'_{minf}, transition shear strength, τ_t, and transition effective stress, $(\sigma - u_w)_t$. The transition net stress is the value of net or effective stress of which lesser than the envelope has non-linear form and beyond that it has the linear form. The transition shear strength is the vertical height representing the strength that corresponds to transition net stress. The minimum friction angle is the inclination of the linear envelope from horizontal.

4.5.1 Stress–strain response and derivation of mobilised shear strength envelopes for tropical sedimentary residual sandy soil

Figure 4.43 shows the stress–strain relationship and the maximum value of deviator stress at failure. The maximum value of deviator stress recorded from the graph is 762 kPa and the corresponding axial strain is 11.13%. The values of effective stress and maximum deviator stress recorded at failure are used to draw Mohr circles and the failure envelopes. The curvilinear shear strength line is well fitted to the Mohr circle as shown in Figure 4.44 using CSESSSM. The application of the curvilinear envelope manages to avoid any underestimation or overestimation of the shear strength over

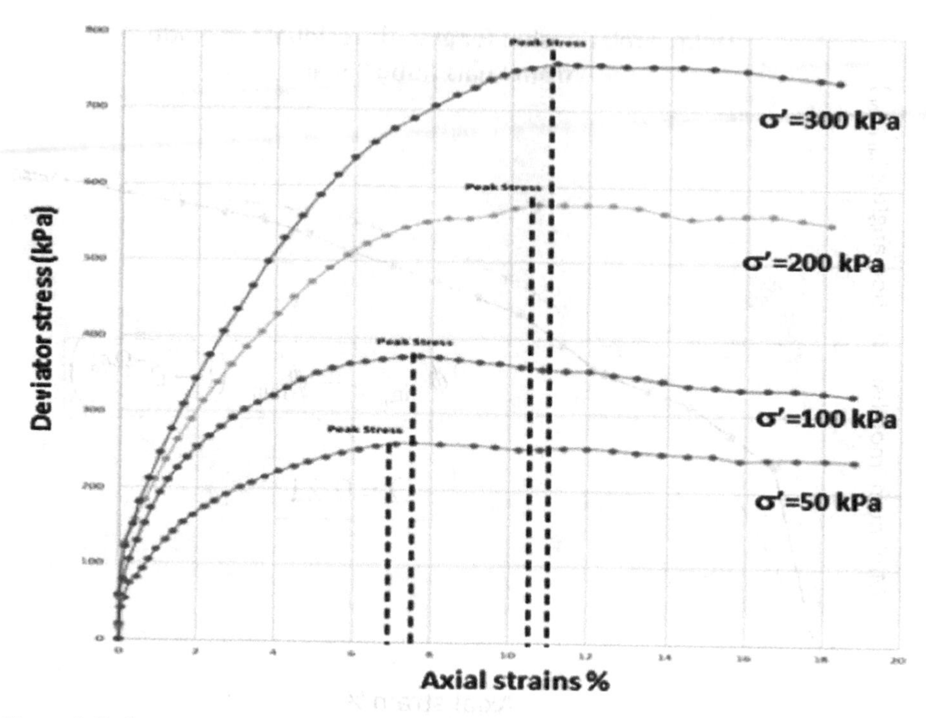

Figure 4.43 Stress strain for saturated specimens at 50, 100, 200 and 300 kPa effective stresses (Alias et al., 2019).

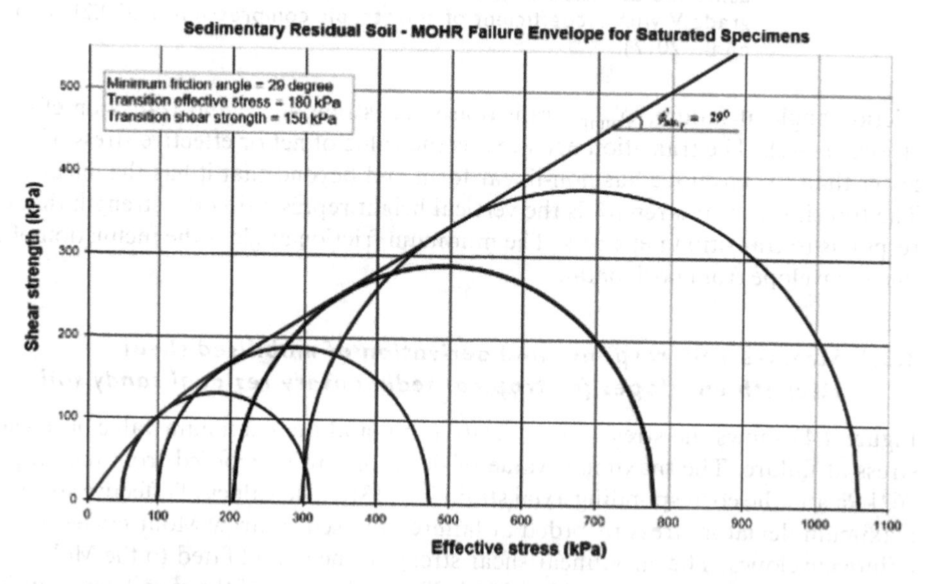

Figure 4.44 Non-linear shear strength failure envelope for saturated specimens (Alias et al., 2019).

Table 4.7 Effective shear stress parameters at failure for non-linear failure envelope at saturation (Alias et al., 2019)

Effective stress (kPa)	*Condition at failure*			*Shear strength parameters*		
	Deviator stress (kPa)	*Pore water pressure (kPa)*	*Cell pressure (kPa)*	ϕ'_{minf}	τ_t *(kPa)*	$(\sigma-u_w)_t$ *(kPa)*
50	262	351	400	29.4°	158	180
100	376	352	450			
200	576	346	550			
300	763	352	650			

the whole stress range (Md Noor, 2016). The values of ϕ'_{minf} , τ_t and $(\sigma-u_w)_t$ were determined and are presented in Table 4.7, where the effective internal friction angle at failure, ϕ'_{minf}, is 29.4° with the transition shear strength τ_t of 158 kPa and transition effective stress $(\sigma-u_w)_t$ of 180 kPa.

Thence, to determine the mobilised shear strength envelopes of the soil, the stress–strain curves need to be normalised so that all the curves will have the same axial strain at failure. The strains for each curve will be normalised so that all the stress–strain curves have a common axial strain at failure by multiplying by a normalised strain factor. The normalised axial strain at failure must follow the curve of maximum axial strain at failure, which is the topmost stress–strain curve representing an effective confining pressure of 300 kPa. The normalised stress–strain curves are shown in Figure 4.45.

Normalised axial strains of 1%–10% with an increment of 1% have been selected for the determination of the mobilised shear strength envelopes. Then the magnitude of deviator stresses corresponding to each of the strain values will be extracted from the stress–strain curves. This is followed by the drawing of the respective Mohr circles and the drawing of the curvilinear shear strength envelopes. The minimum mobilised internal friction angle $\phi'_{min\,mob}$ will be determined for each envelope. The Mohr circles are plotted and the mobilised shear strength envelopes are defined as shown in Figure 4.46. Table 4.8 presents the normalised axial strains and the corresponding parameters used to define the mobilised shear strength envelopes.

The unique relationship of this soil is plotted in Figure 4.47. This is the relationship between the minimum mobilised friction angle, $\phi'_{min_{mob}}$, and the normalised axial strains, ε_a, as presented in Table 4.8. Essentially, the coefficient of the soil anisotropic compression, Ω, is 0.4.

4.5.2 *Prediction of stress–strain response using the NSRMYSF for tropical sedimentary residual sandy soil*

Once the mobilised shear strength envelopes have been obtained for the soils as in Figure 4.46a and b for axial strains of 1%–10%, then the stress–strain curves at any effective stress can be determined. In other words, the stress–strain response of the

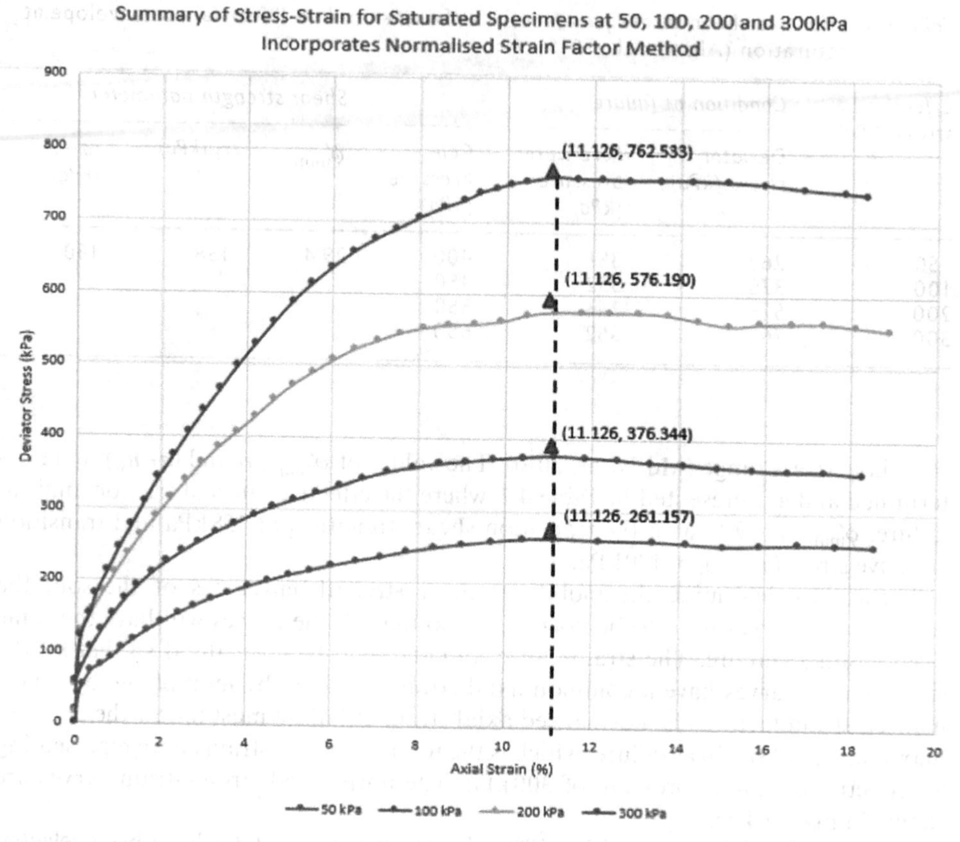

Figure 4.45 The normalised stress–strain curves have a common axial strain at failure, which is 11.13% following the curve of effective stress 300 kPa (Alias et al., 2019).

soil at any depth (i.e., at any effective confining pressure) can be determined. This can avoid the hassle of conducting laboratory tests at various depths to obtain the consolidation coefficients at those depths when the soil formation is discretised into many layers for the settlement determination. In this verification process, the stress prediction is conducted at the same effective stresses with which consolidated drained triaxial tests have been conducted. Then the prediction can be compared with the laboratory test results. Figures 4.48–4.51 show the construction of the predicted Mohr circles that just touches the specific mobilised shear strength envelope for effective stresses of 50, 100, 200 and 300 kPa. The predicted deviator stresses at every considered normalised axial strain are extracted from the diameter of the predicted Mohr circles and are presented in Table 4.9. Figure 4.52 shows the plot of the predicted deviator stresses versus normalised axial strains. To get the actual axial strain values, the values of normalised axial strains will be multiplied by the normalisation inverse factor. This is then followed by the plotting of the predicted stress–strain curves, which gives the graphs of predicted deviator stresses versus actual axial strains. The predicted stress–strain graphs plotted superimposed with the laboratory stress–strain curves

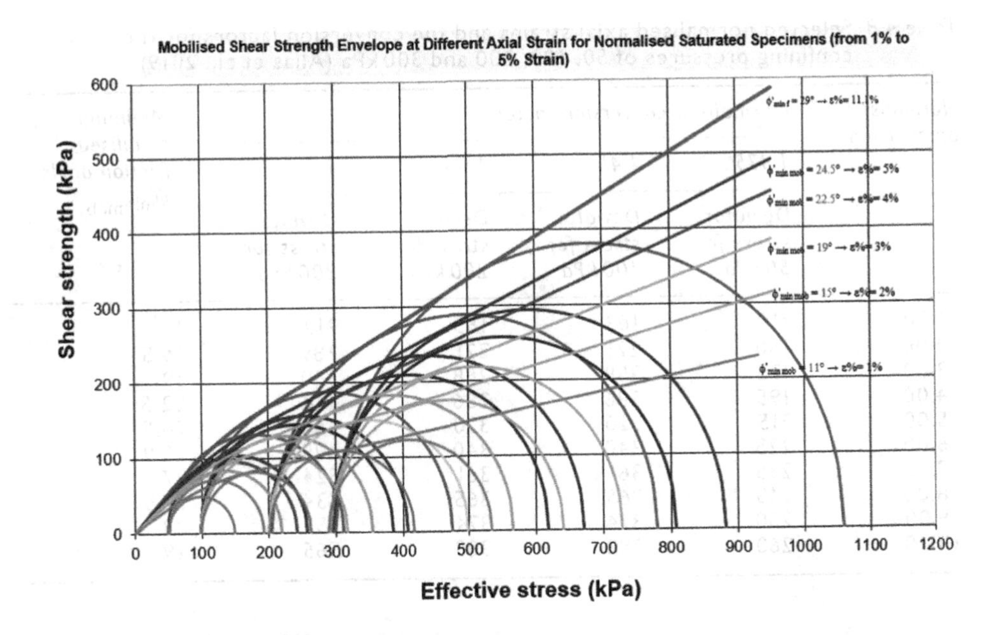

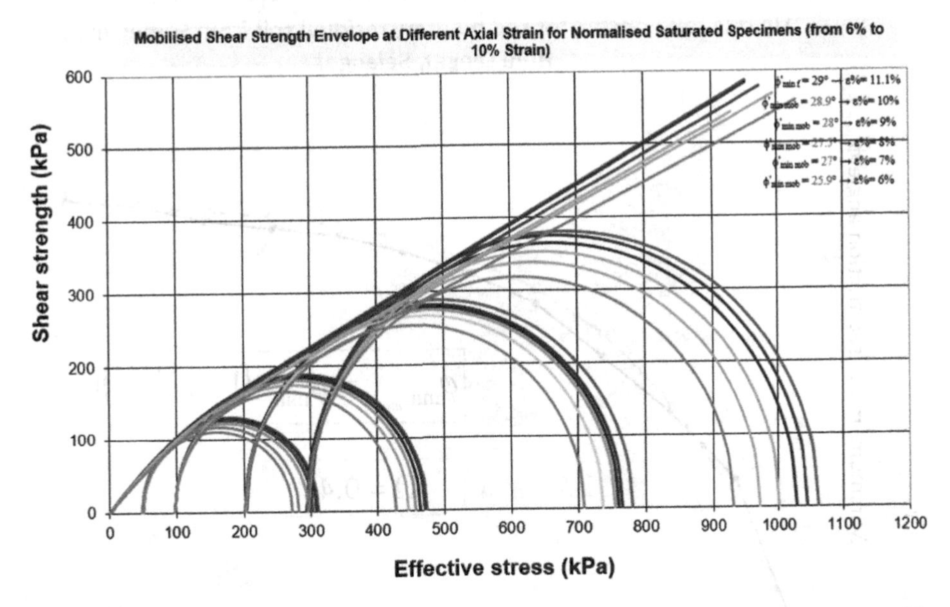

Figure 4.46 Mobilised shear strength envelopes for saturated specimens at 50, 100, 200 and 300 kPa effective confining pressures (Alias et al., 2019). (a) Mobilised shear strength envelopes for 1%–5% axial strains (Alias et al., 2019). (b) Mobilised shear strength envelopes for 6%–11% axial strains.

Table 4.8 Selected normalised axial strains and the conversion factors for the effective confining pressures of 50, 100, 200 and 300 kPa (Alias et al., 2019)

Normalised axial strain, %	*Normalised conversion factor*				*Minimum mobilised friction angle* ϕ'_{minmob}
	1.475	*1.45*	*1.04*	*1*	
	Deviator stress for 50 kPa	Deviator stress for 100 kPa	Deviator stress for 200 kPa	Deviator stress for 300 kPa	
1.00	110	163	163	212	11
2.00	150	221	221	289	15.5
3.00	175	258	258	360	19
4.00	195	286	286	418	22.5
5.00	215	320	320	460	24.5
6.00	225	340	340	495	25.9
7.00	235	361	361	524	27
8.00	245	365	365	534	27.5
9.00	250	374	374	555	28
10.00	260	383	383	565	29

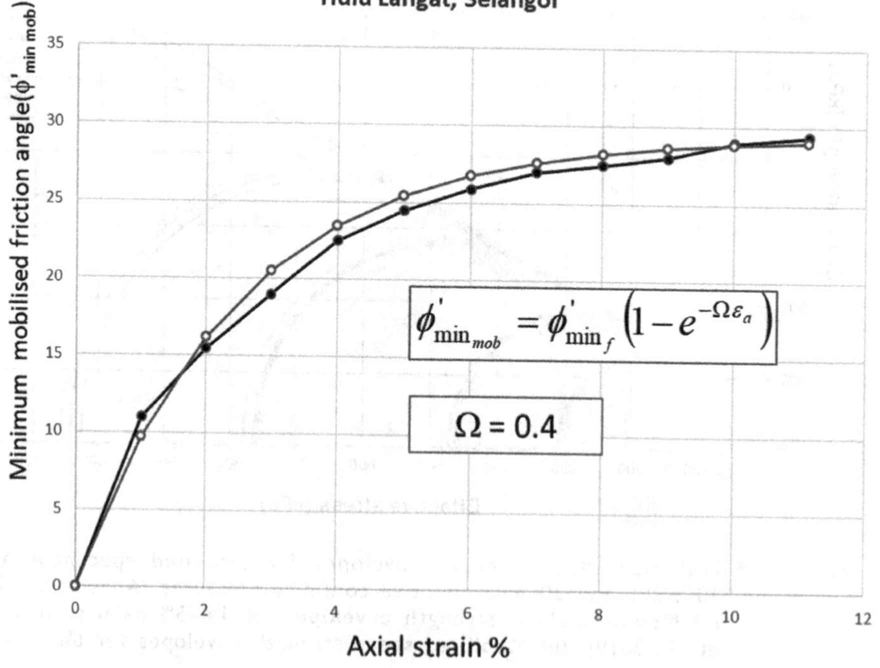

Figure 4.47 Unique relationship for well-graded SAND from Semenyih, Hulu Langat, Selangor, Malaysia (Alias et al., 2019).

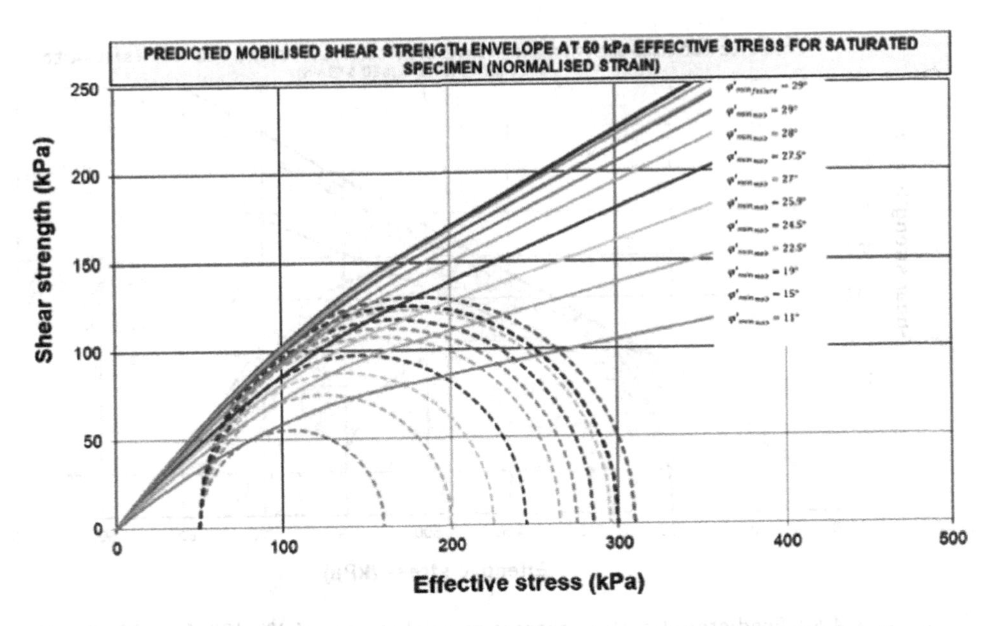

Figure 4.48 Predicted deviator stresses at axial strains of 1%–10% for 50 kPa effective stress (Alias et al., 2019).

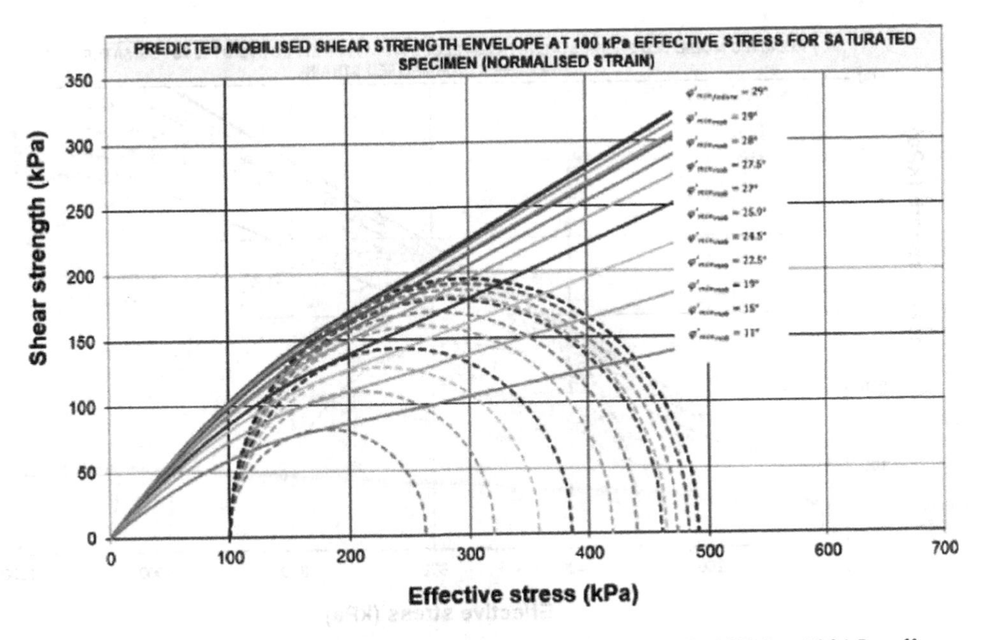

Figure 4.49 Predicted deviator stresses at axial strains of 1%–10% for 100 kPa effective stress (Alias et al., 2019).

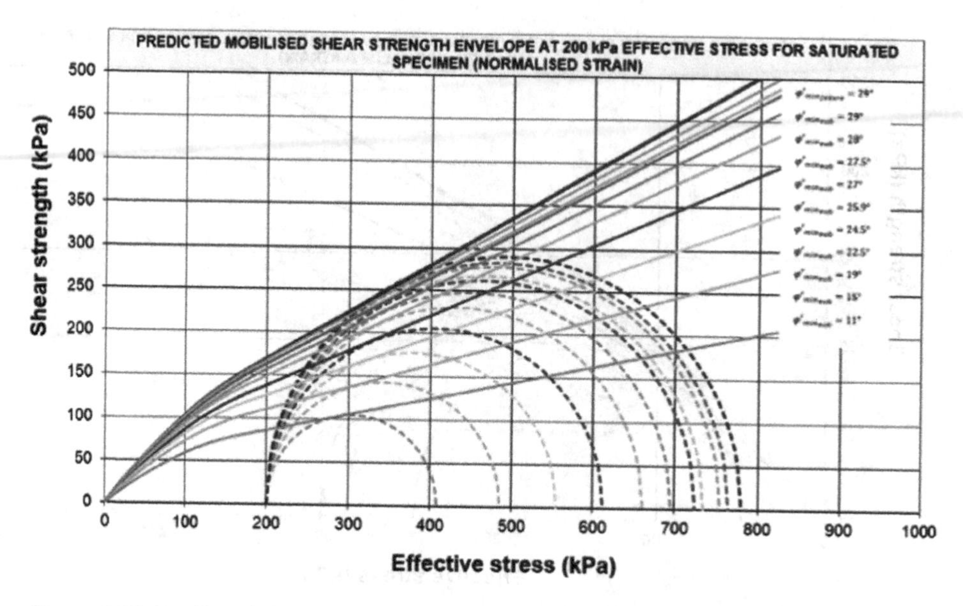

Figure 4.50 Predicted deviator stresses at axial strains of 1%–10% for 200 kPa effective stress (Alias et al., 2019).

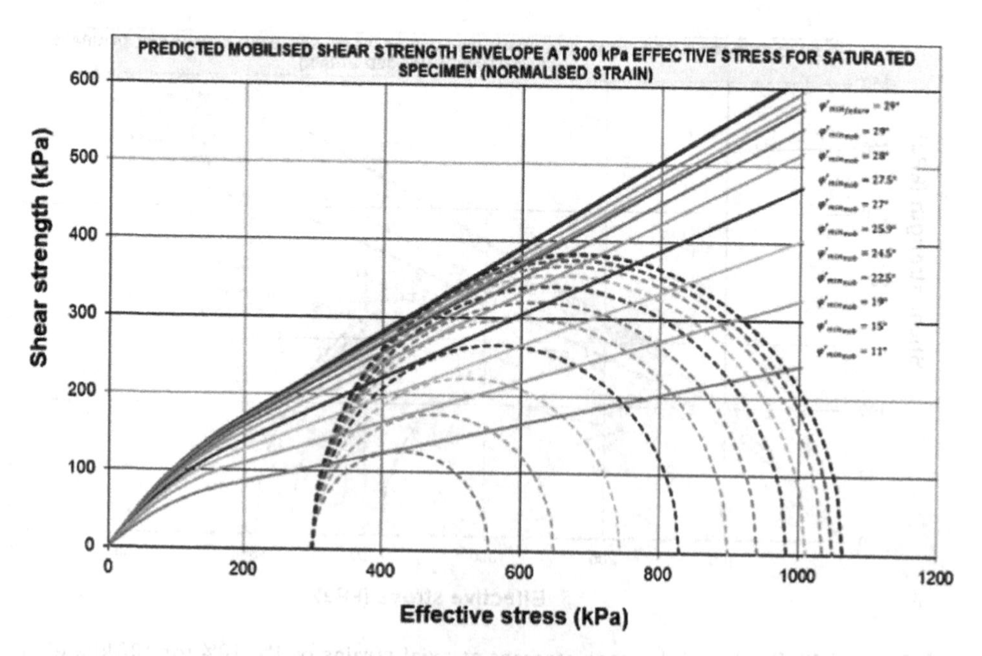

Figure 4.51 Predicted deviator stresses at axial strains of 1%–10% for 300 kPa effective stress (Alias et al., 2019).

Table 4.9 Normalised axial strain with inverse factor and the predicted deviator stress at effective stresses of 50, 100, 200 and 300 kPa

Normalised axial strain, %	*Actual axial strain for 50 kPa*	*Predicted deviator stress for 50 kPa*	*Actual axial strain for 100 kPa*	*Predicted deviator stress for 100 kPa*	*Actual axial strain for 200 kPa*	*Predicted deviator stress for 200 kPa*	*Actual axial strain for 300 kPa*	*Predicted deviator stress for 300 kPa*
	Inverse factor							
	0.678		0.690		0.962		1.00	
1.00	0.678	110	0.690	160	0.962	207	1	212
2.00	1.356	150	1.380	217	1.923	285	2	289
3.00	2.034	175	2.070	258	2.885	355	3	360
4.00	2.712	195	2.760	286	3.846	417	4	418
5.00	3.390	215	3.450	320	4.808	460	5	460
6.00	4.068	225	4.140	340	5.769	495	6	495
7.00	4.745	235	4.830	361	6.731	524	7	524
8.00	5.423	245	5.519	365	7.692	534	8	534
9.00	6.101	250	6.209	374	8.654	555	9	555
10.00	6.779	260	6.899	383	9.615	565	10	565

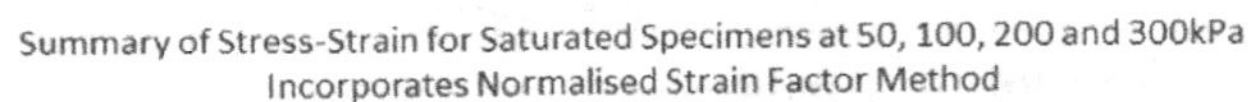

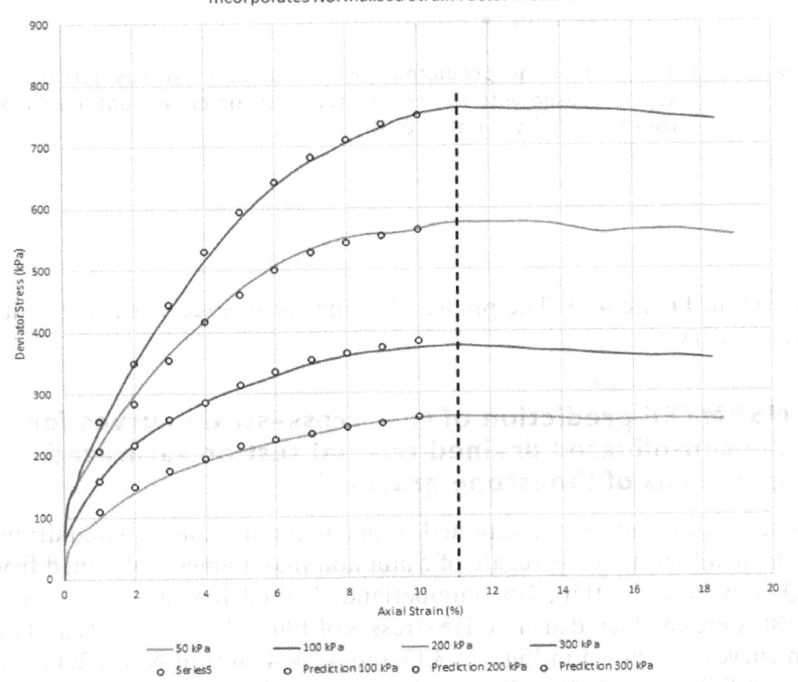

Figure 4.52 The plot of the predicted deviator stresses versus normalised axial strains superimposed with the normalised stress–strain curves obtained from laboratory tests (Alias et al., 2019).

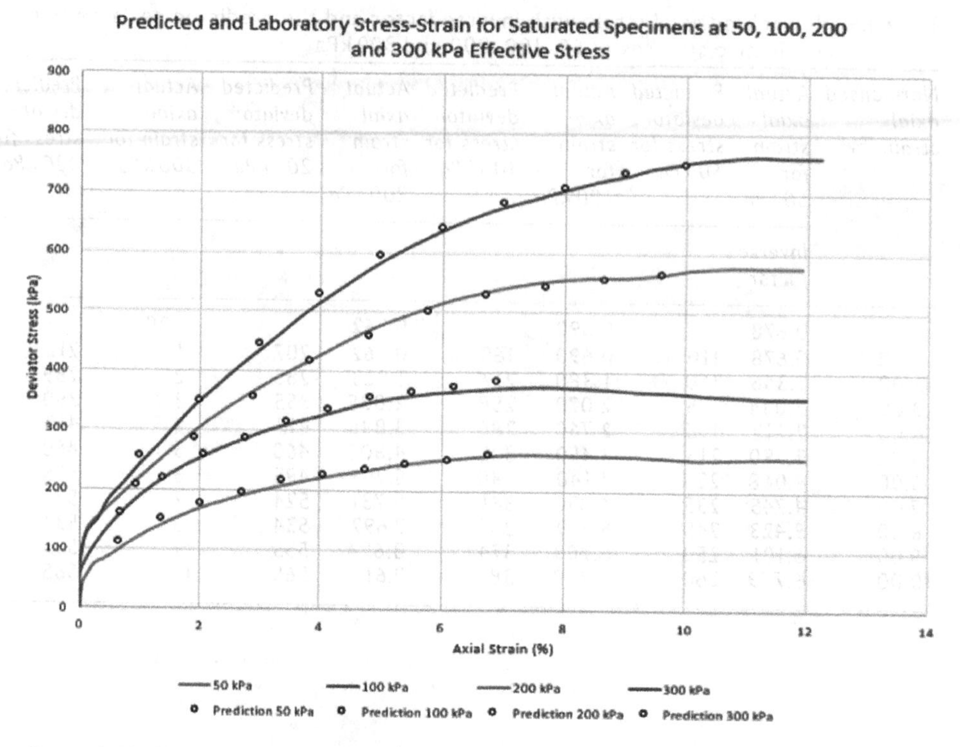

Figure 4.53 The plot of the predicted deviator stresses versus actual axial strains superimposed with the actual stress–strain curves obtained from laboratory tests (Alias et al., 2019).

are shown in Figure 4.53. The predicted graphs show excellent agreement with the laboratory curves.

4.6 NSRMYSF prediction of the stress–strain curves for the consolidated drained triaxial test on saturated specimens of limestone gravel

The stress–strain curves are obtained from conducting consolidated drained triaxial tests on a limestone aggregate of 5 mm nominal diameter obtained from Mootlaw Quarry, near Matfan, Northumberland, United Kingdom (Md Noor, 2006). The tests were conducted at effective stresses of 100, 200 and 300 kPa and the stress–strain curves are shown in Figures 3.13 and 4.54–4.56 (Md Noor, 2006). The axial strains at failure are 1.1%, 1.2% and 1.4%, respectively, and the stress–strain predictions will be conducted accordingly to these axial strains at failure using the NSRMYSF.

(a)

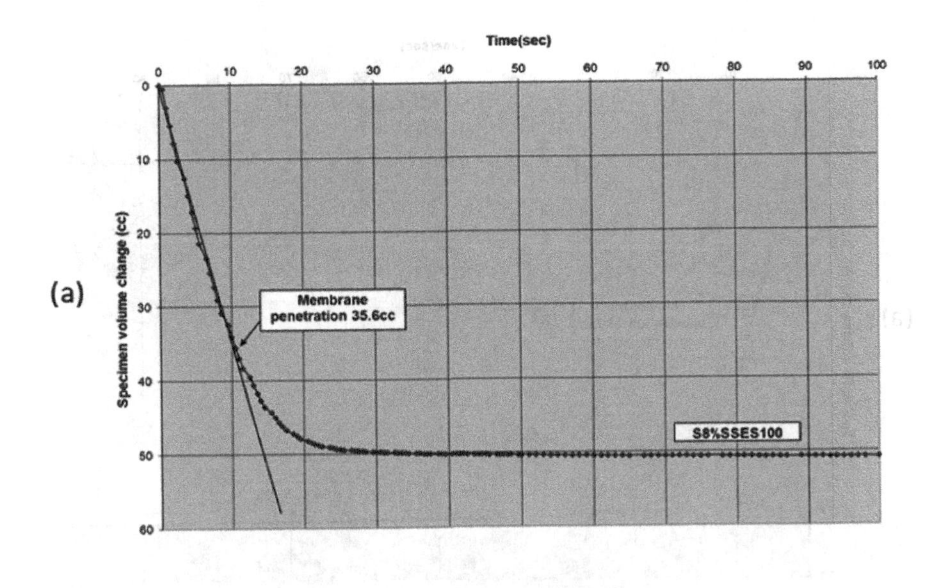

(b)

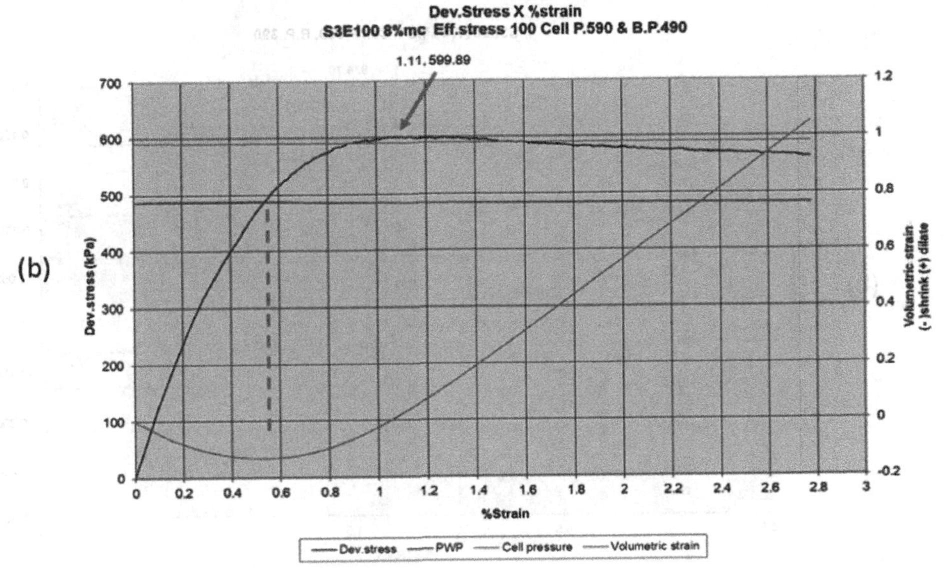

Figure 4.54 Consolidated drained triaxial tests at an effective stress of 100 kPa. (a) Consolidation curve. (b) Stress–strain curve (Md Noor, 2006).

The data for the three consolidated drained triaxial tests are shown in Figures 4.54–4.56 at effective stresses of 100, 200 and 300 kPa, respectively. Figure 4.57 shows the normalised stress–strain curves where the curve has a common axial strain at failure, which is 1.4%. From the normalised stress–strain curves, the mobilised shear strength envelopes are derived as shown in Figure 4.58a and b.

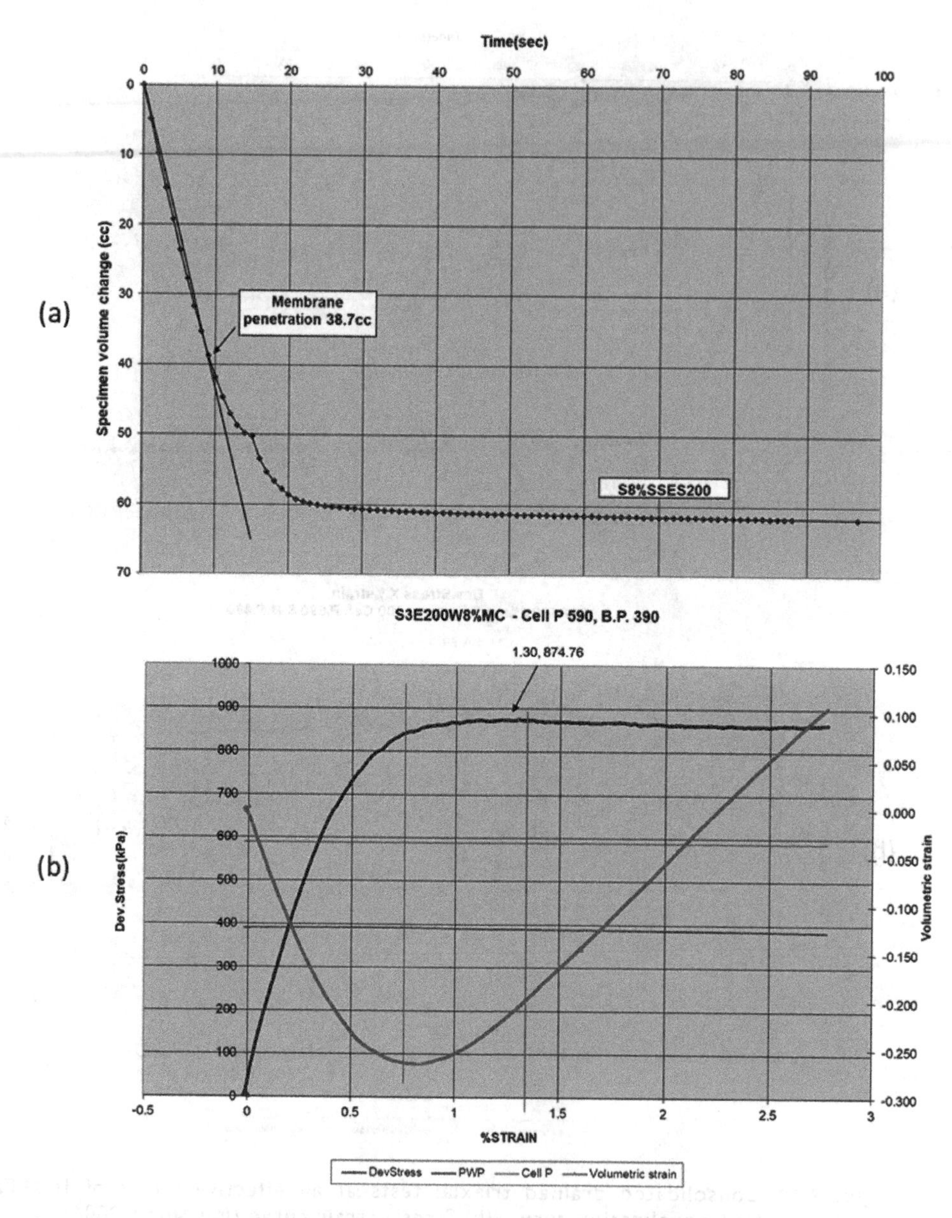

Figure 4.55 Consolidated drained triaxial tests at an effective stress of 200 kPa. (a) Consolidation curve. (b) Stress–strain curve (Md Noor, 2006).

Figures 4.59–4.61 show the predicted Mohr circles of which the diameter represents the predicted deviator stress. Then the stress–strain curves are plotted for the predicted deviator stress against the corresponding axial strain for the effective stresses under considerations which are 100, 200 and 300 kPa. The predicted data

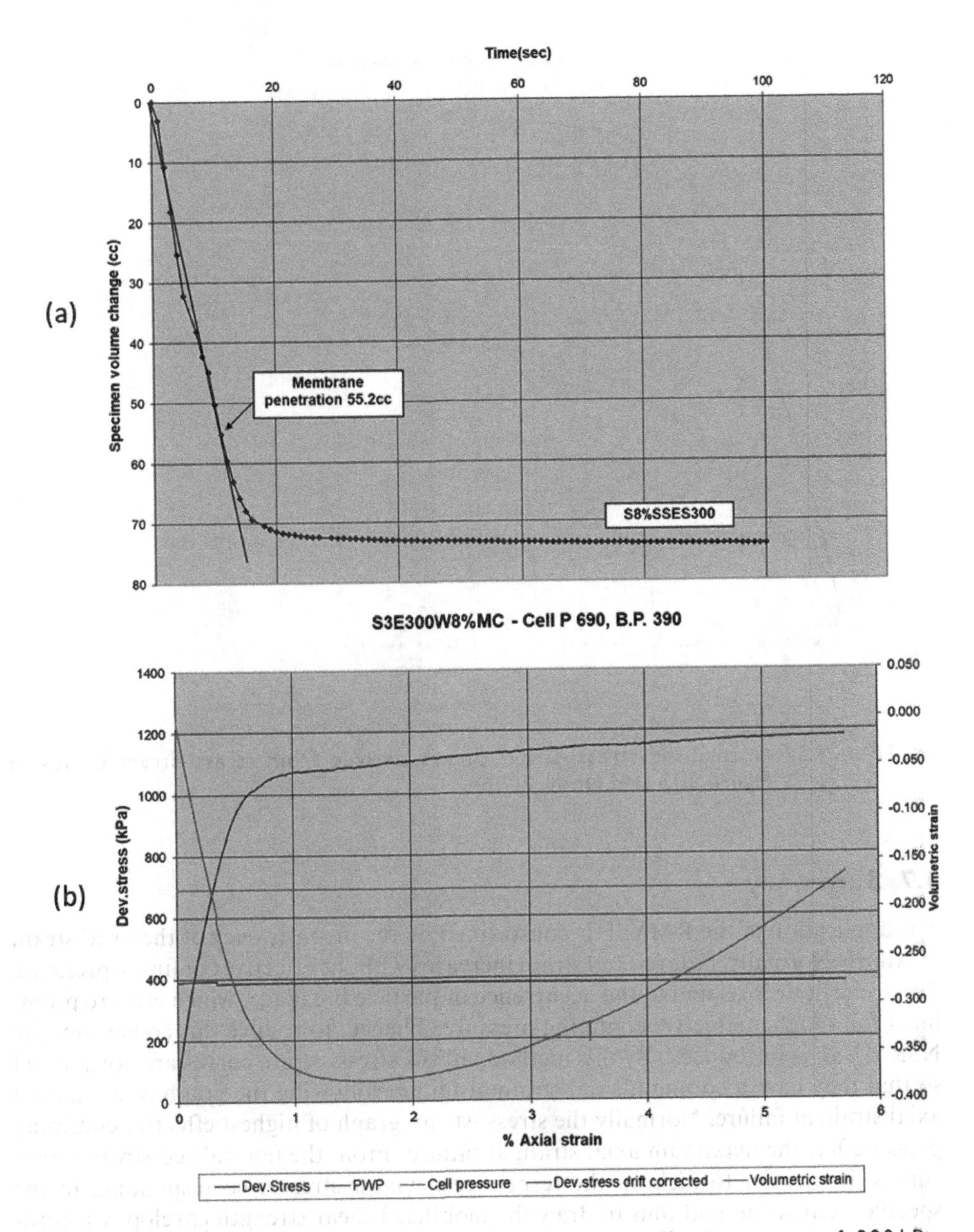

Figure 4.56 Consolidated drained triaxial tests at an effective stress of 300 kPa. (a) Consolidation curve. (b) Stress–strain curve (Md Noor, 2006).

points are presented in Table 4.10. The data points are then superimposed against the laboratory curves as shown in Figure 4.62. There is an excellent agreement between the predicted and the stress–strain curves obtained from the laboratory tests.

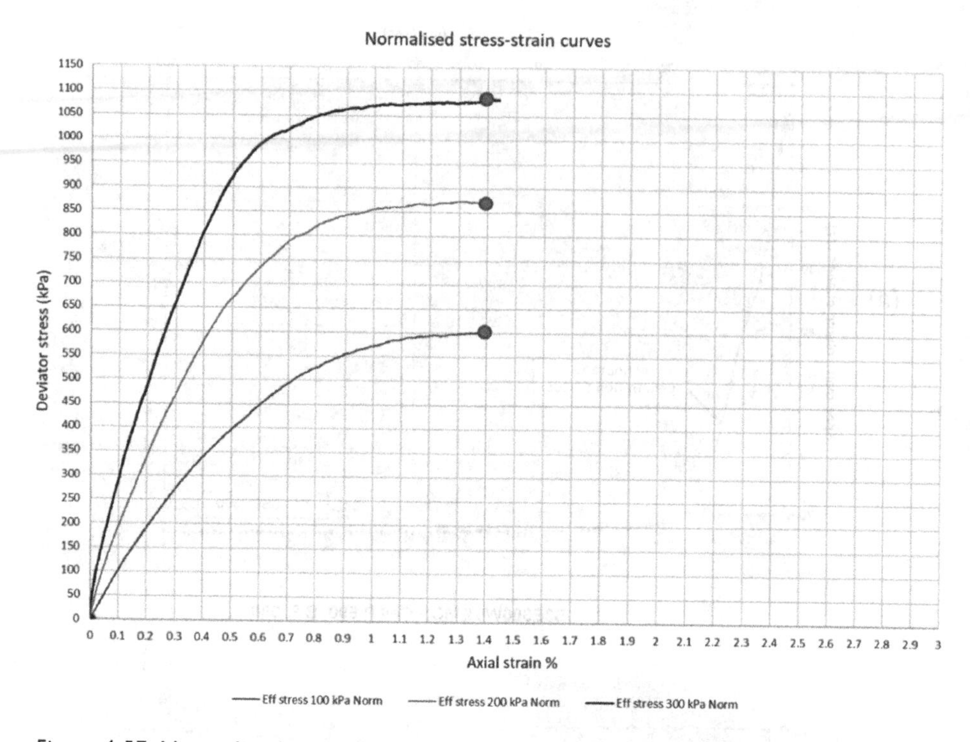

Figure 4.57 Normalised stress–strain curves derived from stress–strain curves in Figure 3.13 (Md Noor, 2006).

4.7 Summary

The application of the RMYSF is constrained by the inconsistency of the axial strain at failure. Normally, failure axial strain increases with the effective confining pressure. This behaviour is related to the occurrence of particle breakage, which is more prominent for a higher effective confining pressure. Thence, to resolve this constraint, the NSRMYSF is introduced. By this method, all the stress–strain curves are normalised so that they have a common axial strain at failure following the graph of maximum axial strain at failure. Normally the stress–strain graph of highest effective confining pressure has the maximum axial strain at failure. From the normalised stress–strain curves, it is easier to pick up the respective deviator stresses corresponding to the specific axial strain and thus to draw the mobilised shear strength envelopes according to the normalised axial strains. By this method, all the stress–strain curves are considered up to their deviator stress at failure, which is the maximum deviator stress. Finally, the normalised strain will be multiplied by the inverse factor to return the values to their actual strain. Then the predicted stress–strain curves will be superimposed against the actual stress–strain curves obtained from laboratory triaxial tests.

The NSRMYSF has been applied to characterise the stress–strain behaviour of various soils including Auckland clay, remoulded granitic residual soil grade V, remoulded sedimentary residual well-graded SAND and limestone gravel from

(a)

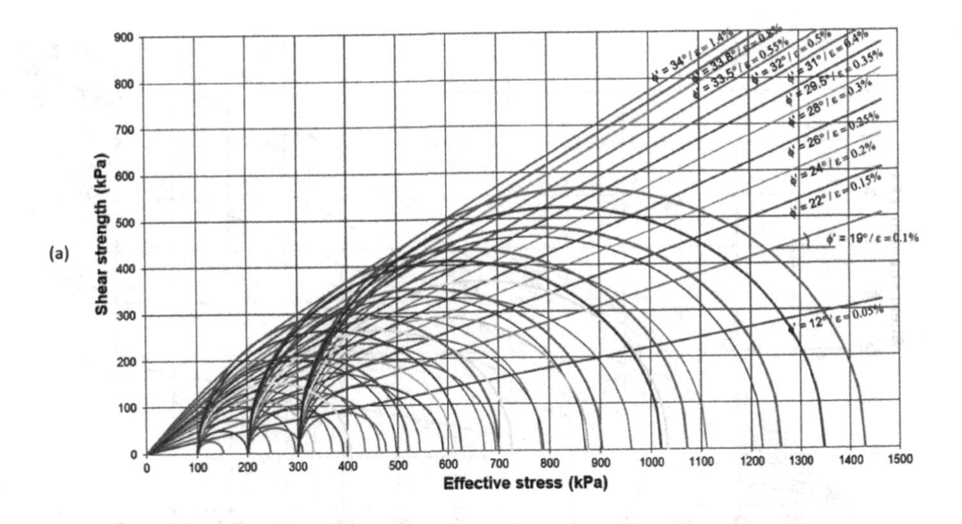

(b)

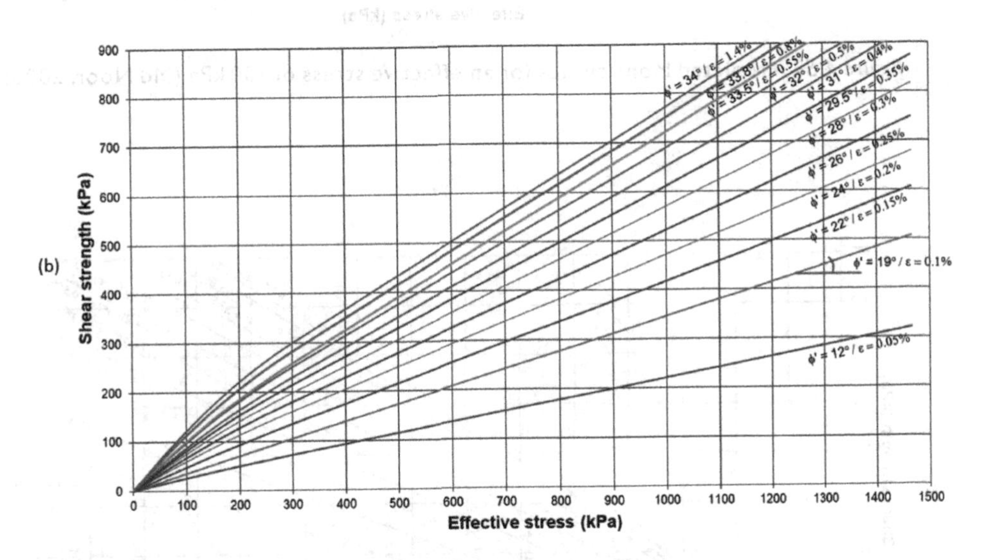

Figure 4.58 Mobilised shear strength envelopes derived from normalised stress–strain curves. (a) Mobilised shear strength envelopes with the derivative Mohr circles. (b) The mobilised shear strength envelopes as the soil intrinsic property (Md Noor, 2006).

Northumberland, United Kingdom. The prediction by the NSRMYSF showed a closer agreement compared to the prediction by the RMYSF, and in addition, it can predict up to the maximum axial strain at failure for each stress–strain curve in contrary to the RMYSF, where it can only predict up to the minimum axial strain at failure among the stress–strain curves. This also has verified the applicability of the NSRMYSF for the stress–strain prediction ranges from fine-grained soil to coarse-grained soils.

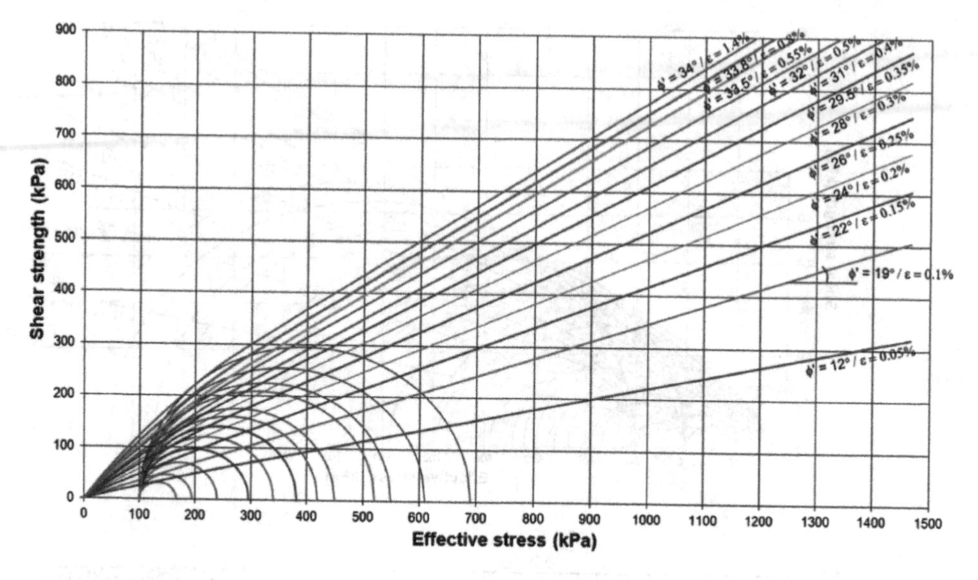

Figure 4.59 Predicted Mohr circles for an effective stress of 100 kPa (Md Noor, 2006).

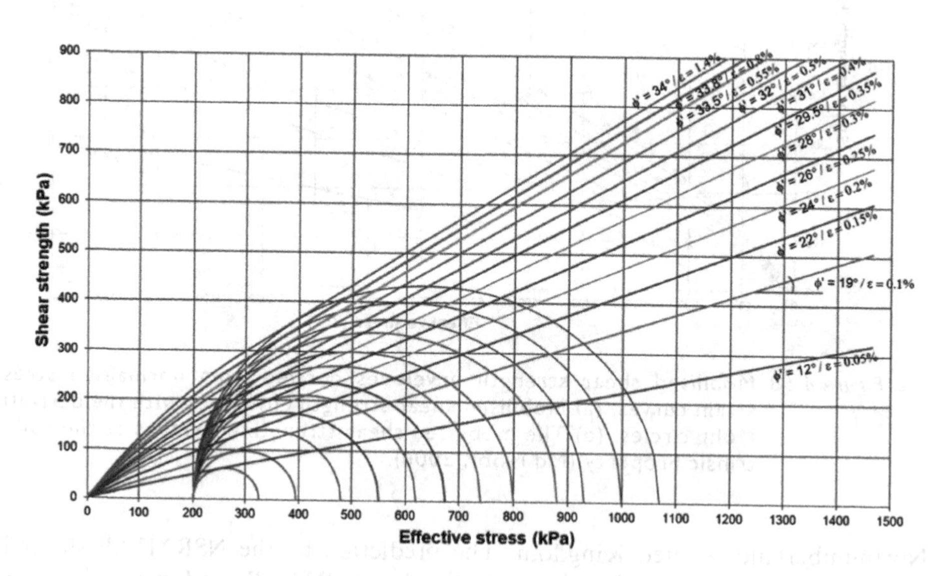

Figure 4.60 Predicted Mohr circles for an effective stress of 200 kPa (Md Noor, 2006).

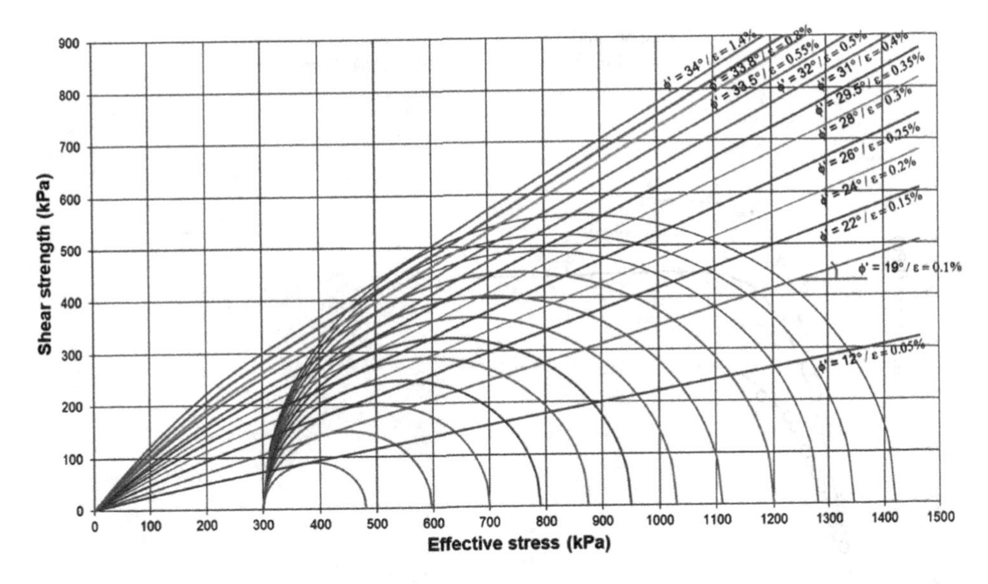

Figure 4.61 Predicted Mohr circles for an effective stress of 300 kPa (Md Noor, 2006).

Table 4.10 Predicted stress–strain data points for effective stresses of 100, 200 and 300 kPa

	Effective stress 100 kPa		*Effective stress 200 kPa*		*Effective stress 300 kPa*	
Inversion factor	1.1/1.4		1.2/1.4		1.4/1.4	
Normalised strain (%)	Actual strain (%)	Predicted deviator stress (kPa)	Actual strain (%)	Predicted deviator stress (kPa)	Actual strain (%)	Predicted deviator stress (kPa)
0	0.00	0.0	0.0	0.0	0.0	0.0
0.05	0.04	65.0	0.0	125.0	0.1	180.0
0.1	0.08	94.0	0.1	195.0	0.1	297.0
0.15	0.12	140.0	0.1	280.0	0.2	400.0
0.2	0.16	195.0	0.2	350.0	0.2	490.0
0.25	0.20	240.0	0.2	420.0	0.3	575.0
0.3	0.24	282.0	0.3	475.0	0.3	650.0
0.35	0.28	320.0	0.3	540.0	0.4	730.0
0.4	0.31	350.0	0.3	600.0	0.4	810.0
0.5	0.39	420.0	0.4	680.0	0.5	900.0
0.55	0.43	450.0	0.5	730.0	0.6	980.0
0.8	0.63	510.0	0.7	800.0	0.8	1045.0
1.4	1.10	590.0	1.2	870.0	1.4	1120.0

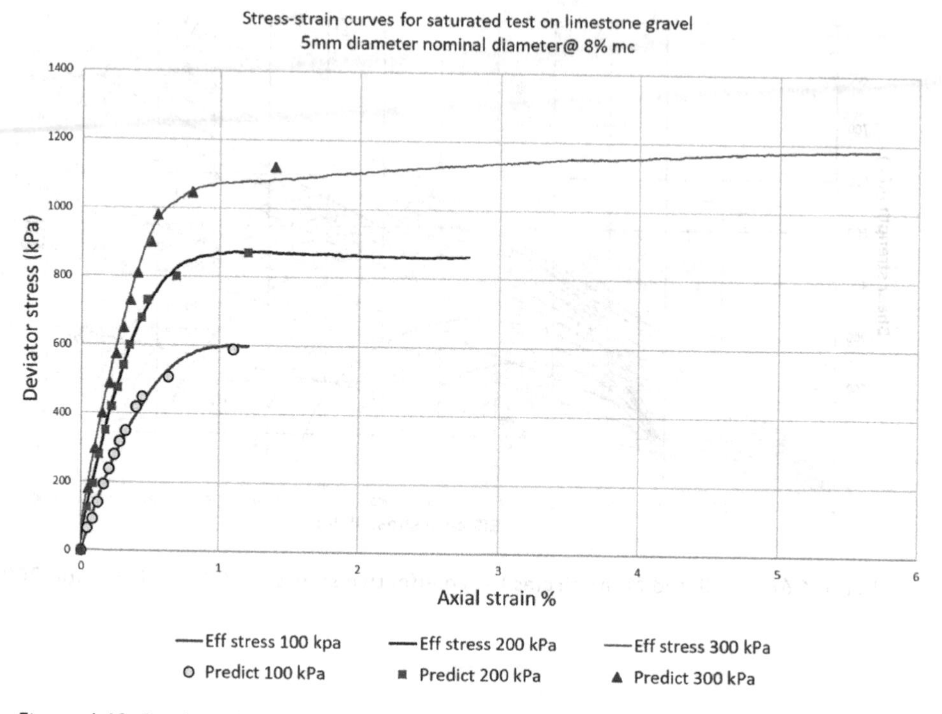

Figure 4.62 Predicted stress–strain data points plotted against the laboratory curves (Md Noor, 2006).

References

Alias, A., Md Noor, M. J., and Mohamad Jais, I. B. (2019). "Soil anisotropic stress-strain prediction using Normalised Rotational Multiple Yield Surface Framework for compacted tropical residual sandy soil." *11th International Conference on Geotechnical Engineering in Tropical Regions (GEOTROPIKA 2019)*, IOP Conference Series: Materials Science and Engineering, Vol. 527, IOP Publishing, Kuala Lumpur, Malaysia.

Bishop, A. W. (1966). "The strength of soils as engineering materials." *Geotechnique*, 16(2), 91–130.

Futai, M. M., and Almeida, M. S. S. (2005). "An experimental investigation of the mechanical behaviour of an unsaturated gneiss residual soil." *Geotechnique*, 55(3), 201–213.

Indraratna, B., Wijewardena, L. S. S., and Balasubramaniam, A. S. (1993). "Large-scale triaxial testing of greywacke rockfill." *Geotechnique*, 43(1), 37–51.

Jais, I. B. M., and Md. Noor, M. J. (2009) "Establishing a unique relationship between minimum mobilised friction angle and axial strain for anisotropic soil settlement model." *4th Asia Pacific Conference on Unsaturated Soils, Newcastle, Australia, 23th–25th November 2009.* Theoretical and Numerical Advances in Unsaturated Soil Mechanics, pp 775–781, CRC Press/ Balkema, Taylor and Francis Group, London, UK; ISBN 978-0-415-87716-9, eBook ISBN 978-0-203-85609-3, Indexed by ISI THOMPSON).

Md Noor, M. J. (2006). "Shear strength and volume change behaviour of saturated and unsaturated soils." Ph.D. Thesis, University of Sheffield.

Md. Noor, M. J., (2016) "Landslide and the hidden roles of shear strength." © UiTM Press, UiTM Malaysia ISBN 978-967-363-385-2.

Md Noor, M. J., Ibrahim, A., and Rahman, A. S. A. (2017). "Normalized rotational multiple yield surface framework (NRMYSF) stress-strain curve prediction method based on small strain triaxial test data on undisturbed Auckland residual clay soils." *4th International Conference on Civil and Environmental Engineering for Sustainability*, Langkawi, Malaysia.

Rahman, A. S. A., Md Noor, M. J., Jais, I. B. M., and Ibrahim, A. (2017). "Prediction of soil anisotropic stress-strain behaviour incorporating shear strength using improvise normalised stress-strain method." *4th International Conference on Civil and Environmental Engineering for Sustainability*, Langkawi Malaysia.

Chapter 5

Modelling inundation settlement and loading collapse settlement using RMYSF

5.1 Earlier studies on soil settlement behaviour

The effective stress concept was introduced by Terzaghi (1943) and it has been the most important concept in soil mechanics. The concept characterises soil settlement based on the increase in the effective stress. This is based on the quote by Terzaghi (1943) which said, "All the measurable effects of a change in stress, such as compression, distortion, and a change in shearing resistance, are exclusively due to changes in effective stress σ_1', σ_2' and σ_3'." The concept has been very successful in describing the settlement behaviour of saturated soils. The expression for effective stress σ' is as in Equation 5.1, where σ is the total stress and u_w is the pore water pressure.

$$\sigma' = \sigma - u_w \tag{5.1}$$

Effective stress is the average stress between the soil grains over a cross-section. In other words, it is the grain-to-grain stress. The existing soil settlement models are developed based on the effective stress concept where settlement is calculated based on the increase in the applied stress. These are like the models of Steinbrenner (1934), Terzaghi (1943) and Janbu et al. (1956) that predict settlement in clays and the models of De Beer and Martens (1951) and Schertmann et al. (1978) that predict settlement in sand. The settlement equations are summarised in Equations 5.2–5.6, respectively.

$$s_i = \frac{qB}{E}\left(1 - v^2\right)I_p$$ (5.2) (Steinbrenner, 1934)

$$S = \frac{C_c H}{1 + e_0}\log\left(\frac{p_0 + \Delta p}{p_0}\right)$$ (5.3) (Terzaghi, 1943)

$$S_i \text{ or } \rho = \frac{\mu_o \mu_1 q B}{E_u}\left(1 - v^2\right)$$ (5.4) (Janbu et al., 1956)

$$s = \frac{H}{C}\text{In}\frac{\sigma_o + \Delta\sigma}{\sigma_o}$$ (5.5) (De Beer and Martens, 1951)

$$S_i = C_1 C_2 q_n \sum \frac{I_z}{E}\Delta Z$$ (5.6) (Schertmann et al., 1978)

These settlement equations or models calculate settlement based on the stress increase in the form of either Δp or $\Delta\sigma$ or q. In other words, the models cannot justify settlement when there is no stress increase. This is like in inundation settlement or wetting collapse where settlement occurs under constant applied stress. These models are developed based on the effective stress concept where settlement is always associated with an increase in effective stress. In addition, effective stress will be increased when there is an increase in the applied load or the lowering of the groundwater table. Thus, a more comprehensive soil settlement model is required to encompass the soil complex behaviours.

Equation 5.3 is based on the theory of consolidation introduced by Terzaghi (1943). This is considering the occurrence of settlement as the pore water pressure is dissipated, which results in the increase in the effective stress Δp. Effective stress is also increased when water table recedes and this can also trigger settlement; a good example for this is the massive soil subsidence reported in Mexico City by Strozzi and Wegmuller (1999) caused by the depletion of groundwater level when it is being pumped for domestic consumption.

However, the concept of effective stress increase with settlement is in contrary for settlement, which occurs due to inundation. Jennings and Burland (1962), Blanchfield and Anderson (2000) and Tadepalli et al. (1992) reported that effective stress decreases when the ground is inundated with subsequent settlement. This is because the effect of inundation diminishes the effect of suction (Bishop, 1959) provided by the surface tension force that pulls two soil particles together when the condition changed from partially saturated to fully saturated. The reduction in the suction force in turn reduces the effective stress. Thence, inundation settlement takes place under effective stress decrease (Md Noor et al., 2008). From this fact, settlement cannot be viewed solely from the standpoint of effective stress.

It was later realised that settlement does occur under effective stress decrease like the inundation settlement or wetting collapse as reported by Blanchfield and Anderson (2000) and Mohamed Jais and Md Noor (2019). This puzzled the geotechnical researchers. Nevertheless, this gives a clear indication that it is not solely the effective stress that governs the soil volume change behaviour. The role of shear strength needs to be incorporated into the volume change framework. The problem is the difficulty to formulate such volume change framework. The effect of yield surface changing with the soil compression makes it more complex. Nevertheless, the involvement of multiple yield surfaces is inevitable and need to be incorporated in the soil volume change framework.

Among the effort to unveil this is the introduction of effective stress equation for unsaturated soils by Bishop (1959) as in Equation 5.7.

$$\sigma' = (\sigma - u_a) + \chi(u_a - u_w) \tag{5.7}$$

However, later it was confirmed by the works of Jennings and Burland (1962), Bishop and Blight (1963) and Burland (1965) that even this effective stress equation shows that the inundation settlement occurs under effective stress decrease. Thence, it is inevitable to incorporate the role of soil shear strength in a soil volume change framework to achieve a comprehensive formulation. Besides, this effective stress equation cannot be applied as a single stress state variable like the Terzaghi (1943) effective stress equation for saturated soil. This is because the term χ should be independent of the external factor, but in this case, χ varies with moisture content. Thence finally, the two independent stress state variables are applied to characterise the mechanical behaviours of unsaturated soils, which are net stress, $(\sigma - u_a)$, and suction, $(u_a - u_w)$.

5.2 Wetting collapse in the Rowe cell inundation test

In the process of understanding the soil inundation settlement or wetting collapse behaviour, inundation tests were conducted in a large Rowe cell of diameter 254mm and the specimen initial thickness is maintained at 74mm in every test (Md Noor et al., 2008). The test material used is an ex-mining sandy soil. The particle size distribution is shown in Figure 5.1 and it is a well-graded SAND. The soil specimen is placed on the lower compartment. A diaphragm was used to separate the water in the upper compartment from the soil specimen as shown in Figure 5.2. The vertical pressure applied to the specimen is provided by pressurising the upper compartment. The air pressure in the specimen is exposed to the atmosphere by opening the outlet that goes through the piston central spindle.

An oven-dried soil sample is used in every test using the same dry weight of soil of 5136g. The initial thickness of the specimen is maintained at 74mm in every test by rotating the special cut wooden plate as shown in Figure 5.3, so that the top specimen level is consistent in every test. In this way, all the tests were started with the same void ratio. Five tests were conducted which are named as TA, TB, TC, TD and TE. The load increments in every test are 0, 50, 100, 200, 400, 800 and 1600kPa. The specimen was not subjected to any inundation in test TA. Thus, in this test, the specimen was compressed under dry condition throughout. In tests TB, TC, TD and TE, the inundation was applied by gradually injecting water from the Rowe cell bottom line until overflow is encountered at the outlet of the pore air pressure line that goes through the piston spindle. In test TB, the specimen was inundated before the application of the first vertical stress of 50kPa; in other words, this is a saturated settlement test. In this case, the settlement encountered upon the application of 50kPa vertical stress must be representing the wetting collapse for an effective stress of 50kPa.

In tests TC, TD and TE, the specimens were inundated when the vertical stress reached 200, 400 and 800kPa, respectively. In every pressure increment, the settlement was allowed to be fully stabilised before the next pressure increment was applied. The graphs of the void ratio against the log of net stress (i.e., e versus $\log(\sigma - u_a)$) for the tests are shown in Figure 5.4. The magnitude of wetting collapse under net stresses of 200, 400 and 800kPa cannot be easily noticed in Figure 5.4 and thus the graphs that show wetting collapse at net stress 200 and 400kPa are enlarged in Figure 5.5. The amounts of inundation settlement, ΔH, which is the drop in the specimen height in the tests are shown in Table 5.1. It is indicating that the magnitude of the inundation settlement decreases as the net stress becomes higher. In other words, the inundation settlement is bigger under low net stress. This is another soil peculiar volume change behaviour that needs an engineering explanation.

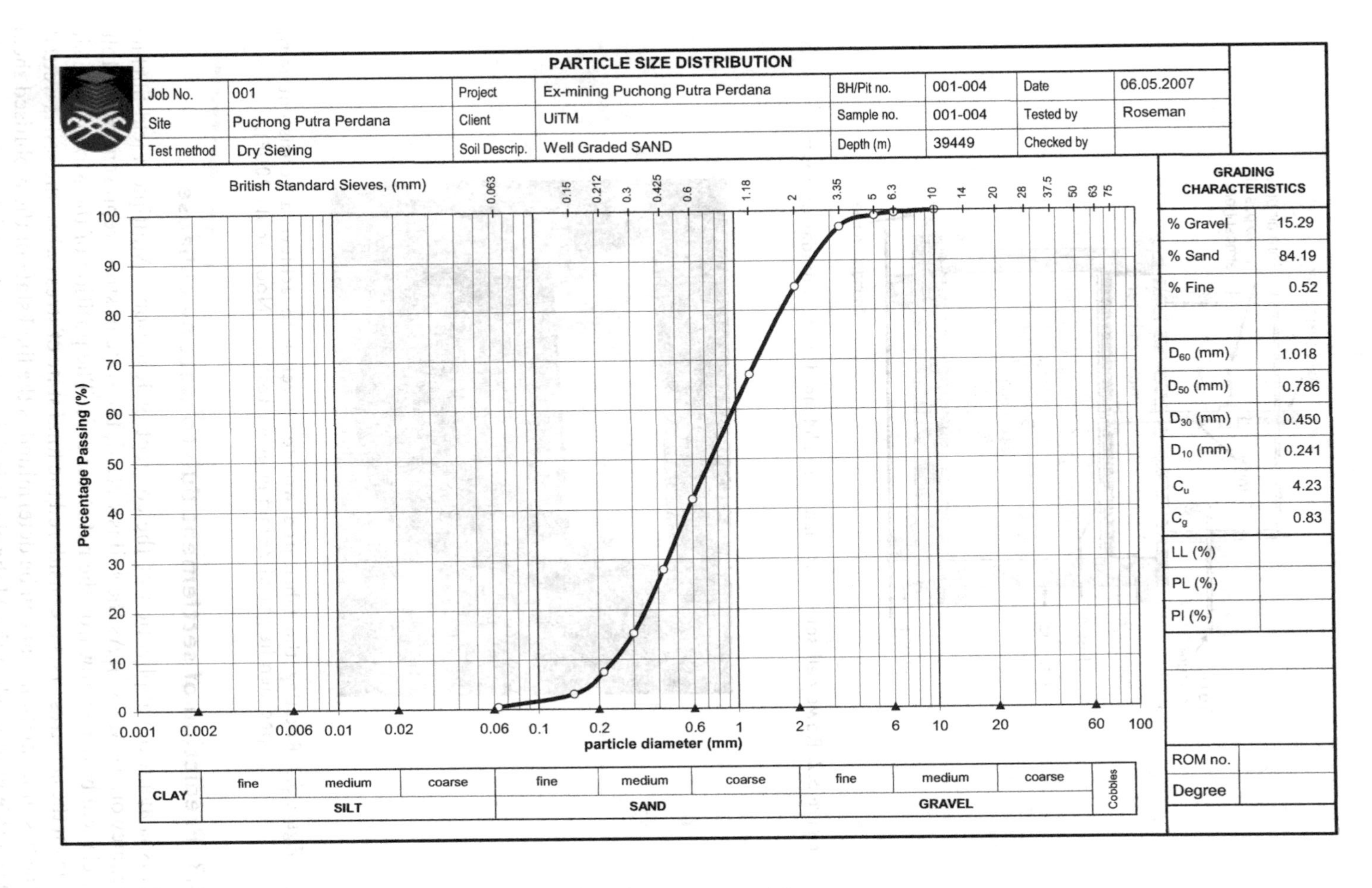

Figure 5.1 Particle size distribution of ex-mining sand as the test material.

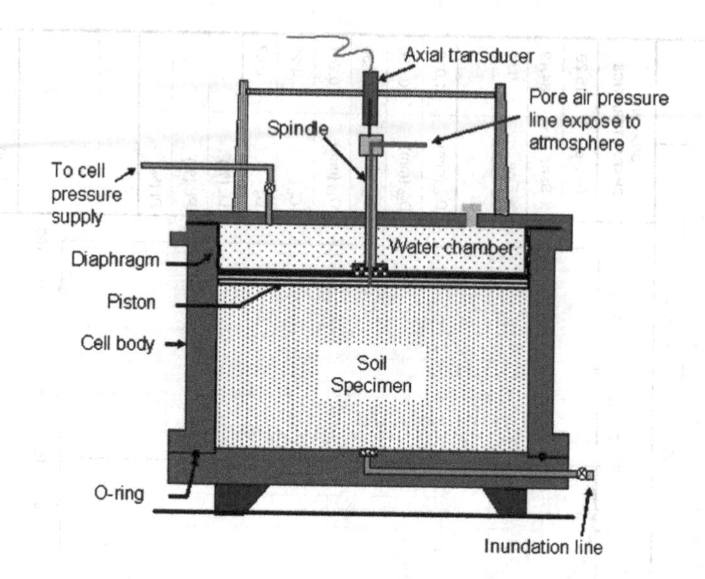

Figure 5.2 Rowe cell set-up of diameter 254 mm for the inundation tests.

Figure 5.3 Rotating the fabricated wooden leveller to achieve a consistent initial specimen height of 74 mm in every test (Md Noor et al., 2008).

5.3 Prediction of settlement due to wetting collapse

This is again the application of the concept of Rotational Multiple Yield Surface Framework (RMYSF). As the soil is wetted, the state of stress representative Mohr circle will be driven towards the net stress axis. The position of the Mohr circle sitting on the net stress axis is considered the final state of stress when the soil is wetted. The first information that can be determined is the final state of the mobilised shear strength envelope when the Mohr circle is driven to sit on top of the net tress axis.

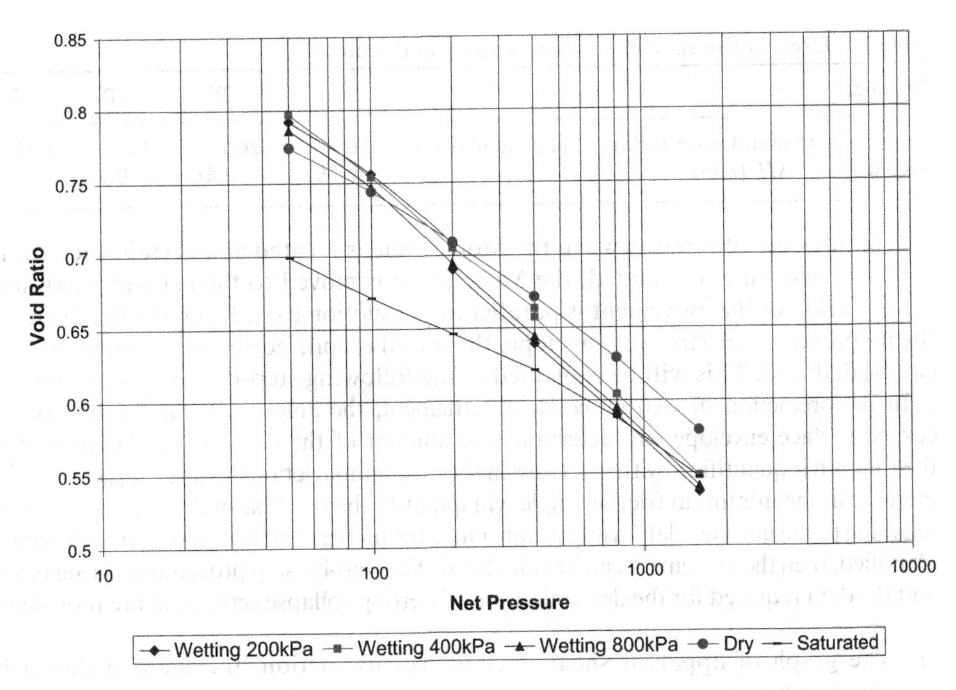

Figure 5.4 Graphs of void ratio versus log net stress (Md Noor et al., 2008).

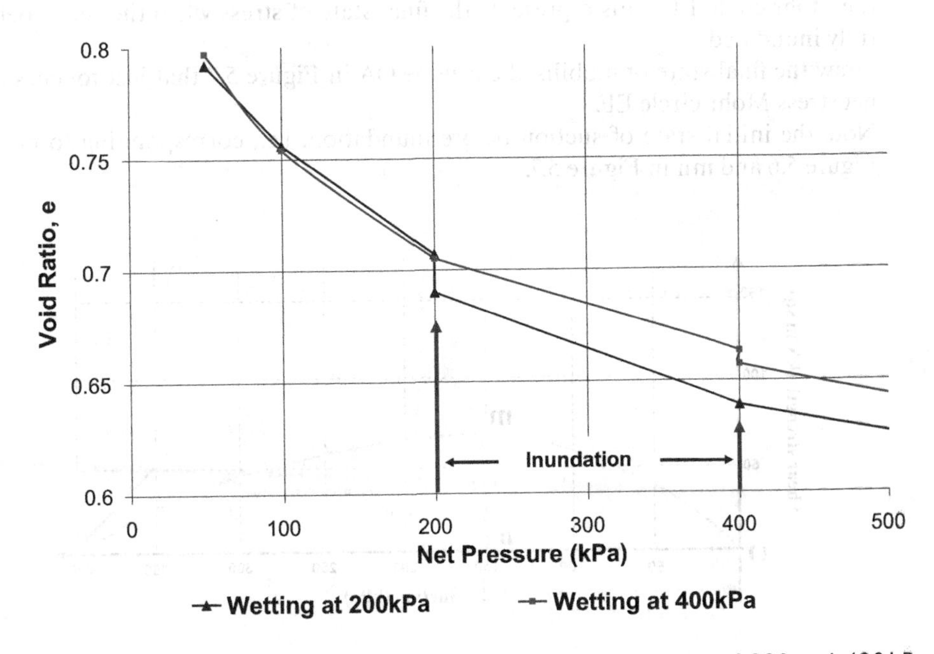

Figure 5.5 Wetting collapse settlement induced at net stresses of 200 and 400 kPa (Md Noor et al., 2008).

Table 5.1 Drop in the specimen height upon inundation

Test code	*TA*	*TB*	*TC*	*TD*	*TE*
Net stress at inundation (kPa)	No inundation	50	200	400	800
Height drop, ΔH (mm)	–	7.76	0.66	0.25	0.01

The initial state of stress is when the Mohr circle is located at a certain suction. In addition, when it is inundated, the Mohr circle is moved to the net stress axis and the direction of the movement is parallel to the suction axis. From the final state of the mobilised shear strength envelope, the initial mobilised shear strength envelope can be deduced. This will be explained in the following step-by-step procedure.

In the prediction of wetting collapse settlement, the amount of the rotation of the curved surface envelope will determine the amount of the settlement. The amount of the rotation is quantified by the increase in the minimum friction angle, $\phi'_{\min mob}$. Then the increase in the minimum friction angle will quantify the increase in the % axial strain by referring to the unique relationship graph. Once the increase in the % axial strain has been identified, then the settlement can be calculated. The step-by-step procedure is given below and the data required for the determination of wetting collapse settlement are provided;

1. The graph of apparent shear strength versus suction like the one shown in Figure 5.6.
2. Determine the initial state of net stress like the Mohr circle CD as shown in Figure 5.7 and then draw the Mohr circle directly sitting on the net stress axis like the Mohr circle EF. This represents the final state of stress when the soil is being fully inundated.
3. Draw the final state of mobilised envelope OA in Figure 5.7 that just touches the net stress Mohr circle EF.
4. Note the initial state of suction before inundation, i.e., corresponding to mn in Figure 5.6 and mn in Figure 5.7.

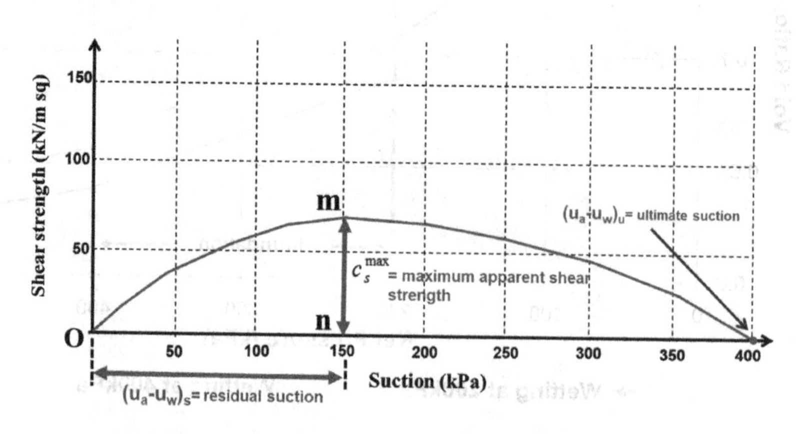

Figure 5.6 An example of a variation of apparent shear strength with respect to suction (Md Noor, 2006).

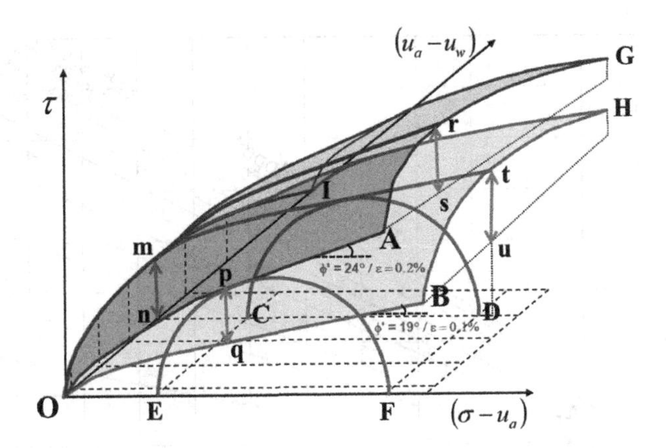

Figure 5.7 Rotation of the mobilised curved surface envelope upon wetting when suction drops from n to O and the Mohr circle shifted from CD to EF.

5. Note the final state of suction. The final suction is zero if the soil is inundated to full saturation as represented by the Mohr circle EF sitting directly on the net stress axis as in Figure 5.7.
6. Determine the drop in the apparent shear strength due to the suction drop caused by the inundation from the graph of apparent shear strength versus suction in (1), i.e., the initial apparent shear strength is mn of 70 kN/m^2 and the final suction is zero (see Figures 5.6 and 5.7).
7. That drop in the apparent shear strength, i.e., mn to zero in Figures 5.6 and 5.7 is equivalent to the rise of the envelope pq in Figure 5.7 at the specific net stress. This is because the curvatures Om, Ar and Bt are the same.
8. Note the drop in the apparent shear strength and make a vertical stick pq according to scale. Map the vertical stick pq vertically down below the contact point between the net stress Mohr circle and the mobilised envelope as in Figure 5.7.
9. Draw the initial state of the mobilised envelope represented by OB in Figure 5.7 before inundation by just touching the bottom bit of the stick.
10. Determine the mobilised shear strength envelopes of the soil like in Figure 5.8.
11. Note the rise of the envelope from OB to OA as in Figures 5.7 and 5.8, as that specific net stress is causing the rotation of the envelope.
12. Determine the increase in the minimum mobilised friction angle $\phi'_{\text{min mob}}$ from 19° to 24° as shown in Figures 5.7 and 5.8.
13. From the unique relationship graph in Figure 5.9, determine the change in the axial strain, $\Delta\varepsilon\%$, that corresponds to the change in the minimum mobilised friction angle, $\phi'_{\text{min mob}}$. From $\Delta\varepsilon\% = 0.2\% - 0.1\% = 0.1\%$.
14. Thence, the collapse settlement ρ_{collapse} is calculated as

$$\rho_{\text{collapse}} = \Delta\varepsilon\% \times H,$$

where H is the thickness of the formation.

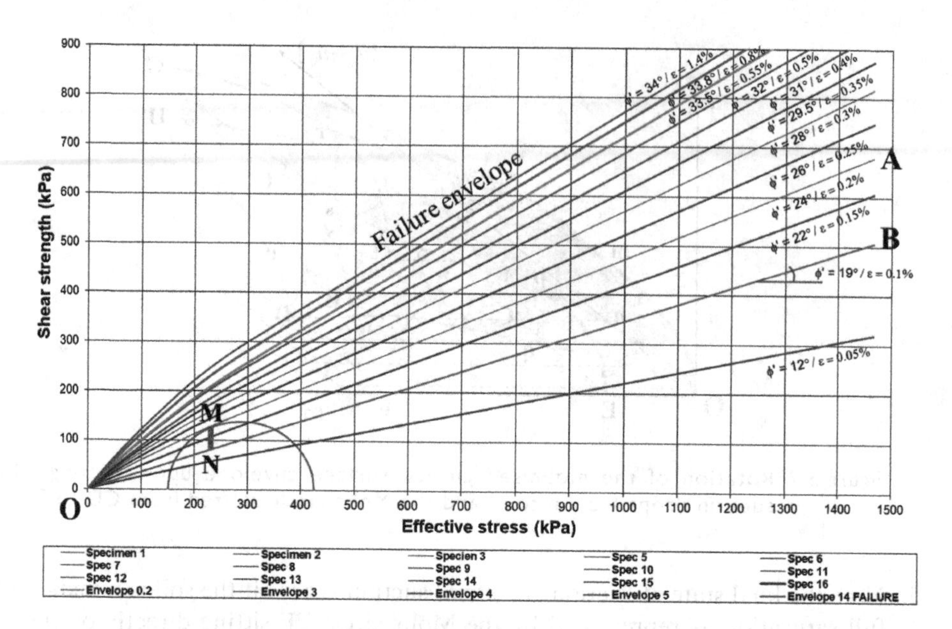

Figure 5.8 Mapping the final stress state after the soil being inundated on to the plot of mobilised shear strength envelopes directly sitting on the net stress axis corresponding to zero suction to locate the initial and the final mobilised shear strength envelopes (Md Noor, 2006).

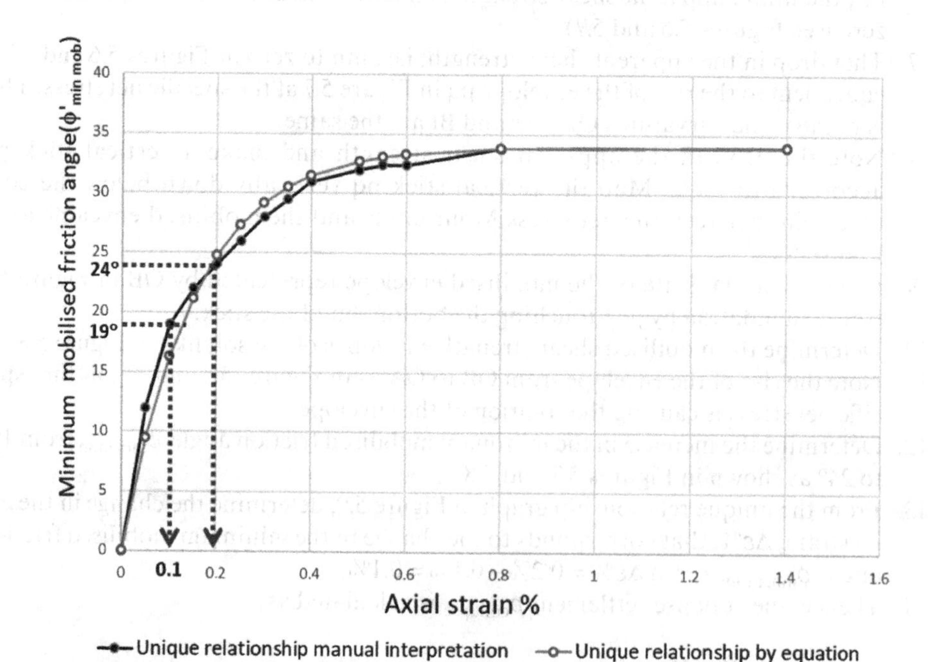

Figure 5.9 Example of a unique relationship graph of the soil under consideration (Md Noor, 2006).

Note that this determination of the wetting collapse is based on the mobilised shear strength envelopes at saturation as in Figure 5.8. This is because the distribution of the mobilised shear strength envelopes is the same for saturated and unsaturated conditions. Except that when the conditions become unsaturated, the whole set of the mobilised shear strength envelopes are raised by the magnitude of the apparent shear strength for the suction under consideration. This prediction of the wetting collapse settlement can be done by using either the RMYSF or the NSRMYSF. However, the application of the NSRMYSF is of better accuracy.

5.4 Modelling of wetting collapse under low and high applied net stress

The inundation settlement that takes place under effective stress decreases. This contradicts the effective stress concept of Terzaghi (1943) and thence, it is a soil volume change behaviour that is difficult to characterise. Nevertheless, the framework manages to model it qualitatively as in Figure 2.31 in Section 2.10. Then this bigger inundation settlement under low net stress compared to higher net stress as presented in Section 5.2 will be a more difficult soil volume change behaviour to characterise. This is an abnormal soil volume change behaviour because the lesser settlement is encountered when the loading is high. This section will substantiate that the framework can qualitatively model this weird soil volume change behaviour.

Consider the graph of apparent shear strength versus suction at net stress of zero for limestone gravel from the United Kingdom (Md Noor, 2006) as shown in Figure 5.10. In the τ-suction plane, the Mohr circle at a suction of 10 kPa is seen as a vertical line of height 25 kN/m^2. In other words, the apparent shear strength of this gravel at the suction of 10 kPa is 25 kN/m^2. Upon inundation, the Mohr circle will be driven laterally along the suction axis to the left to a suction of zero. Simultaneously, the Mohr circle will push the curved surface envelope upwards by the height of 25 kN/m^2 as illustrated in Figure 5.11. Thence, the envelope will be raised by the height of the apparent shear strength when the suction is reduced to zero. The rise of the envelope depends on the magnitude of the apparent shear strength, irrespective of the net stress value. This is demonstrated in Figures 5.12 and 5.13 where the rise of the envelope is the same, i.e., 25 kN/m^2 whether the inundation occurs at a net stress of 50 or 150 kPa. Notice that the Mohr circle at a net stress of 150 kPa is taller than the one at a net stress 50 kPa

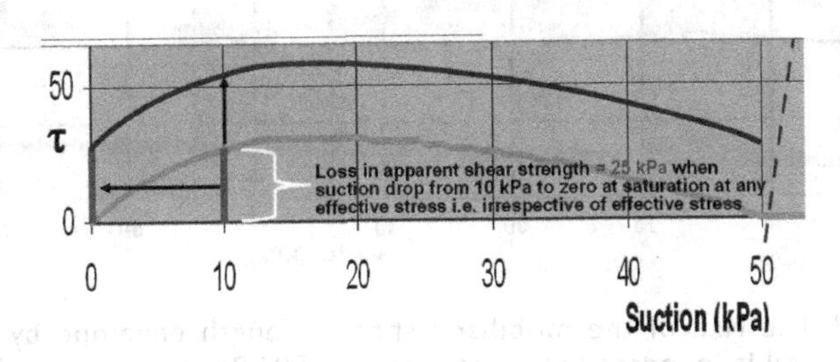

Figure 5.10 Graph of apparent shear strength versus suction for limestone gravel from the United Kingdom (Md Noor, 2006).

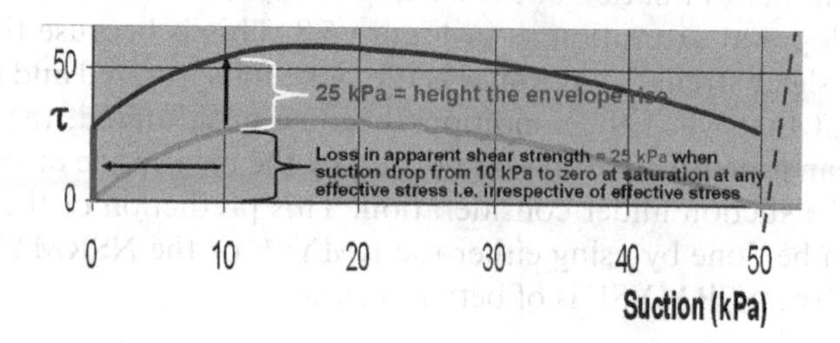

Figure 5.11 Upon wetting to suction zero, the curved surface envelope is raised by 25 kPa.

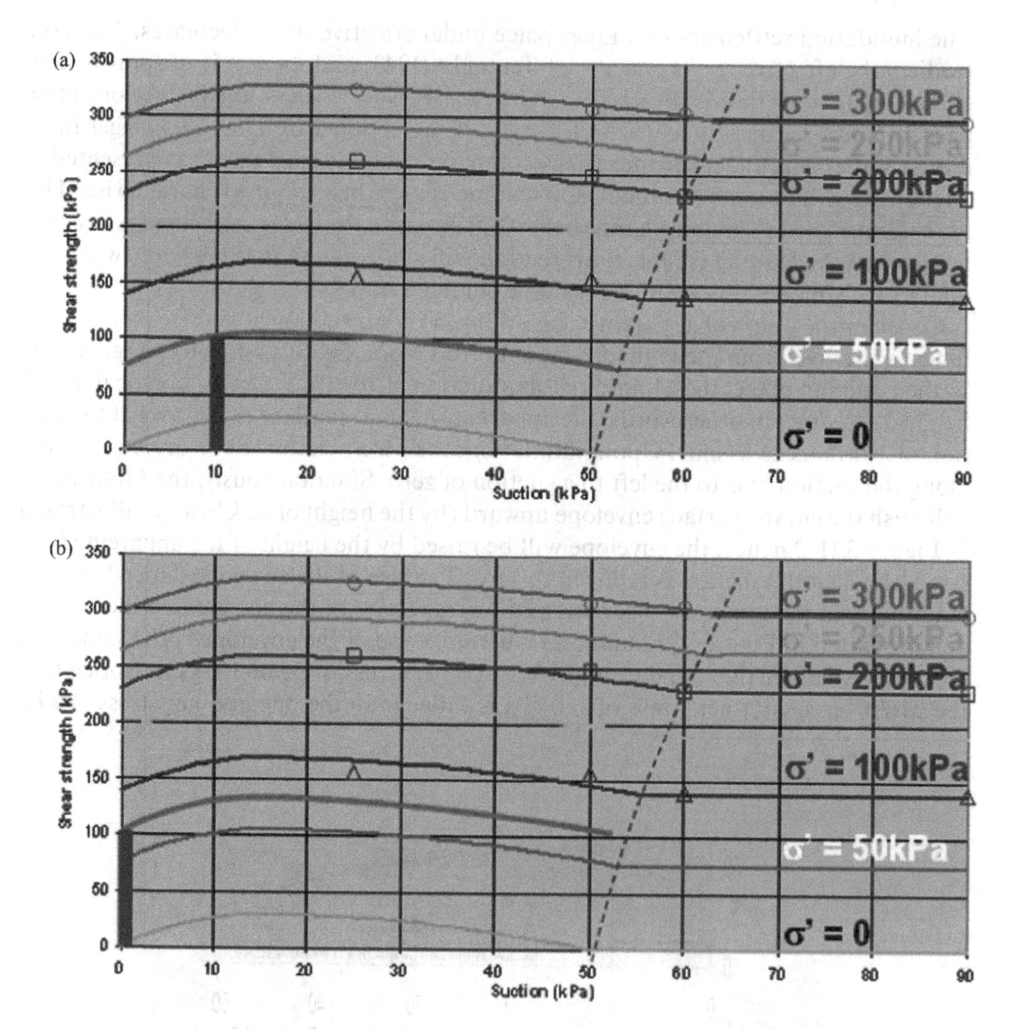

Figure 5.12 The rise of the mobilised shear strength envelope by 25 kPa when the soil is inundated at a net stress of 50 kPa at a suction of 10 kPa. (a) Mohr circle at a suction of 10 kPa. (b) Mohr circle at a suction of zero with the envelope raised (Md Noor, 2006).

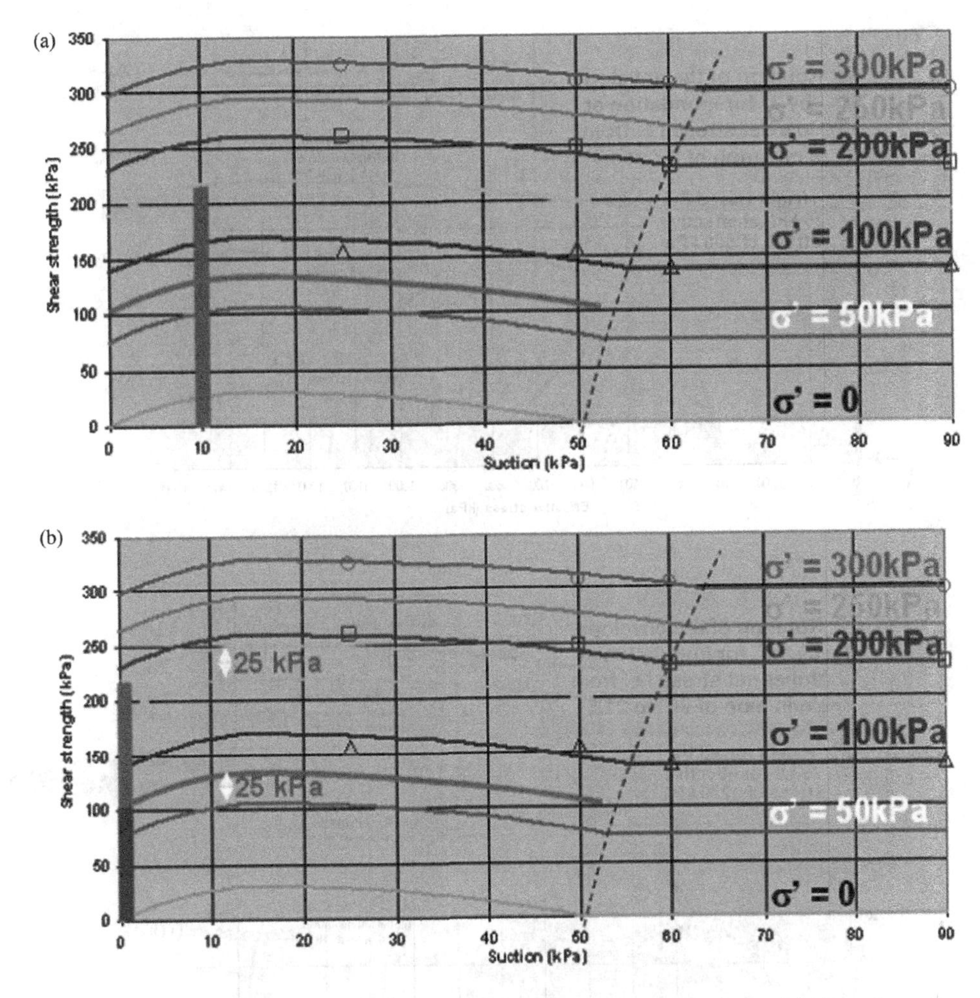

Figure 5.13 The rise of the mobilised shear strength envelope by 25 kPa when the soil is inundated at a net stress of 150 kPa at a suction of 10 kPa. (a) Mohr circle at a suction of 10 kPa. (b) Mohr circle at a suction of zero with the envelope raised (Md Noor, 2006).

due to the greater σ_3' and σ_1', i.e., the applied stress is higher for a net stress of 150 kPa. However, this equal rise in the envelope does not produce equal rotation of the envelope as illustrated in Figure 5.14a and b. This is the trickiest part of the modelling. The inundation at low net stress causes a rotation of 3°, i.e., the increase of $\Delta\phi'_{\min\text{mob}}$ of 3° compared to a rotation of 1.5°, i.e., the increase of $\Delta\phi_{\min_{\text{mob}}}$ of 1.5° at a higher net stress. Thence, the framework can model this weird soil volume change behaviour. Note that the amount of the increase in the $\Delta\phi'_{\min\text{mob}}$ represents the magnitude of the settlement.

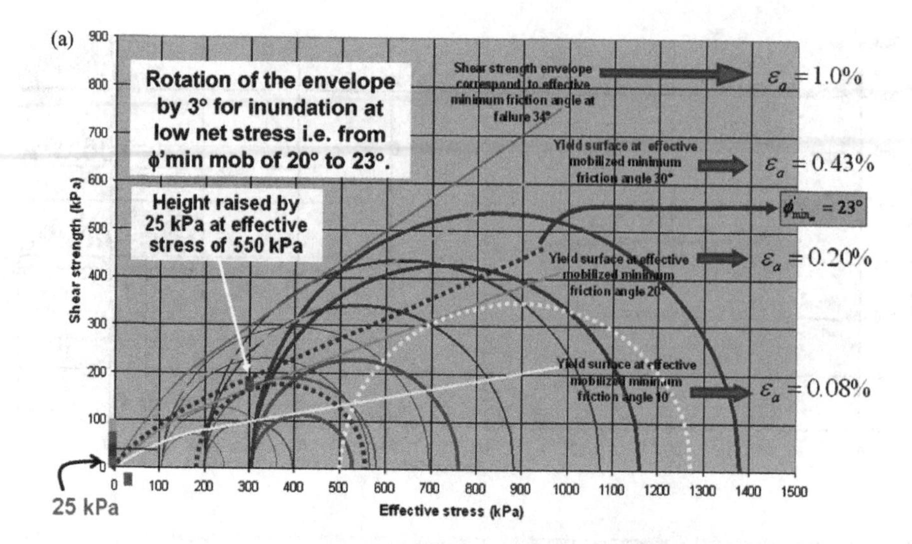

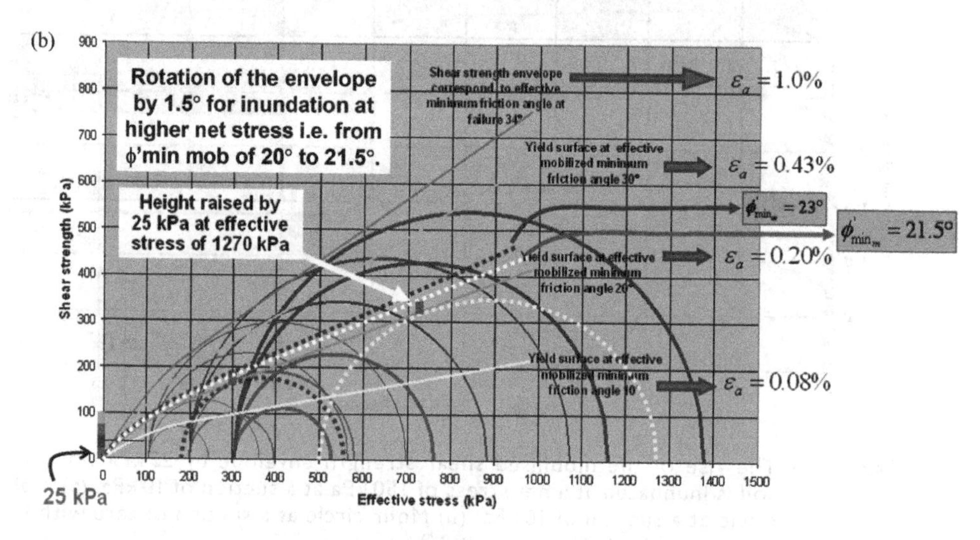

Figure 5.14 Equal rise of the envelope at different net stresses due to inundation causes a bigger rotation at low net stress compared to the higher net stress. (a) Rotation of 3° for low net stress. (b) Rotation of 1.5° at higher net stress.

5.5 Prediction of settlement due to loading collapse

Consider a pad footing foundation for a low-rise building. Normally, a footing is constructed at 2.0 m below the formation level or the ground surface (Figure 5.15). Assume that the friction angle of the soil is 34° and the bulk unit weight of the soil is 18 kN/m^3 and the groundwater table is far below the ground surface. Then the coefficient

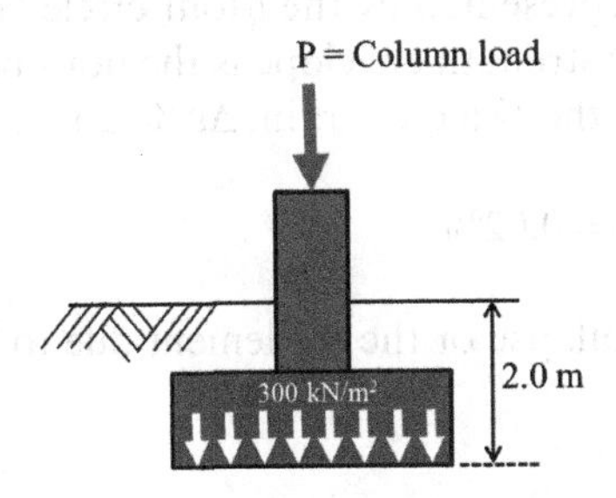

Figure 5.15 A pad footing imposing a uniform pressure of 300 kN/m^2 to the ground.

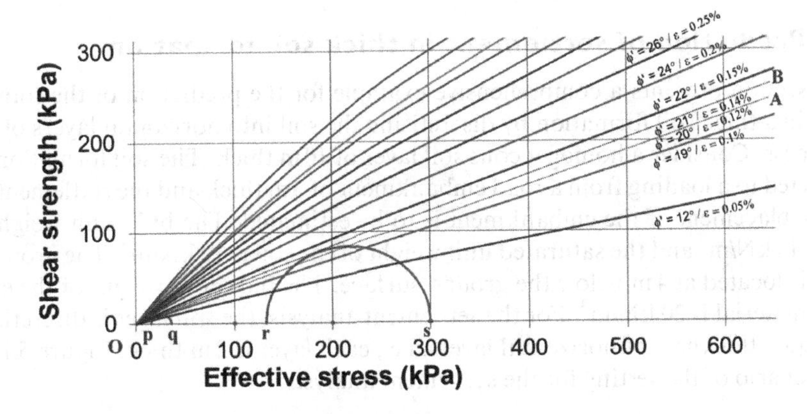

Figure 5.16 Mobilised shear strength envelopes of the soil under consideration.

of earth pressure at rest according to Jaky's equation (1948) is as in Equation 5.8, as follows:

$$k_o = 1 - \sin\phi \tag{5.8}$$

i.e., $k_o = 1 - \sin\phi = 1 - \sin 34° = 1 - 0.559 = 0.44$

Therefore, the initial minor and major principal stresses are;

$$\sigma_1' = \gamma D = 18 \times 2.0 = 36\text{kN/m}^2$$

$$\sigma_3' = k_o \gamma D = 0.44 \times 36 = 15.84\text{kN/m}^2$$

This initial stress state is represented by the Mohr circle "pq" in Figure 5.16. The corresponding mobilised shear strength envelope is the non-linear line OA.

Then, consider when the footing load of 300 kN/m^2 is applied, it will represent the final stress state. Thence

$$\sigma_1' = 300\text{kN/m}^2$$
$$\sigma_3' = k_o \times 300 = 0.44 \times 300 = 132\text{kN/m}^2$$

This final stress state is represented by the Mohr circle "rs" in Figure 5.16. The corresponding mobilised shear strength envelope is the non-linear line OB.

Thence, the increase in the % axial strain, $\Delta\varepsilon\%$, due to the footing load is

$$\Delta\varepsilon\% = 0.14\% - 0.12\% = 0.02\%$$

From here, the loading collapse or the settlement due to loading can be calculated as Equation 5.9;

$$\rho_{\text{collapse}} = \Delta\varepsilon\% \times H$$

where H is the thickness of the discrete soil layer under consideration.

5.6 Prediction of settlement in thick soil formation

This section presents a comprehensive example for the prediction of the total settlement in a thick soil formation by discretising the soil into horizontal layers of smaller thickness. Consider a homogeneous soil layer of 10 m thick. The soil formation will be subjected to a loading from a road embankment of 3 m thick and the settlement caused by the placement of the embankment is to be estimated. The bulk unit weight of the soil is 18 kN/m^3 and the saturated unit weight of the soil is 19 kN/m^3. The groundwater table is located at 4 m below the ground surface. The bulk unit weight of the embankment material is 20 kN/m^3. For the settlement analysis, the soil layer is discretised into five equal thickness of horizontal layers, i.e., each layer is 2 m thick. Figure 5.17 shows the scenario of the setting for the settlement analysis.

5.6.1 Characterising the soil mobilised shear strength envelopes of the soil

The stress–strain curves of the soil formation at effective stresses of 50, 100, 200 and 300 kPa obtained from conducting consolidated drained triaxial tests are shown in Figure 5.18. The axial strains at failure for effective stresses of 50, 100, 200 and 300 kPa are 0.05%, 0.055%, 0.058% and 0.062%, respectively. The analysis adopts the method of RMYSF. The common % axial strain at failure is taken at 0.05%. Then the mobilised

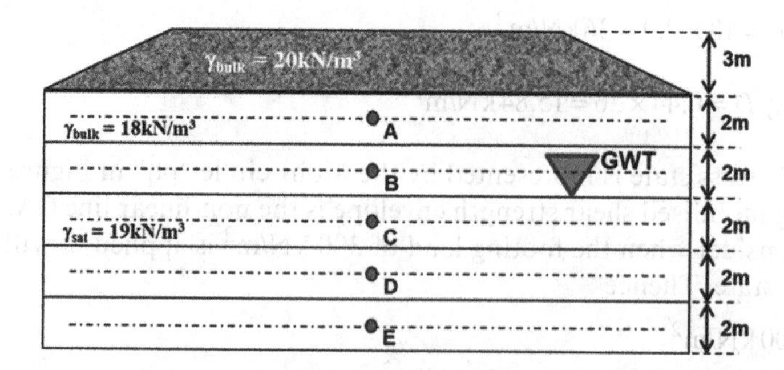

Figure 5.17 The 10 m thickness of soil is subjected to an embankment loading.

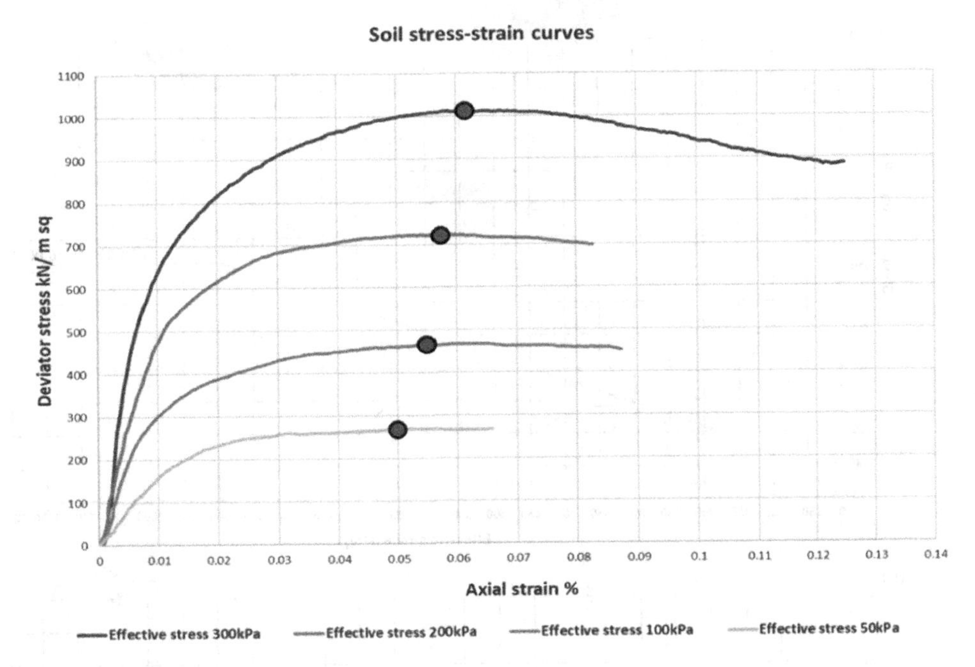

Figure 5.18 Stress–strain curves of the soil at effective stresses of 50, 100, 200 and 300 kPa.

shear strength envelopes at axial strains of 0.005%, 0.01%, 0.02%, 0.03%, 0.04% and 0.05% are determined by drawing the respective Mohr circles representing each stress–strain curve, i.e., curves at effective stresses of 50, 100, 200 and 300 kPa. The interpreted mobilised shear strength envelopes are shown in Figure 5.19a–e. Figure 5.20 shows the superimposed mobilised shear strength envelopes of the soil at % axial strains of 0.005%, 0.01%, 0.02%, 0.03%, 0.04% and 0.05% superimposed with the related Mohr circles.

These mobilised shear strength envelopes are the intrinsic property of the soil itself, which is directly related to the % axial strain, resulted from anisotropic volume change. It governs the volume change behaviour of the soil from the standpoint of the mobilised shear strength developed within the soil mass when it is being deformed. Note that the mobilised shear strength envelope applies to the whole range of effective stress. This means that the stress–strain response at any effective stress or any depth in the ground can be predicted reliably using this soil intrinsic or inherent property. It is interpreted from the soil response when the soil is subjected to anisotropic volume change under anisotropic stress condition in a triaxial cell. This laboratory characterisation of the anisotropic soil volume change behaviour replicates the true soil volume change behaviour in the ground. This intrinsic soil property dictates the soil stress–strain behaviour at any effective stress in the ground. In other words, this is the ultimate goal wanted by geotechnical engineers where the soil stress–strain response at any depth can be accurately determined. This is very useful for settlement prediction and soil volume change modelling.

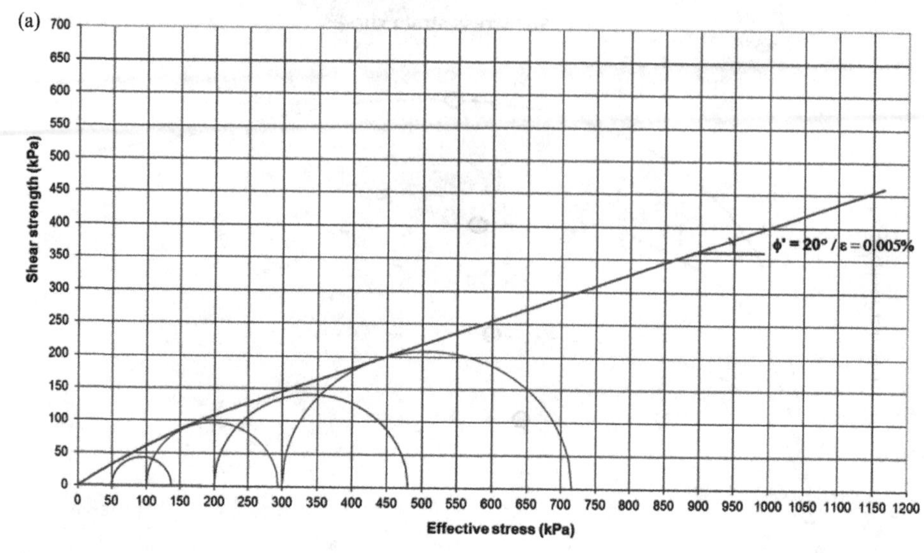

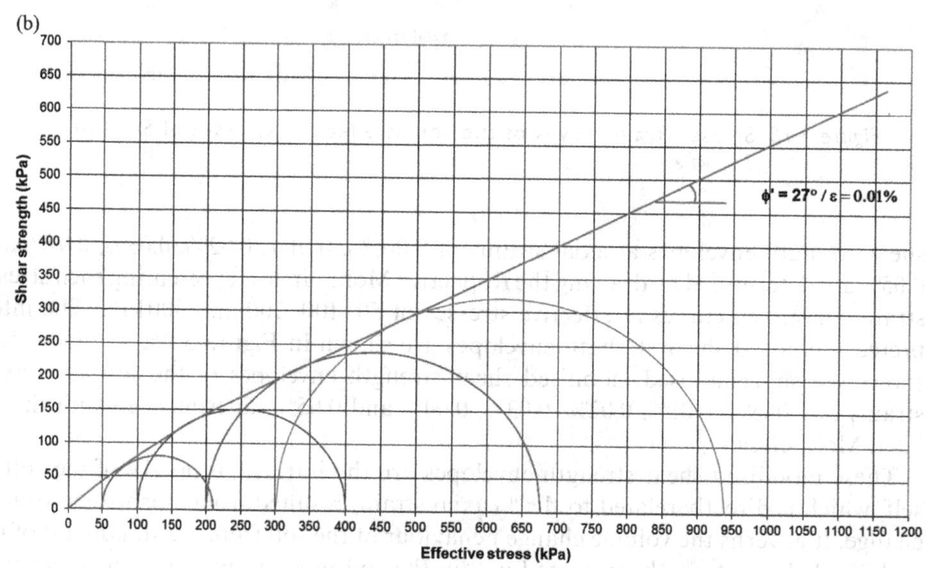

Figure 5.19 Mobilised shear strength envelopes at % axial strains of (a) 0.005%, (b) 0.01%, (c) 0.02%, (d) 0.03%, (e) 0.04% and (f) 0.05% of the soil.

(Continued)

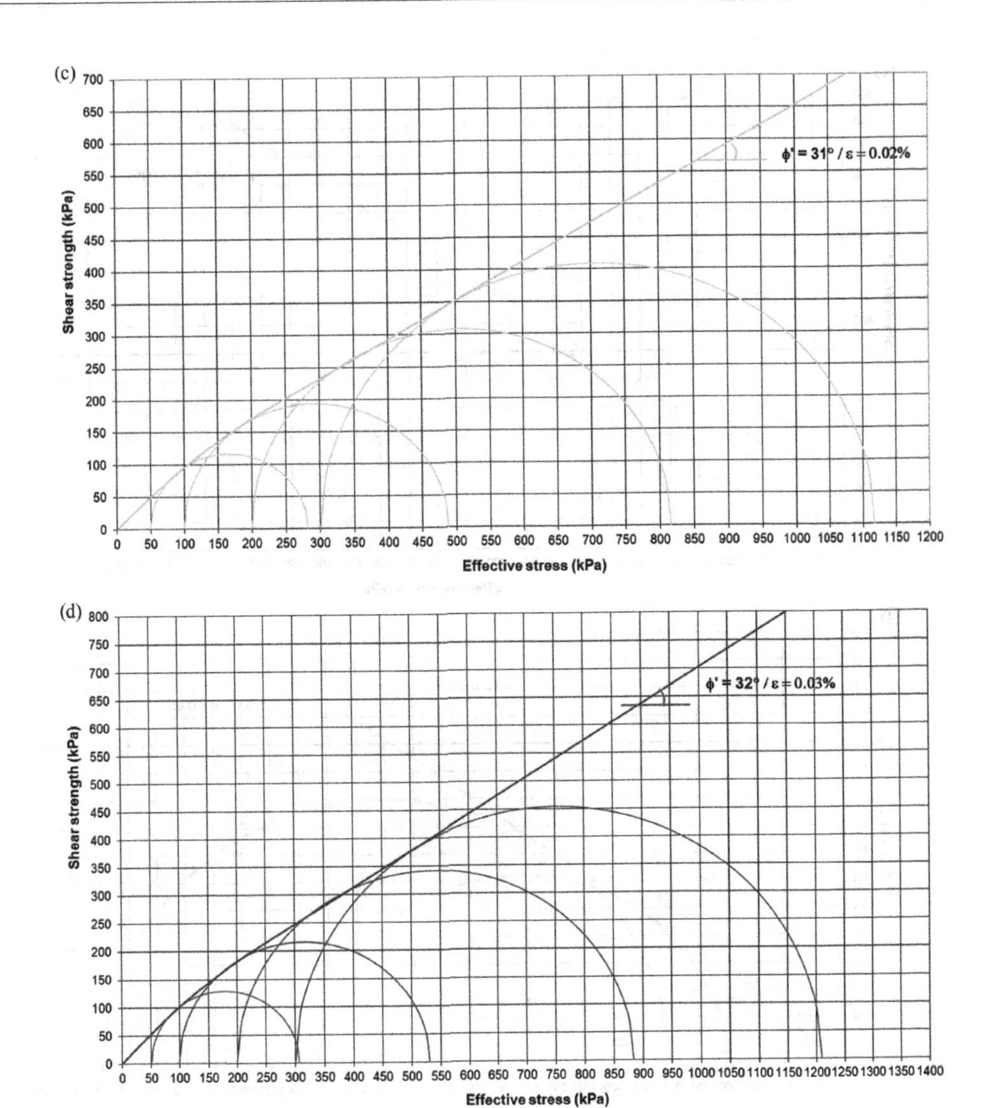

Figure 5.19 (Continued) Mobilised shear strength envelopes at % axial strains of (a) 0.005%, (b) 0.01%, (c) 0.02%, (d) 0.03%, (e) 0.04% and (f) 0.05% of the soil.

(Continued)

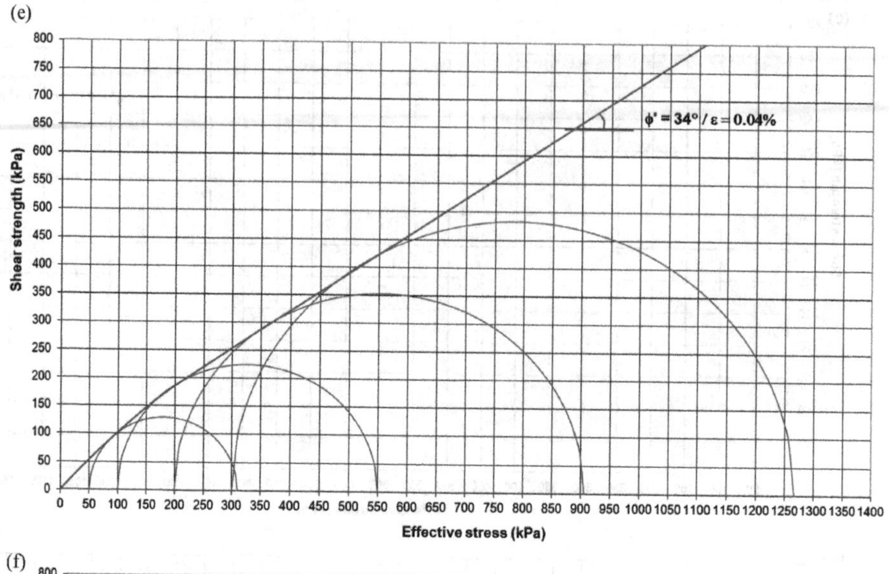

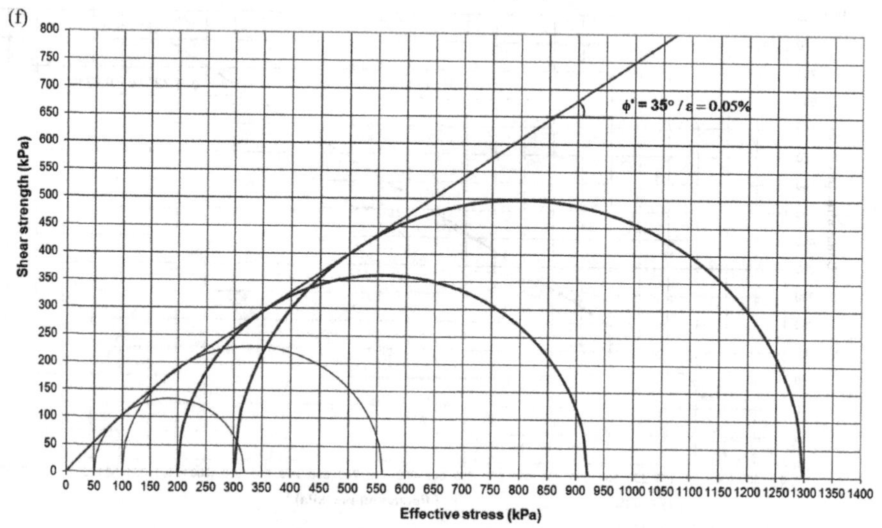

Figure 5.19 (Continued) Mobilised shear strength envelopes at % axial strains of (a) 0.005%, (b) 0.01%, (c) 0.02%, (d) 0.03%, (e) 0.04% and (f) 0.05% of the soil.

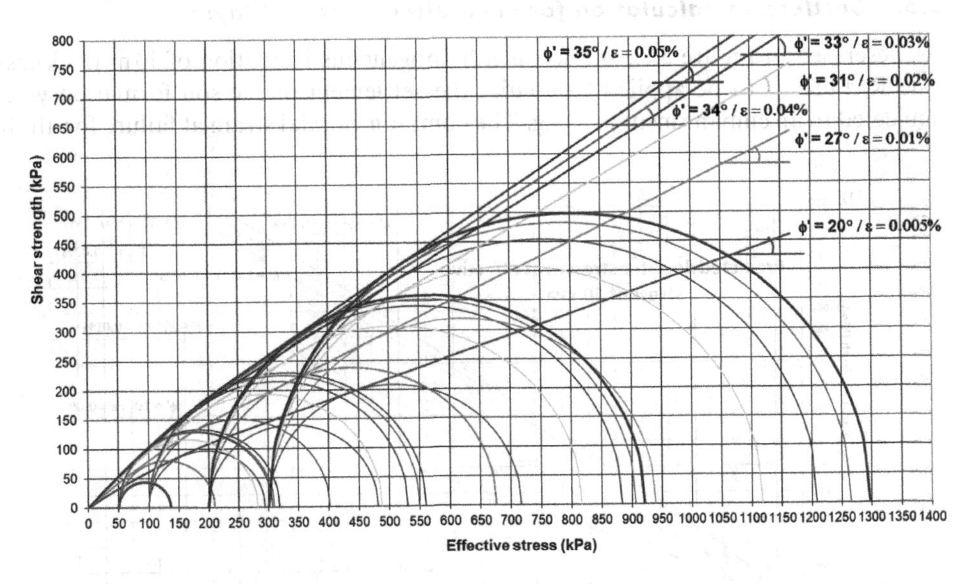

Figure 5.20 Mobilised shear strength envelopes of the soil at % axial strains of 0.005%, 0.01%, 0.02%, 0.03%, 0.04% and 0.05% superimposed with the related Mohr circles.

5.6.2 Verifying the accuracy of RMYSF

Nevertheless, the accuracy of the stress–strain prediction using RMYSF in this particular case is the first need to be tested. The mobilised shear strength envelopes and the shear strength envelope of the soil are applied to predict the stress–strain response at the four magnitudes of the effective stress applied in the consolidated drained triaxial tests conducted in the laboratory. The prediction of the deviator stresses at a specific axial strain for effective stresses of 50, 100, 200 and 300 kPa is carried out by drawing the respective Mohr circles to touch the respective mobilised shear strength envelopes. The diameter of the Mohr circles will represent the magnitude of the deviator stress and the touched envelope represents a specific % axial strain. The prediction of the stress–strain responses at effective stresses of 50, 100, 200 and 300 kPa is shown in Figure 5.21a–d. The predicted deviator stresses and the corresponding % axial strains are presented in Table 5.2 and the predicted data points are mapped on to the stress–strain curves obtained from the laboratory consolidated drained triaxial tests as shown in Figure 5.22. Essentially, the prediction shows an excellent resemblance with the laboratory test curves. Thence, this verified that the technique of RMYSF could be reliably applied to predict settlement in such soil formation when the soil is subjected to a loading because the prediction of the stress–strain response is of a high degree of accuracy.

5.6.3 Settlement calculation for each discretised soil layer

As described in Figure 5.17, the soil is a homogeneous formation of 10 m thickness. The RMYSF is to be applied to predict the settlement of the soil formation when subjected to an embankment loading. The common % axial strain at failure for all the

(a)

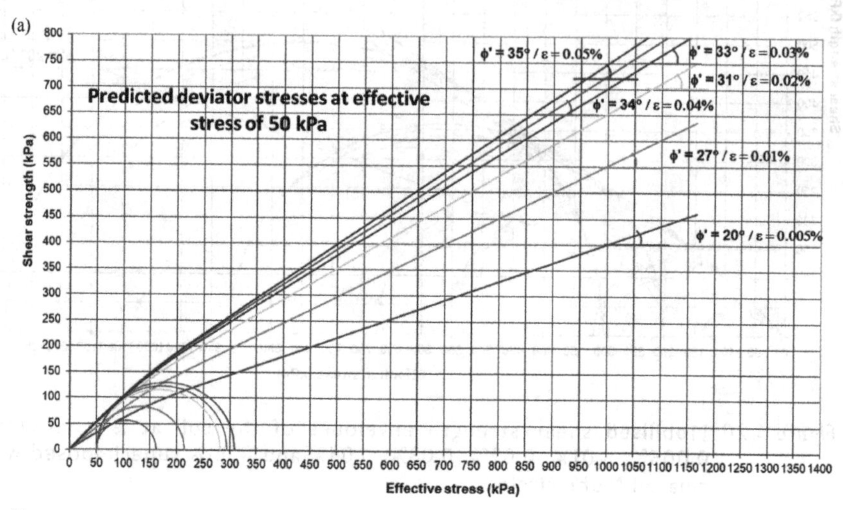

(b)

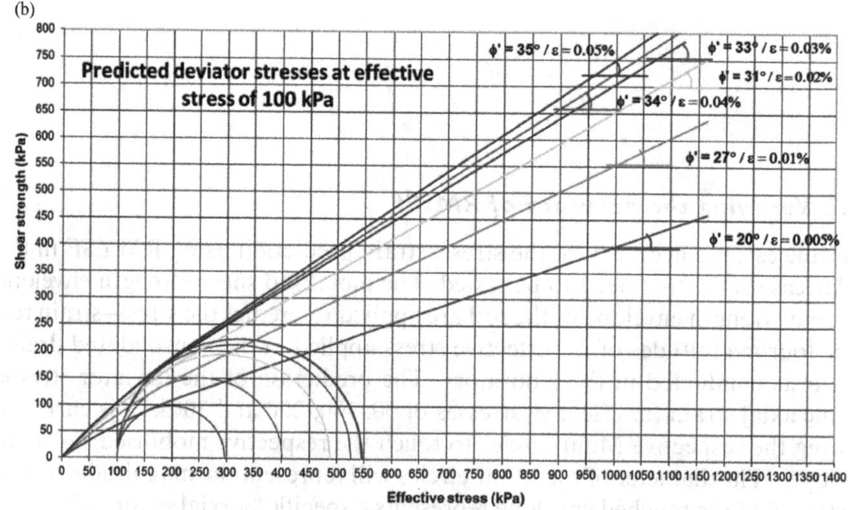

Figure 5.21 Drawing the Mohr circles touching the respective mobilised shear strength envelopes and the shear strength envelope at failure to determine the predicted deviator stresses corresponding to the specific axial strain represented by the envelopes. (a) Prediction of deviator stresses at an effective stress of 50 kPa by taking the minor principal stress at 50 kPa. (b) Prediction of deviator stresses at an effective stress of 100 kPa by taking the minor principal stress at 100 kPa. (c) Prediction of deviator stresses at an effective stress of 200 kPa by taking the minor principal stress at 200 kPa. (d) Prediction of deviator stresses at an effective stress of 300 kPa by taking the minor principal stress, at 300 kPa.

(Continued)

(c)

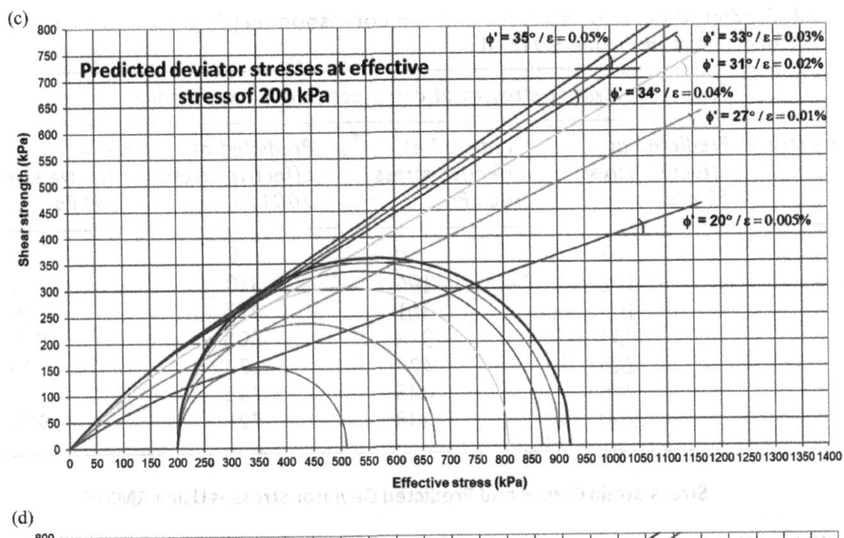

(d)

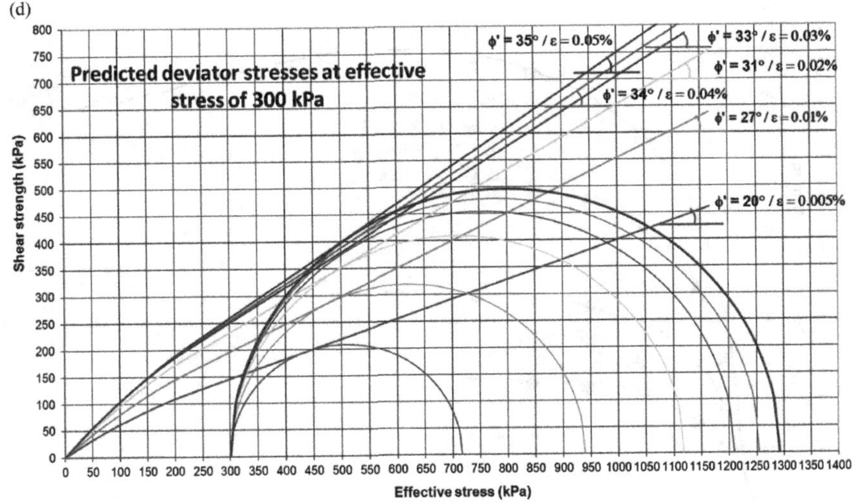

Figure 5.21 (Continued) Drawing the Mohr circles touching the respective mobilised shear strength envelopes and the shear strength envelope at failure to determine the predicted deviator stresses corresponding to the specific axial strain represented by the envelopes. (a) Prediction of deviator stresses at an effective stress of 50 kPa by taking the minor principal stress at 50 kPa. (b) Prediction of deviator stresses at an effective stress of 100 kPa by taking the minor principal stress at 100 kPa. (c) Prediction of deviator stresses at an effective stress of 200 kPa by taking the minor principal stress at 200 kPa. (d) Prediction of deviator stresses at an effective stress of 300 kPa by taking the minor principal stress, at 300 kPa.

four stress–strain curves is taken as 0.05%. The bulk unit weight of the soil is 18 kN/m^3 and the saturated unit weight of the soil is 19 kN/m^3. The groundwater table is at 4 m below the ground surface. The layer was backfilled with a soil of 3 m thickness with a bulk unit weight of 20 kN/m^3. Divide the soil layer into five equal thickness horizontal layers, i.e., each layer is 2 m thick.

Table 5.2 Predicted deviator stresses and the corresponding % axial strains as presented in Figure 5.21a–d

	Predicted deviator stresses at corresponding % axial strains			
% axial strain	*Predicted at effective stress 50 kPa*	*Predicted at effective stress 100 kPa*	*Predicted at effective stress 200 kPa*	*Predicted at effective stress 300 kPa*
0	0	0	0	0
0.005	112	196	310	416
0.01	165	301	473	639
0.02	232	388	608	817
0.03	245	426	670	909
0.04	258	445	702	955
0.05	260	448	721	992

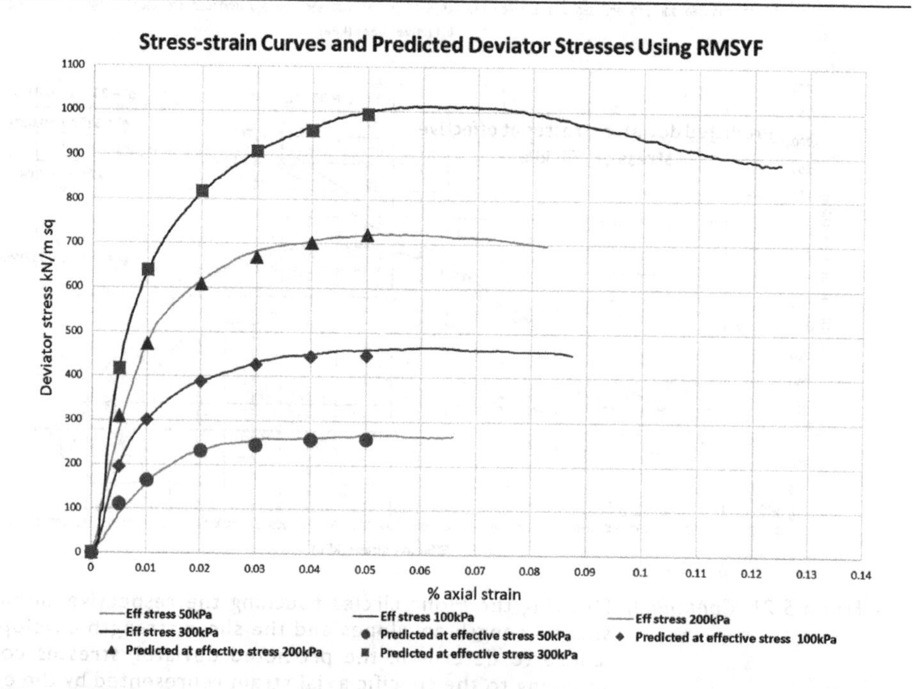

Figure 5.22 Data points of the predicted deviator stresses plotted superimposed with the stress–strain curves obtained from the laboratory consolidated drained triaxial tests showed an excellent replication.

The first task is to determine the state of effective stress at the midheight of each soil layer, i.e., at points *A*, *B*, *C*, *D* and *E* before the embankment load is applied. The stresses at these points will be the initial state of effective stresses and the final state of stresses will be when the embankment load has been applied. Therefore, it is required to predict the state of soil stress–strain response under the initial stress conditions. In other words, it is required to predict the stress–strain curves at the midpoints of each soil layers under the initial state of effective stresses.

The initial state of effective stress p_0 at each midpoint of the soil layers is calculated as follows:

p_0=initial effective stress at the middle of each layer after backfill

$$p_0 = H_{\text{Dry Sand}} \times \gamma_{\text{dry(sand)}} + H_{\text{Wet Sand}}\left[\gamma_{\text{sat(sand)}} - \gamma_{\text{w}}\right]$$

p_f=final effective stress at the middle of each layer after backfill

$$p_f = H_{\text{Backfill}} \times \gamma_{\text{backfill}} + H_{\text{Dry Sand}} \times \gamma_{\text{dry(sand)}} + H_{\text{Wet Sand}}\left[\gamma_{\text{sat(sand)}} - \gamma_{\text{w}}\right]$$

Initial effective stresses at midpoints A, B, C, D and E are as follows.

$$p_{0A} = 18 \times 1 = 18\,\text{kN/m}^2$$

$$p_{0B} = 18 \times 3 = 54\,\text{kN/m}^2$$

$$p_{0C} = 18 \times 4 + (19 - 9.81)1 = 81.19\,\text{kN/m}^2$$

$$p_{0D} = 18 \times 4 + (19 - 9.81)3 = 99.57\,\text{kN/m}^2$$

$$p_{0E} = 18 \times 4 + (19 - 9.81)5 = 117.95\,\text{kN/m}^2$$

Final effective stresses at midpoints A, B, C, D and E are as follows.

$$p_{fA} = 20 \times 3 + 18 \times 1 = 78\,\text{kN/m}^2$$

$$p_{fB} = 20 \times 3 + 18 \times 3 = 114\,\text{kN/m}^2$$

$$p_{fC} = 20 \times 3 + 18 \times 4 + (19 - 9.81)1 = 141.19\,\text{kN/m}^2$$

$$p_{fD} = 20 \times 3 + 18 \times 4 + (19 - 9.81)3 = 159.57\,\text{kN/m}^2$$

$$p_{fE} = 20 \times 3 + 18 \times 4 + (19 - 9.81)5 = 177.95\,\text{kN/m}^2$$

The next task is to predict the soil stress–strain response under the initial state of effective stress condition before the embankment load is placed. Thence, this is to predict the stress–strain curves at effective stresses of 18, 54, 81.19, 99.57 and 117.95 kN/m^2, which are the state of initial effective stresses at points *A*, *B*, *C*, *D* and *E* respectively. The predicted deviator stresses are conducted as shown in Figure 5.23a–e, respectively, and the values are presented in Table 5.3.

The predicted stress–strain curves at the midheight of each soil layer are plotted superimposed with the laboratory stress–strain curves obtained from conducting the consolidated drained triaxial tests as shown in Figure 5.24a–e, respectively. Figure 5.25 superimposed all the predicted stress–strain curves at the midheight of each soil layer at points *A*, *B*, *C*, *D* and *E*.

The next task is to determine the increase in the % axial strain for each soil layer due to the increase in the effective stress when the embankment loading is placed. The initial and the final effective stresses at midheight of each soil layer are summarised in Table 5.4. The predicted stress–strain curves at the midheight of each soil layer under

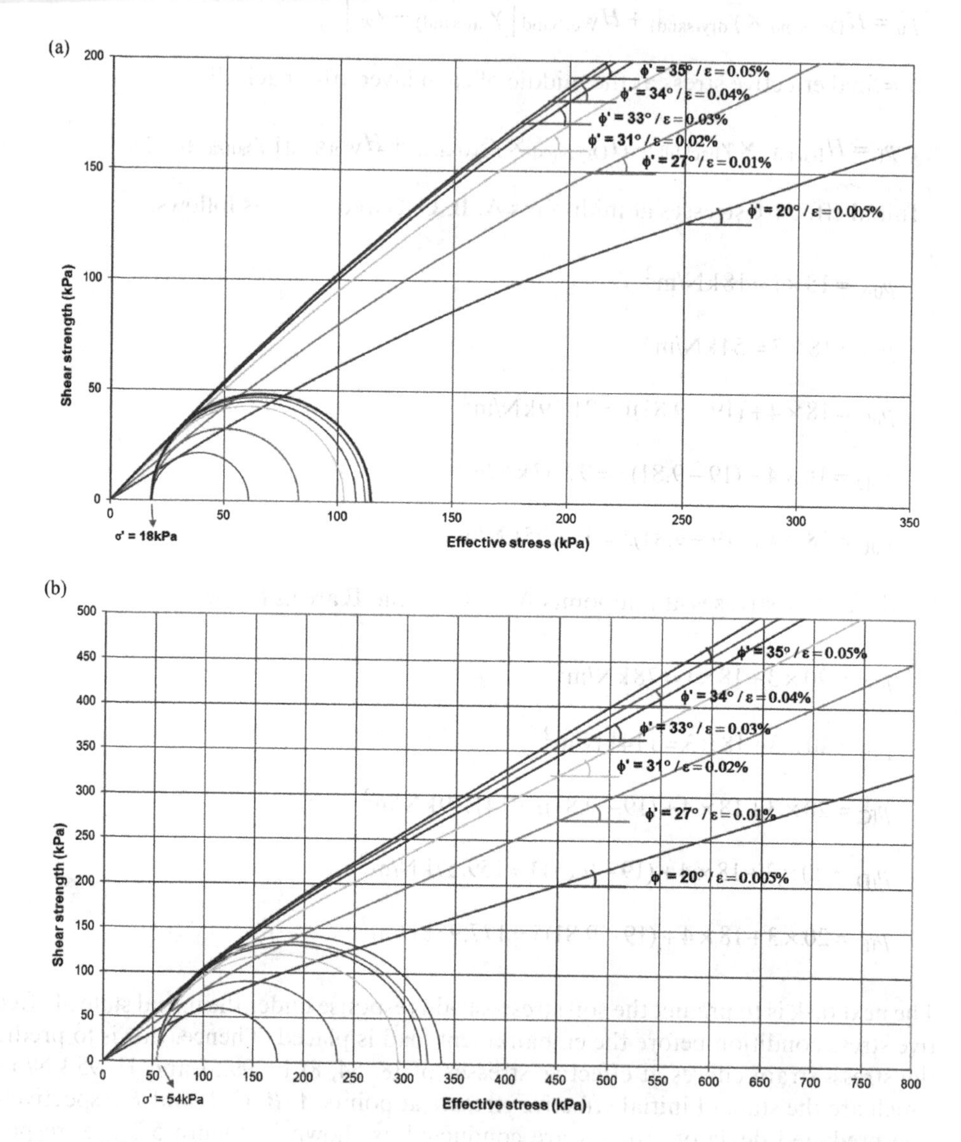

Figure 5.23 Prediction of stress–strain curves at midpoints of each soil layers. (a) Prediction of stress–strain curve at point A for layer 1. (b) Prediction of stress–strain curve at point B for layer 2. (c) Prediction of stress–strain curve at point C for layer 3. (d) Prediction of stress–strain curve at point D for layer 4. (e) Prediction of stress–strain curve at point E for layer 5.

(Continued)

initial effective stress condition are used to determine the increase in strain due to the embankment loading. The initial effective stress state will show the corresponding initial state of % axial strain while the final effective stress state will show the corresponding final state of % axial strain. Thence, the increase in the % axial strain for each soil layer can be determined. The magnitudes of the strain increased due to the

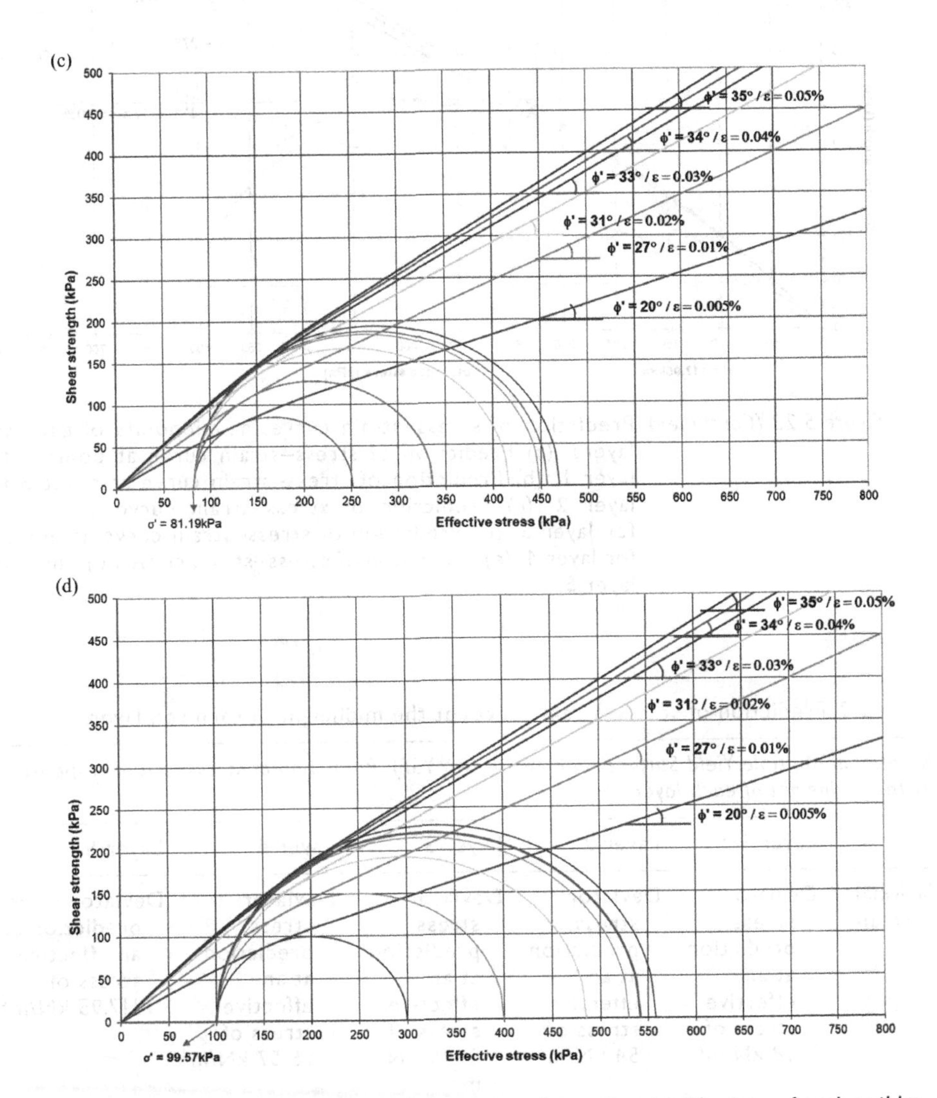

Figure 5.23 (Continued) Prediction of stress–strain curves at midpoints of each soil layers. (a) Prediction of stress–strain curve at point A for layer 1. (b) Prediction of stress–strain curve at point B for layer 2. (c) Prediction of stress–strain curve at point C for layer 3. (d) Prediction of stress–strain curve at point D for layer 4. (e) Prediction of stress–strain curve at point E for layer 5.

(Continued)

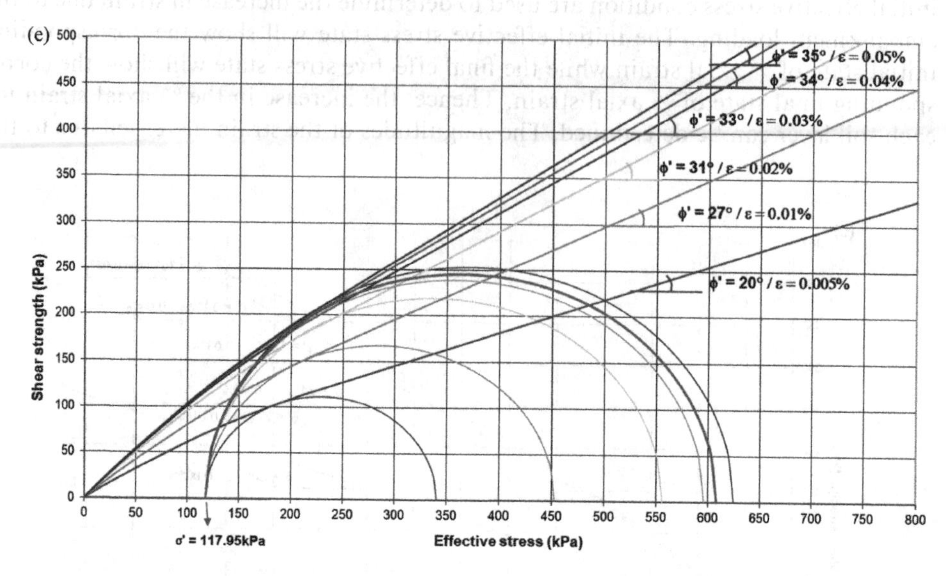

Figure 5.23 (Continued) Prediction of stress–strain curves at midpoints of each soil layers. (a) Prediction of stress–strain curve at point A for layer 1. (b) Prediction of stress–strain curve at point B for layer 2. (c) Prediction of stress–strain curve at point C for layer 3. (d) Prediction of stress–strain curve at point D for layer 4. (e) Prediction of stress–strain curve at point E for layer 5.

Table 5.3 Prediction of stress–strain curves at the midheight of each soil layer

Rotational Multiple Yield Surface Framework (RMYSF): Prediction of stress–strain response at the midheight of each layer

	Layer 1	Layer 2	Layer 3	Layer 4	Layer 5
% Axial strain	Deviator stress prediction at an effective stress of 18 kN/m^2	Deviator stress prediction at an effective stress of 54 kN/m^2	Deviator stress prediction at an effective stress of 81.19 kN/m^2	Deviator stress prediction at an effective stress of 95.57 kN/m^2	Deviator stress prediction at an effective stress of 117.95 kN/m^2
0	0	0	0	0	0
0.005	40	120	170	200	223
0.01	60	180	255	300	335
0.02	85	240	335	385	438
0.03	90	263	367	430	478
0.04	94	270	375	442	490
0.05	96	283	387	458	506

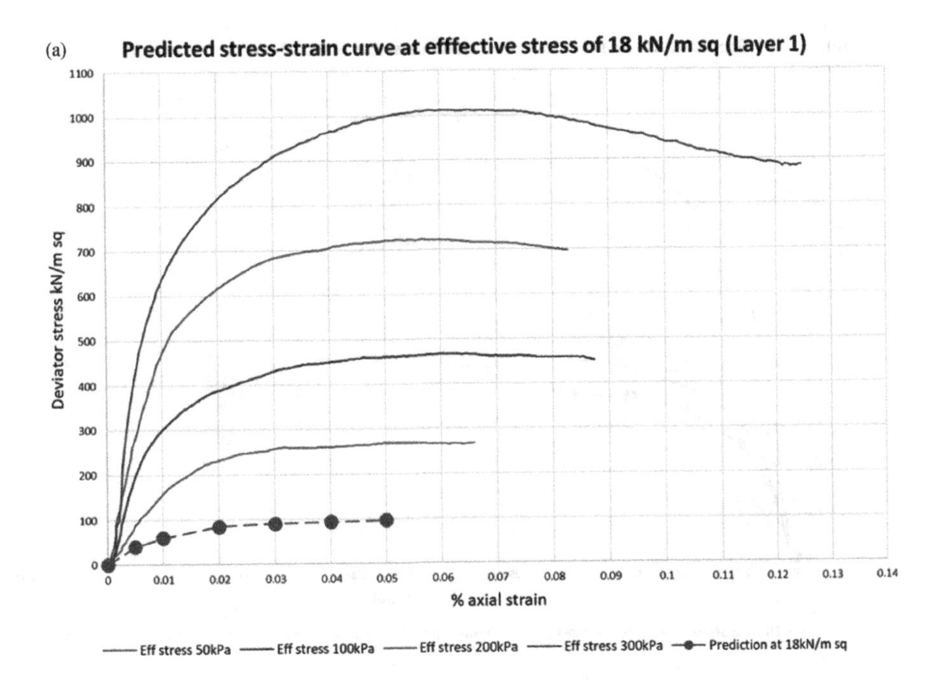

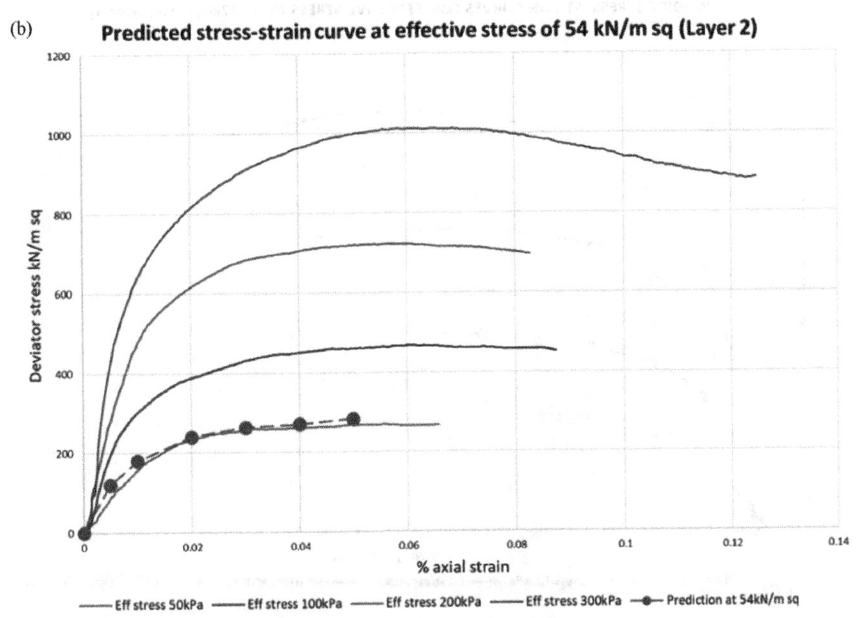

Figure 5.24 Predicted stress–strain curves at the midheight of each soil layer plotted superimposed with the laboratory stress–strain curves. (a) Predicted stress–strain curve at the midheight of layer 1. (b) Predicted stress–strain curve at the midheight of layer 2. (c) Predicted stress–strain curve at the midheight of layer 3. (d) Predicted stress–strain curve at the midheight of layer 4. (e) Predicted stress–strain curves at the midheight of layer 5.

(Continued)

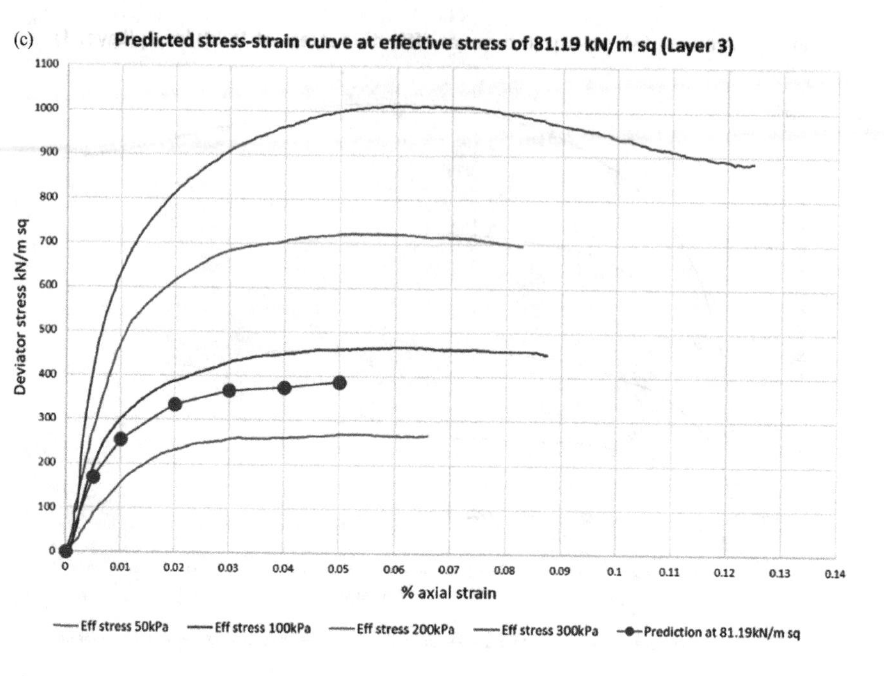

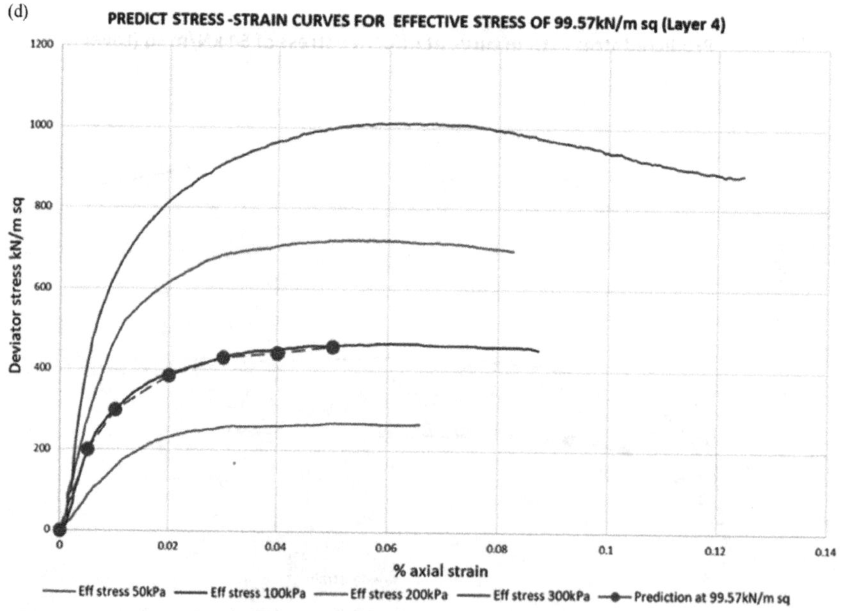

Figure 5.24 (Continued) Predicted stress–strain curves at the midheight of each soil layer plotted superimposed with the laboratory stress–strain curves. (a) Predicted stress–strain curve at the midheight of layer 1. (b) Predicted stress–strain curve at the midheight of layer 2. (c) Predicted stress–strain curve at the midheight of layer 3. (d) Predicted stress–strain curve at the midheight of layer 4. (e) Predicted stress–strain curves at the midheight of layer 5.

(Continued)

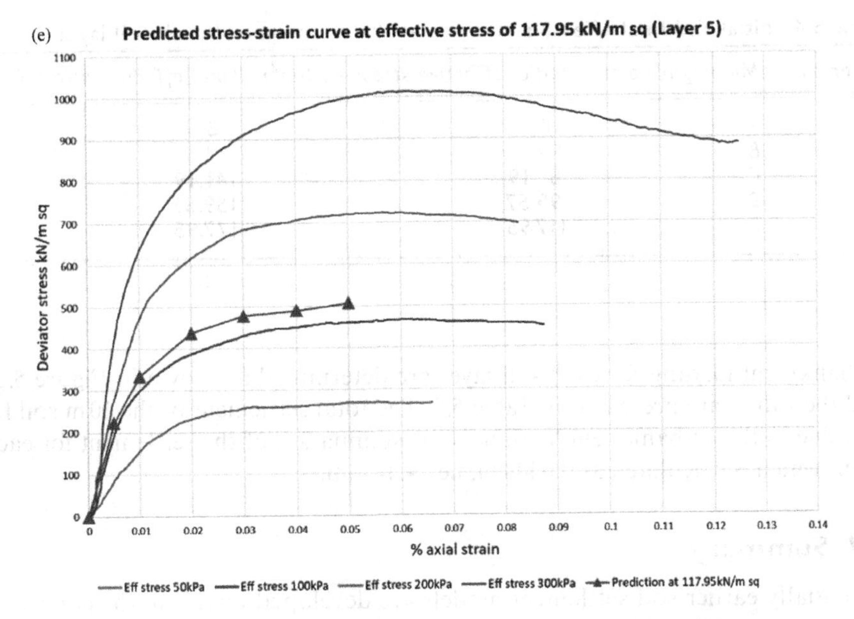

Figure 5.24 (Continued) Predicted stress–strain curves at the midheight of each soil layer plotted superimposed with the laboratory stress–strain curves. (a) Predicted stress–strain curve at the midheight of layer 1. (b) Predicted stress–strain curve at the midheight of layer 2. (c) Predicted stress–strain curve at the midheight of layer 3. (d) Predicted stress–strain curve at the midheight of layer 4. (e) Predicted stress–strain curves at the midheight of layer 5.

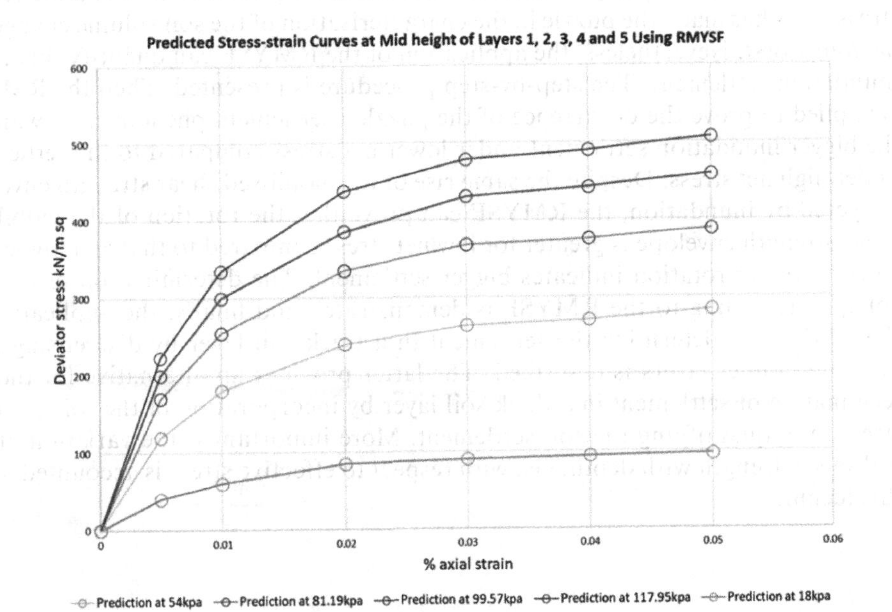

Figure 5.25 All the predicted stress–strain curves at the midheight of each soil layer.

Table 5.4 Initial and final effective stresses at the midheight of each soil layer

Layer no.	*Midheight point*	*Initial effective stress (kN/m²)*	*Final effective stress (kN/m²)*
1	A	18	78
2	B	54	114
3	C	81.19	141.19
4	D	99.57	159.57
5	E	117.95	177.95

embankment loading for each soil layer are determined as shown in Figure 5.26a–e, and the values are presented in Table 5.5. The total settlement of the 10 m soil formation due to the embankment loading is the summation of the settlement for each soil layer, which is calculated as 0.0444 m, i.e., 4.44 mm.

5.7 Summary

Essentially earlier soil settlement models are developed based on the concept of effective stress of Terzaghi (1943). The models determine the magnitude of settlement based on either the stress increased or the magnitude of the load applied. However, the occurrence of inundation settlement has contradicted the concept of effective stress when the settlement is happening under effective stress decrease. The laboratory Rowe cell inundation tests proved that the settlement takes place under a constant load or constant net stress when the soil is inundated. The Rowe cell inundation tests also revealed another complex soil settlement behaviour whereby inundation settlement is bigger under low net stress compared to that at a higher net stress. This has made the puzzle in the characterisation of the soil volume change behaviour worst. Nevertheless, the application of the RMYSF can quantify this weird inundation settlement. The step-by-step procedure is presented. Then the RMYSF is applied to prove the occurrence of the puzzled settlement phenomenon, which is the bigger inundation settlement under lower net stress compared to the settlement under high net stress. Despite the same rise of the mobilised shear strength envelope triggered by inundation, the RMYSF can prove that the rotation of the mobilised shear strength envelope is greater for low net stress compared to that at a higher net stress. Greater rotation indicates bigger settlement. The determination of loading collapse according to the RMYSF is demonstrated and finally, the application of the RMYSF to determine the settlement in a thick soil layer by discretising it to multiple thinner layers is described. The latter provides an alternative for the determination of settlement in a thick soil layer by incorporation of the role of shear strength in quantifying the soil settlement. More importantly, the variation of the soil shear strength with depth, i.e., with respect to effective stress is accounted for in this technique.

(a)

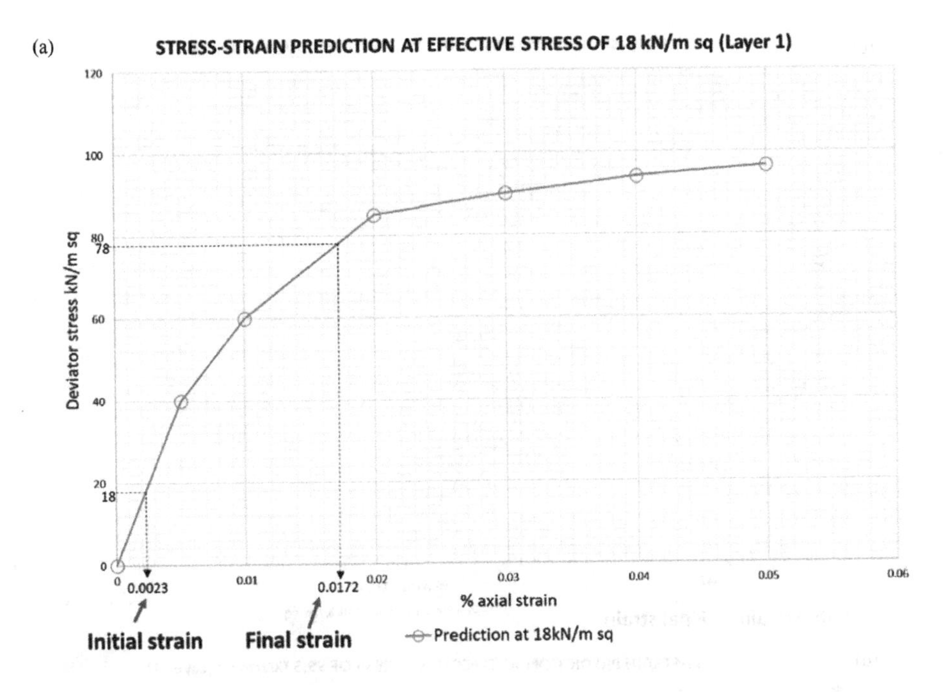

(b)

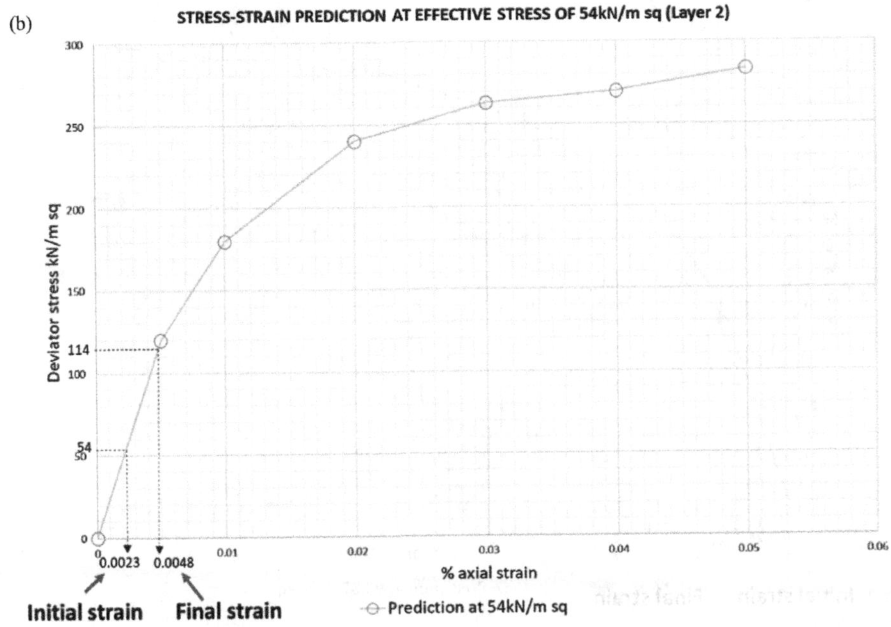

Figure 5.26 Determination of strain increased at midheight of each soil layer due to the embankment loading. (a) Determination of strain increased at the midheight of layer 1. (b) Determination of strain increased at the midheight of layer 2. (c) Determination of strain increased at the midheight of layer 3. (d) Determination of strain increased at the midheight of layer 4. (e) Determination of strain increased at the midheight of layer 5.

(Continued)

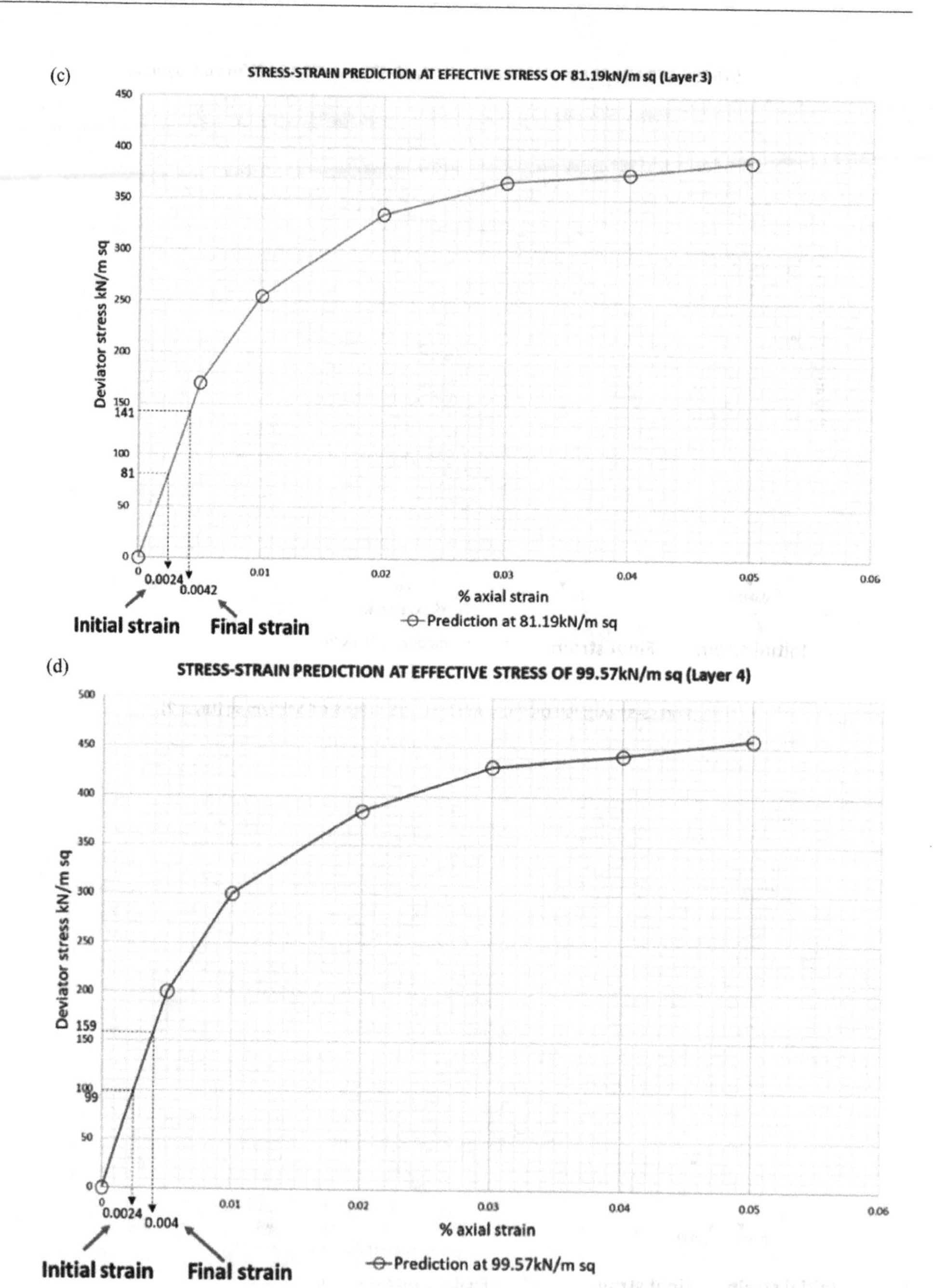

Figure 5.26 (Continued) Determination of strain increased at midheight of each soil layer due to the embankment loading. (a) Determination of strain increased at the midheight of layer 1. (b) Determination of strain increased at the midheight of layer 2. (c) Determination of strain increased at the midheight of layer 3. (d) Determination of strain increased at the midheight of layer 4. (e) Determination of strain increased at the midheight of layer 5.

(Continued)

(e)

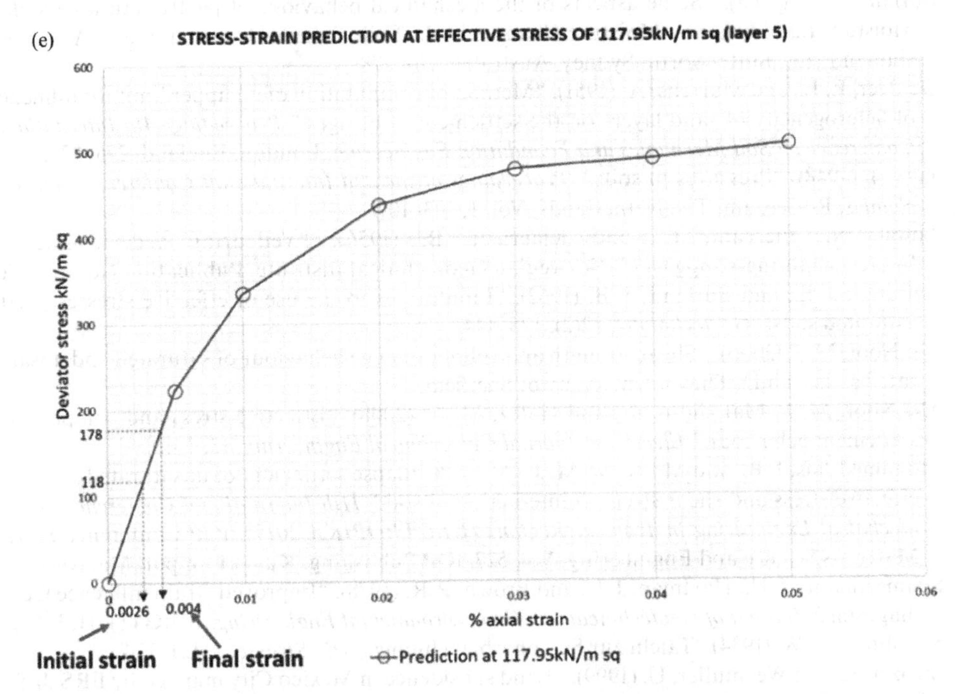

Figure 5.26 (Continued) Determination of strain increased at midheight of each soil layer due to the embankment loading. (a) Determination of strain increased at the midheight of layer 1. (b) Determination of strain increased at the midheight of layer 2. (c) Determination of strain increased at the midheight of layer 3. (d) Determination of strain increased at the midheight of layer 4. (e) Determination of strain increased at the midheight of layer 5.

Table 5.5 Settlement calculation for each soil layer

Layer no.	*Midpoint*	*Initial % axial strain*	*Final % axial strain*	*Increase in % axial strain* $\Delta\varepsilon\%$	*Layer thickness H (m)*	*Settlement (m) =* $\Delta\varepsilon\% \times H$
1	A	0.0023	0.0172	0.0149	2	0.0298
2	B	0.0023	0.0048	0.0025	2	0.005
3	C	0.0024	0.0042	0.0018	2	0.0036
4	D	0.0024	0.004	0.0016	2	0.0032
5	E	0.0026	0.004	0.0014	2	0.0028
				Total settlement		0.0444

References

Bishop, A. W. (1959). "The principle of effective stress." *Teknisk oeblad*, 106(39), 859–863.

Bishop, A. W., and Blight, G. E. (1963). "Some aspects of effective stress in saturated and unsaturated soils." *Geotechnique*, 13(3), 177–197.

Blanchfield, R., and Anderson, W. F. (2000). "Wetting collapse in opencast coalmine backfill." *Geotechnical Engineering*, 143(3), 139–149.

Burland, J. B. (1965). "Some aspects of the mechanical behaviour of partly saturated soils." Moisture Equilibria and Moisture Changes in the Soils Beneath Covered Areas, A Symposium in Print, Butterworth, Sydney, Australia, 270–278.

De Beer, E. E., and Martens, A. (1951). "Method of computation of an upper limit for influence of heterogeneity of sand layers on the settlement of bridges." *Proceedings 4th International Conference on Soil Mechanics and Foundation Engineering*, London, England, 275–278.

Jaky, J. (1948). "Pressures in soils." *Proceedings of the 2nd International Conference Soil Mechanics*, Rotterdam, The Netherlands, Vol. 1, 103–107.

Janbu, N., Bjerrum, L., and Kjaernsli, B. (1956). "Veiledring ved losning av fundermentering-soppgaver." Norwegian Geotechnical Institute, Publication No. 16, Oslo.

Jennings, J. E., and Burland, J. B. (1962). "Limitations to the use of effective stresses partly saturated soils." *Geotechnique*, 12(2), 125–144.

Md Noor, M. J. (2006). "Shear strength and volume change behaviour of saturated and unsaturated soils." Ph.D. Thesis, University of Sheffield.

Md Noor, M. J., Mat. Jidin, R., and Hafez, M. A. (2008). "Effective stress and complex soil settlement behaviour." *Electronic Journal Geotechnical Engineering*, 13, 1–13.

Mohamed Jais, I. B., and Md Noor, M. J. (2019). "Collapse settlement of unsaturated soil from effective stress and shear strength interaction of soil." *11th International Conference on Geotechnical Engineering in Tropical Regions (GEOTROPIKA 2019)*, IOP Conference Series: Materials Science and Engineering, Vol. 527, IOP Publishing, Kuala Lumpur, Malaysia.

Schmertmann, J. H., Hartman, J. P., and Brown, P. R. (1978). "Improved strain influence factor diagrams." *Journal of Geotechnical and Geoenvironmental Engineering*, 104(GT8), 1131–1135.

Steinbrenner, W. (1934) "Tafeln zur Setzungsberechnung." *Die Strasse*, 1, 121–124.

Strozzi, T., and Wegmuller, U. (1999). "Land subsidence in Mexico City mapped by ERS differential SAR interferometry." *IEEE 1999 International Geoscience and Remote Sensing Symposium*, Hamburg, Germany doi:10.1109/IGARSS.1999.774993.

Tadepalli, R., Rahardjo, H., and Fredlund, D. G. (1992). "Measurement of matric suction and volume change during inundation of collapsible soil." *ASTM Geotechnical Testing Journal*, 15(2), 115–122.

Terzaghi, K. (1943). *Theoretical soil mechanics*, Wiley Publications, New York.

Chapter 6

Anisotropic and elastic–plastic rock deformation model for accurate prediction of intact rock stress–strain response

6.1 Introduction

The study of rock deformation when subjected to stress is very important, because often rock formation normally becomes the foundation-supporting layer for high-rise buildings. Thence, the amount of settlement when the formation is loaded need to be known, so that the magnitude of the deformation is within the limit allowed or the subjected stress does not exceed the rock peak strength under a certain confining pressure. Researchers focussed on the prediction of the rock mass peak strength by developing and improving the rock damage model. Normally, this rock mass breaking point is predicted based on the rock stiffness, *k*, or Young's modulus, *E*. This concept assumes that the rock is an elastic material whereas the true rock response to stress is elastic–plastic. As an example, the foundation of the PETRONAS Twin Towers in Kuala Lumpur, Malaysia is supported on 104 piles. The piles were extended to the depths ranging from 200 to 374 ft to reach the bedrock formation. The load from the twin towers is transferred to the rock formation through the piles. With the piles foundation reaching this depth, PETRONAS Twin Towers become the structure having the deepest-seated foundation in the world. Thence, the prediction of the tower settlement is of very importance, especially because it involved two towers. The settlement prediction for the towers need to be accurate to avoid any differential settlement over the massive foundation platform and the most important factor of all is that it does not exceed the rock peak strength.

Thence, a very accurate rock deformation model is required. The application of the Normalised Strain Rotational Multiple Yield Surface Framework (NSRMYSF) for predicting rock settlement begins with the prediction of rock stress–strain behaviour and the determination of the intact rock peak strength. This deformation model is developed from the rock actual stress–strain behaviour and thus has directly incorporated the non-linear effect of the stress–strain or the degradation of stiffness as strain increases. The non-linear stress–strain curves are obtained from conducting uniaxial compression tests in Hoek's cell (Hoek, 1968) subjected to a certain confining pressure. Thence, the rock specimen is subjected to an anisotropic stress condition. The model can accurately predict not just the intact rock peak strength but also the intact rock stress–strain curves.

6.2 Rock stress–strain behaviour

Rock deformed when subjected to stress. The deformation behaviour of rock is defined by its stress–strain curves under various confining pressures when subjected to uniaxial compression. A typical stress–strain curve for intact rock is shown in Figure 6.1, which is the stress–strain curve for intact Hawkesbury sandstone from Australia at confining pressures of 0, 4 and 8 MPa (Hamzah et al., 2020). All the curves slightly bent downward at initial straining and this is indicating the initial loss in stiffness. However, subsequently, the graph starts to curve upwards indicating the increase in the stiffness at a mid-range strain. Finally, when approaching the peak strength, the graph starts to bend down towards the peak strength to indicate the loss of stiffness again when nearing fracture. Largely, the rock stress–strain behaviour is non-linear. Once the failure plane has developed, the stress–strain curves become horizontal, if there is confining pressure, i.e., the graphs become straining under constant stress as demonstrated by the graphs for 4 and 8 MPa confining pressures in Figure 6.1. Notice that as the confining pressure increases, the rock peak strength also increases. The deviator stresses at failure for this rock at confining pressures of 0, 4 and 8 MPa are 40, 58 and 73 MPa, respectively. Besides, the axial strain at failure also increases with the increase in the confining pressure. The axial strains at failure for this rock are 0.55%, 0.75% and 0.9% for the confining pressures of 0, 4 and 8 MPa, respectively.

6.3 Rock elastic–plastic behaviour under anisotropic stress conditions

The question arises whether rock responds elastic or elastic–plastic when subjected to anisotropic stress or the uniaxial compression. Some thought that rock responds fully elastic before failure because it is considered as a brittle material. Once failed, the rock will fully break. That is why most rock deformation models applied the stiffness, k, that is the best representative gradient of the stress–strain curve to predict the failure stress or the rupture strength. The slight bending downwards of the stress–strain curves at low strain and when approaching peak strength due to stiffness degradation

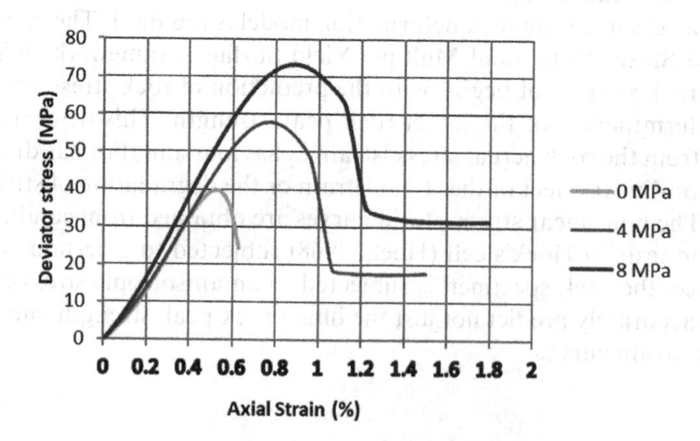

Figure 6.1 Stress–strain curves for Hawkesbury sandstone (Hamzah et al., 2020).

Figure 6.2 The cylindrical core specimens of Hawkesbury sandstone of diameter 42 mm and height 100 mm used in the uniaxial monotonic compressive testing. (a) Fresh cored specimen. (b) Specimen fixed with a vertical strain gauge to monitor axial displacement. (c) Specimen fixed with a horizontal strain gauge to monitor lateral displacement (Hamzah et al., 2020).

is ignored for simplicity. In other words, the non-linearity aspect of the stress–strain curve is neglected in most rock damage models.

The true respond of a rock to the subjected stress can only be understood by carrying out unloading and reloading tests during the rock uniaxial compression test. Rock uniaxial tests were conducted using the INSTRON 1342 machine and rock specimens used are of 42 mm diameter and 100 mm height as shown in Figure 6.2a–c. The specimen is equipped with vertical strain gauge as shown in Figure 6.2b to monitor axial displacement and horizontal strain gauge as shown in Figure 6.2c to monitor lateral displacement.

Figure 6.3 shows the rock unloading and reloading curve. The curve has substantiated that rock is an elastic–plastic material. When the rock specimen was loaded from O at the origin up to point A, the response must be elastic–plastic because when it is being unloaded at A to B, the deformation can be seen to be constituted of elastic, BC, and plastic strain, OB. Note that AB represents the unloading and reloading curves and these curves are the fully elastic curves. In other words, even though the unloading and reloading are repeated many times, the paths AB and BA are maintained. However, upon reaching A and beyond, the curve follows the normal stress–strain behaviour, which is the elastic–plastic response.

Rock response by anisotropic behaviour under uniaxial compression test condition (Barla, 1974). However, he quoted that rock is frequently treated as linearly elastic, homogeneous and isotropic material. Significant errors can be encountered in stress and deformation analysis when assuming anisotropic rock to be isotropic (Berry and

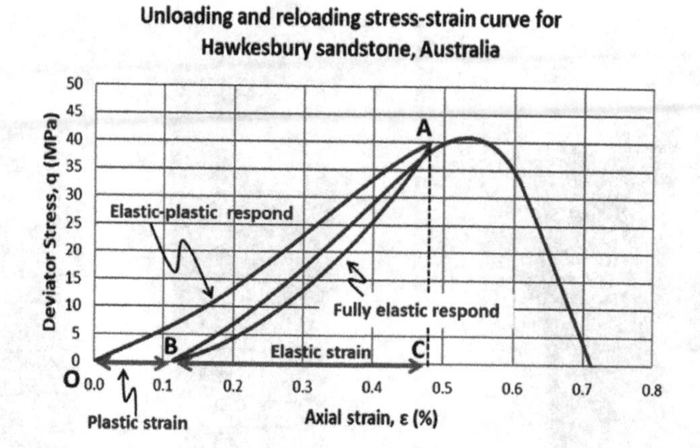

Figure 6.3 Unloading and reloading stress–strain curve for Hawkesbury sandstone from Australia (Hamzah et al., 2020).

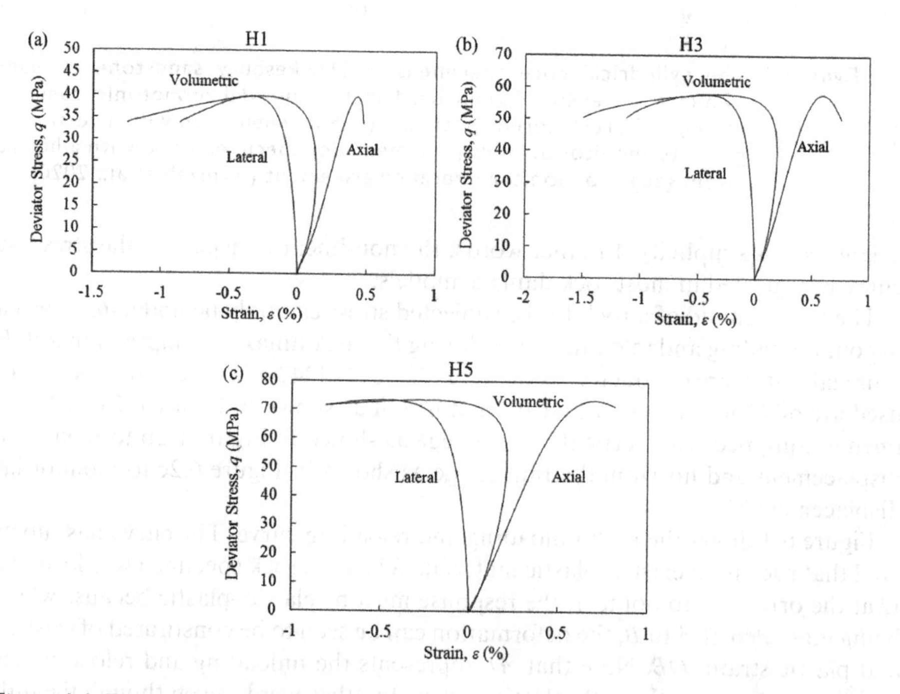

Figure 6.4 The response of the Hawkesbury sandstone based on the axial, lateral and volumetric deformation behaviours when tested using a uniaxial compression machine (Hamzah et al., 2020). (a) 0 MPa confining stress. (b) 4 MPa confining stress. (c) 8 MPa confining stress.

Fairhurst, 1966; Becker and Hooker, 1967; Barla and Wane, 1970). Figures 6.4a–c shows the response of the Hawkesbury sandstone based on the axial, lateral and volumetric deformation behaviours when tested using a uniaxial compression machine.

It is proved that the rock expanded laterally when being compressed and thence substantiated the rock anisotropic behaviour under uniaxial compression. The volumetric strain is calculated based on Equation 6.1;

$$\varepsilon_{vol} = \varepsilon_{axial} + 2\varepsilon_{lateral} \tag{6.1}$$

where

ε_{vol} is the volumetric strain,
ε_{axial} is the average axial strain recorded from two axial strain gauges and
$\varepsilon_{lateral}$ is the average lateral strain recorded from two lateral strain gauges.

The axial strain gauges are responding with increasing positive value indicating the shortening of the specimen when the deviator stress is applied. The lateral strain gauges are responding with an increasing negative value, indicating the specimen is expanding laterally. Thence, the readings of these strain gauges attached to the rock specimens have substantiated that these rock specimens are responding by shortening axially and expanding laterally due to the anisotropic compression stress.

6.4 Previous rock deformation models

The rock deformation models are commonly referred to as rock damage models. These models are designed to predict the rock mass peak strength. The models are based on the laboratory experimental results conducted on intact rock specimens. Most of these models incorporate Young's modulus, E, Poisson ratio, v, and the confining pressure, σ_3, in characterising the rock peak strength. Besides, these models mostly ignored the effect of stiffness degradation, which results in the non-linear behaviour of the stress–strain curves as strain increases during the uniaxial compression.

The non-linear Hoek–Brown failure criterion for rock masses was introduced by Hoek and Brown (1980a, 1980b) to predict the strength of rock mass based on the strength of intact rock. This rock strength model is one of the earliest and has been accepted and applied widely by the rock mechanics community (Hoek and Brown, 1990). It is defined by Equation 6.2.

$$\sigma_1' = \sigma_3' + \sqrt{m\sigma_c\sigma\sigma_3' + s\sigma_c^2} \tag{6.2}$$

where

σ_1' is the major principal stress at failure,
σ_3' is the minor principal stress or the applied confining pressure,
m and s are the material constants and
σ_c is the uniaxial compressive strength of the intact rock at zero confining pressure.

By this equation, the peak deviator stress $\Delta\sigma_f$ or the rock peak strength can be determined when σ_1' and σ_3' are known as

$$\Delta\sigma_f = \sigma_1' - \sigma_3'$$

Then, Hoek et al. (2002) has refined the equation as Equation 6.3.

$$\sigma_1' = \sigma_3' + \sigma_{ci}\left(m\frac{\sigma_3'}{\sigma_{ci}} + s\right)^{0.5} \tag{6.3}$$

where σ_{ci} is the uniaxial compressive strength of the intact rock.

Figure 6.5 shows the change in the Hoek–Brown non-linear failure envelope as a function of the material constant m plotted in shear versus normal stress. Note that larger values of m give a steeper Hoek–Brown failure envelope equivalent to higher friction angles than the lower m values.

Practically all software for soil and rock mechanics are written in terms of the Mohr–Coulomb criterion, i.e., in the form of c' and ϕ'. Thence, there is a need to correlate the Hoek–Brown criterion with the non-linear parameters m and s, and the Mohr–Coulomb criterion with the c' and ϕ'. Hoek et al. (2002) then introduced the equation for the conversion of the rock strength in terms of the principal stresses into the Mohr–Coulomb failure criterion by the estimation of c' and ϕ'. The proposed equations are given in Equations 6.4 and 6.5 respectively.

$$c' = \frac{\sigma_{ci}\left[(1+2a)s + (1-a)m_b\sigma_{3n}'\right](s + m_b\sigma_{3n}')^{a-1}}{(1+a)(2+a)\sqrt{1 + \left(6am_b(s + m_b\sigma_{3n}')^{a-1}\right)/\left((1+a)+(2+a)\right)}} \tag{6.4}$$

$$\phi' = \sin^{-1}\left[\frac{6am_b(s + m_b\sigma_{3n}')^{a-1}}{2(1+a)(2+a) + 6am_b(s + m_b\sigma_{3n}')^{a-1}}\right] \tag{6.5}$$

where $\sigma_{3n} = \dfrac{\sigma_{3\max}'}{\sigma_{ci}}$ and

$$m_b = m_i \exp\left(\frac{\text{GSI} - 100}{28 - 14D}\right) \tag{6.6}$$

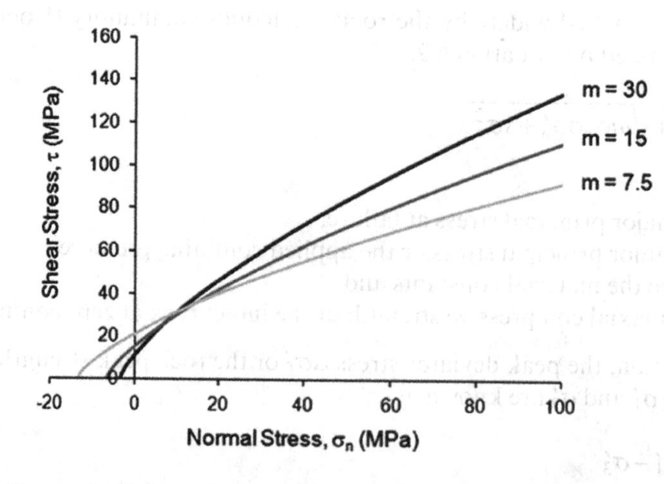

Figure 6.5 Change in the Hoek–Brown failure envelope as a function of the material constant m plotted in shear versus normal stress.

s and a are constants for the rock mass given by the following relationships:

$$s = \exp\left(\frac{GSI - 100}{9 - 3D}\right) \quad (6.7)$$

$$a = \frac{1}{2} + \frac{1}{6}\left(e^{-GSI/15} - e^{-20/3}\right) \quad (6.8)$$

where GSI is the geological strength index introduced by Hoek et al. (1992).

Essentially, the formulae for the estimation of c' and ϕ' as in Equations 6.4 and 6.5 involved various coefficients and certainly, the mathematics of rock mechanics is very complex. This also may indicate the complexities involved in the study of the strength of the rock mass, which is to be derived from the result of the laboratory strength tests on small intact rock specimens.

Figure 6.6 shows the difference between a linear Mohr–Coulomb envelope and a non-linear Hoek–Brown failure envelope plotted against triaxial test data for intact rock (Eberhardt, 2012). The true non-linear Hoek–Brown failure envelope is approximated to the Mohr–Coulomb linear failure envelope where the c' and ϕ' are determined based on Equations 6.4 and 6.5, respectively.

Figure 6.7 shows how the rock mass strength is estimated from the intact rock strength using GSI.

Another effort to estimate the strength of rock mass is the damage model of Eberhardt et al. (1999), which defined damage in terms of axial, lateral and volumetric cumulative strains according to Equations 6.9–6.11 and in addition to the recorded number of acoustic events (ω_{AE}). Eberhardt et al. (1999) accounted for the influence of the acoustic events(ω_{AE}) by performing acoustic emission monitoring during the

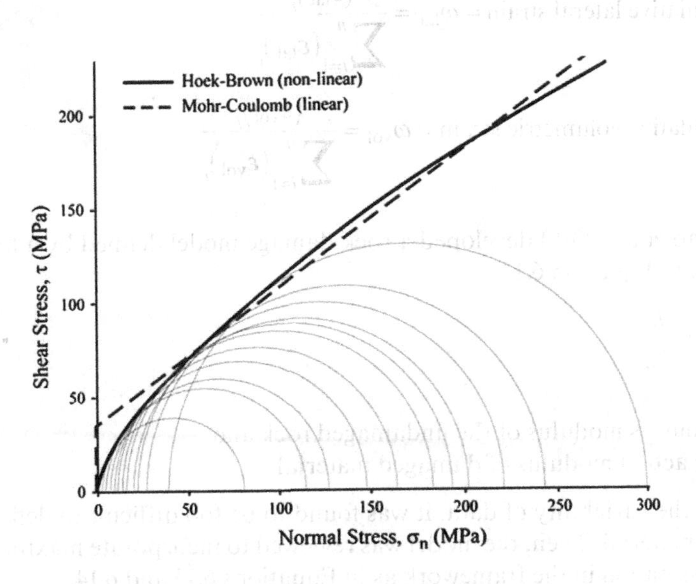

Figure 6.6 The difference between linear Mohr–Coulomb envelope and the non-linear Hoek–Brown failure envelope plotted against triaxial test data for intact rock (Eberhardt, 2012).

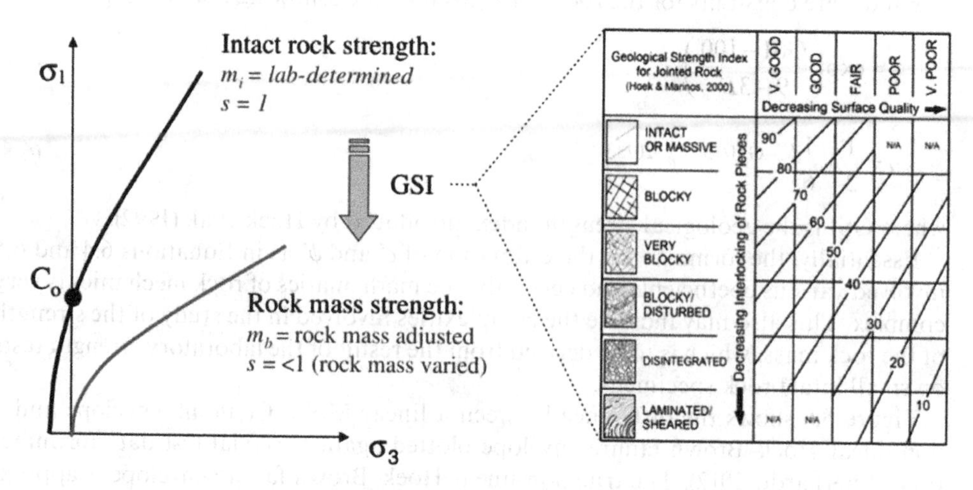

Figure 6.7 Scaling of Hoek–Brown non-linear failure envelope for an intact rock to that for rock mass strength (Eberhardt, 2012).

uniaxial compression. Nevertheless, the model did not consider the confining pressure, peak strength or the number of loading repeatedly applied to the rock specimen.

$$\text{Cumulative axial strain} = \omega_{\text{ax}} = \frac{(\varepsilon_{\text{ax}})_i}{\sum_{i=1}^{n} (\varepsilon_{\text{ax}})_i} \tag{6.9}$$

$$\text{Cumulative lateral strain} = \omega_{\text{lat}} = \frac{(\varepsilon_{\text{lat}})_i}{\sum_{i=1}^{n} (\varepsilon_{\text{lat}})_i} \tag{6.10}$$

$$\text{Cumulative volumetric strain} = \omega_{\text{vol}} = \frac{(\varepsilon_{\text{vol}})_i}{\sum_{i=1}^{n} (\varepsilon_{\text{vol}})_i} \tag{6.11}$$

Then, Xiao et al. (2010) developed a rock damage model defined by a loss in stiffness according to Equation 6.12.

$$D = 1 - \frac{E}{E_o} \tag{6.12}$$

where

E_o is Young's modulus of the undamaged rock and
E is the actual modulus of damaged material.

Owing to the variability of data, it was found to be too difficult to define modulus of damaged material. Then, the model was reviewed to incorporate maximum strain and energy dissipation in the framework as in Equations 6.13 and 6.14.

$$D = \frac{E}{E_{\text{tot}}} \tag{6.13}$$

where
E_{tot} is the total energy dissipation capacity in unit volume and
E is the instantaneous energy dissipation after some cycles.

$$D = \frac{e_{max}^{n} - e_{max}^{o}}{e_{max}^{f} - e_{max}^{o}} \tag{6.14}$$

where
e_{max}^{o} is the initial maximum strain,
e_{max}^{n} is the instantaneous maximum strain after n cycles and
e_{max}^{f} is the ultimate maximum strain.
Then, Chen et al. (2006) developed a rock damage model using Equation 6.15.

$$D = 1 - \exp\left\{ -\left\{ \left[\varepsilon_1 E - \left(\frac{1+\sin\phi}{1-\sin\phi} - 2v \right) \sigma_3 \right] \Big/ E\varepsilon_o \right\}^m \right\} \tag{6.15}$$

where
ϕ is the internal angle of friction,
ε_1 is the axial strain,
σ_3 is the confining pressure,
E is Young's modulus,
v is Poisson's ratio and
ε_o is the scale parameter related to the average uniaxial strength.

Therefore, Chen et al. (2006) has incorporated the peak strength, deformability, confining pressure and the cyclic loading amplitude in the rock damage model.

Note that these rock deformation models only predict the rock peak strength, which is the failure deviator stress, and only the Hoek–Brown model that considers strength in the form of failure envelope. Moreover, most of these rock damage models did not incorporate the shear strength τ that is developed within the rock mass to resist the compression when it is being uniaxially loaded which is according to the Mohr–Coulomb failure criterion. This developed shear strength is known as mobilised shear strength. However, the maximum limit of this mobilised shear strength is the rock peak strength where the rock mass can no longer withstand the mobilised shear strength that leads to the rupture failure. Besides, these models did not incorporate the applied compressive stress that is distributed within the rock mass in the form of Mohr–Coulomb strength criterion described by a Mohr circle. In addition, the application of Young's modulus, E, assumes that the rock responds linearly with respect to stress. In other words, the non-linear aspects of the rock stress–strain behaviour are ignored.

6.5 Development of mobilised shear strength in Hawkesbury sandstone when the rock undergoes anisotropic compression

Rock uniaxial tests were conducted using the INSTRON 1342 machine, which is a servo-controlled testing machine with a loading capacity of 250 kN to demonstrate the development of the mobilised shear strength when the rock specimen is compressed.

Cylindrical specimens of Hawkesbury sandstone were prepared according to ISRM (International Society for Rock Mechanics) standards with approximate dimensions of 100 mm height and 42 mm diameter (Hamzah et al., 2020). Figure 6.2 shows the typical specimen of Hawkesbury sandstone used in the tests. The INSTRON 1342 machine is capable of applying different confining pressures, which can be varied from 0 to 150 MPa. The monotonic loading was applied to every rock specimen with the application of deviator stress at a constant loading rate of 0.01 mm/sec until the failure point was reached. The experimental setup of the uniaxial compression test is shown in Figure 6.8. Three rock uniaxial tests were conducted where the specimens were subjected to three confining pressures, which are 0, 4 and 8 MPa (Hamzah et al., 2020). The stress–strain curves obtained from these tests are shown in Figure 6.1.

During the shearing stage in the uniaxial testing, the specimens underwent axial compression and subsequent lateral expansion. This condition is known as anisotropic stress condition whereby the axial and lateral stresses applied are not equal. The lateral pressure is produced by applying cell pressure while the vertical pressure is the sum of the applied deviator stress and applied cell pressure. The deviator stresses at failure for the tests at confining pressures of 0, 4 and 8 MPa are found to be 40, 58 and 73 MPa, respectively. These stress–strain curves are normalised as shown in Figure 6.9 with a common axial strain at failure of 0.85%. The mobilised shear strength envelopes shown in Figures 6.10 and 6.11 are defined using Equations 5.16–5.18 and determined from Mohr circles derived from the normalised stress–strain curves as in Figure 6.9. The mobilised shear strength envelopes for rock are drawn using curvilinear equations applied for soil shear strength envelopes as in Equations 2.16–2.18 with a slight modification by adding the term c that is the rock cementation. These curvilinear shear strength equations for rock are shown in Equations 6.16–6.18, which are adopted from equations of Md Noor and Anderson (2006).

$$\tau_{\mathrm{satf}} = c + \frac{(\sigma - u_{\mathrm{w}})}{(\sigma - u_{\mathrm{w}})_{\mathrm{t}}}\left[1 + \frac{(\sigma - u_{\mathrm{w}})_{\mathrm{t}} - (\sigma - u_{\mathrm{w}})}{N(\sigma - u_{\mathrm{w}})_{\mathrm{t}}}\right]\tau_{\mathrm{t}} \tag{6.16}$$

Figure 6.8 Experimental setup of the uniaxial compression test using an INSTRON 1342 machine (Hamzah et al., 2020).

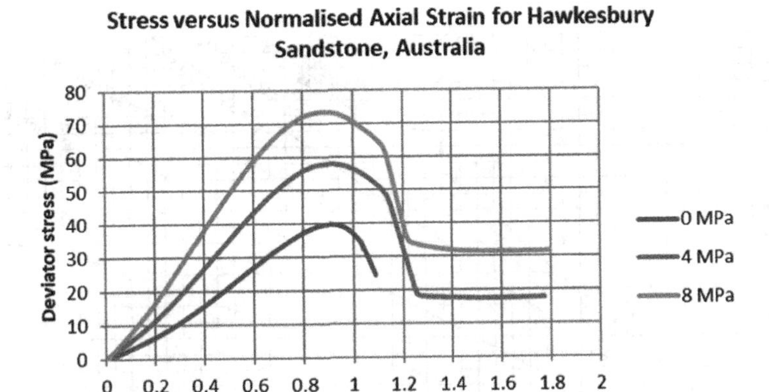

Figure 6.9 Normalised stress–strain curves for Hawkesbury sandstone with common axial strain at failure of 0.89% (Hamzah et al., 2020).

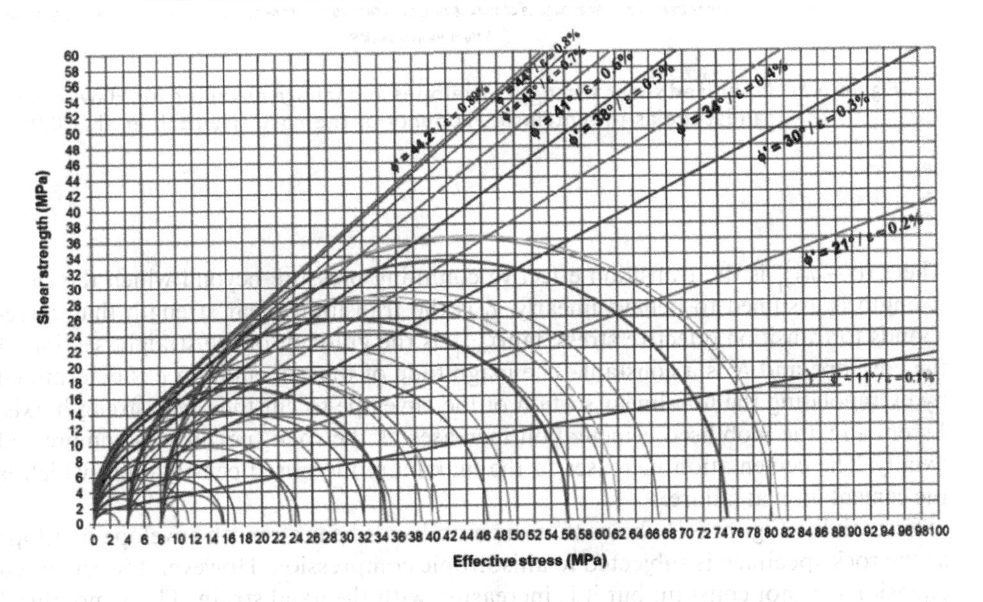

Figure 6.10 Mobilised shear strength envelopes for Hawkesbury sandstone at axial strains of 0.2%, 0.4%, 0.6%, 0.8% and shear strength envelope at failure at an axial strain of 0.89% (Hamzah et al., 2020).

$$\tau_{\text{satf}} = c + (\sigma - u_{\text{w}})\tan\phi'_{\text{minf}} + \left[\tau_{\text{t}} - (\sigma - u_{\text{w}})_{\text{t}} \tan\phi'_{\text{minf}}\right] \tag{6.17}$$

$$N = \frac{1}{1 - \left[(\sigma - u_{\text{w}})_{\text{t}} \dfrac{\tan\phi'_{\text{minf}}}{\tau_{\text{t}}}\right]}, \tag{6.18}$$

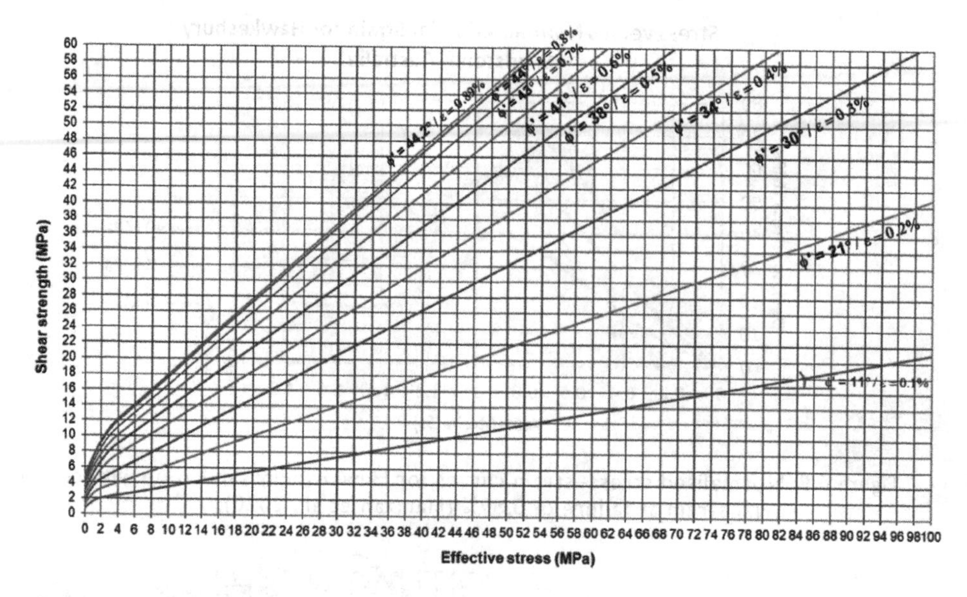

Figure 6.11 Mobilised shear strength envelopes and failure envelope for Hawkesbury sandstone as the intrinsic property of the rock (Hamzah et al., 2020).

where $(\sigma - u_w)_t$ is the transition effective confining pressure beyond which the shear strength is assumed to behave linearly, τ_t is the transition shear strength that corresponds to transition effective stress, $\tan\phi'_{\text{minf}}$ is the gradient of the straight section of the envelope and N is a constant. The magnitude of the cementation c is determined by extrapolating the non-linear section of the envelope to cut the shear strength axis. Essentially, the mobilised cementation increases as the rock specimen is compressed axially. The cementation increases to the maximum mobilised cementation, which is the cementation at failure.

Essentially, Figure 6.11 shows that the mobilised shear strength envelope develops as the rock specimen is subjected to anisotropic compression. However, the cemented cohesion c' is not constant but it is increasing with the axial strain. These mobilised shear strength envelopes are the intrinsic property of the rock similar to the failure envelope.

The mobilised cementation also increases with the amount of axial strain. The greater the specimen is compressed and deformed, the greater the mobilised cementation. The variation of the mobilised cementation shows a linear variation with the increase in the axial strain as shown in Figure 6.12. The maximum mobilised cementation of 5.9 MPa at 0.89% axial strain corresponds to the failure shear strength envelope of the rock (Table 6.1). Once failed, the cementation is destroyed due to the rupture of the rock mass because of its brittle nature.

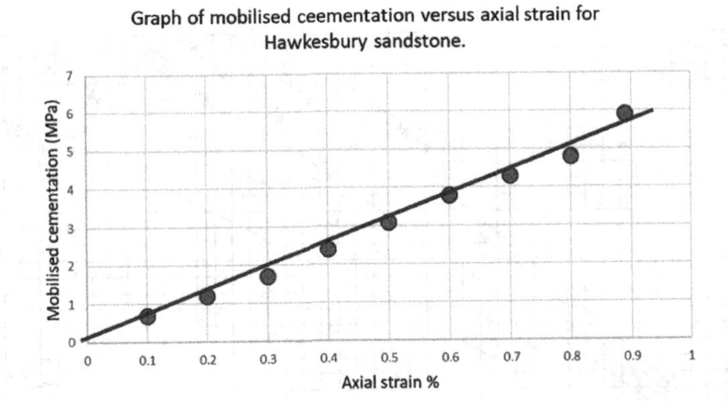

Figure 6.12 Graph showing the linear variation of mobilised cementation with % axial strain for Hawkesbury sandstone, Australia (Hamzah et al., 2020).

Table 6.1 Mobilised shear strength envelopes for various axial strains deduced from the normalised stress–strain curves (Hamzah et al., 2020)

Normalised strain %	*Cementation, c′ (MPa)*	*Transition effective stress, $(\sigma - u_w)_t$ (MPa)*	*Transition shear strength, τ_t (MPa)*	*Minimum mobilised friction angle, ϕ'_{minmob}*
0.1	0.7	2	1.3	11
0.2	1.2	2	2.0	21
0.3	1.7	2	2.7	30
0.4	2.4	4	5.2	34
0.5	3.1	4	5.9	38
0.6	3.8	4	6.3	41
0.7	4.3	4	6.8	43
0.8	4.8	4	7.1	44
0.89	5.9	4	7.1	44.2

6.5.1 Prediction of stress–strain curves for Hawkesbury sandstone using NSRMYSF at confining stresses of 0, 4 and 8 MPa

Based on the determined mobilised shear strength envelopes for the Hawkesbury sandstone as in Figure 6.11, the stress–strain behaviour of the rock at any effective confining pressure σ'_3 can be predicted. This is because the mobilised envelopes are valid for any magnitude of effective confining pressure. For a better prediction of the stress–strain response of the rock, more mobilised shear strength envelopes can be drawn especially on the lower axial strain where the gaps between the mobilised envelopes are greater. Perhaps, the mobilised envelopes for axial strains of 0.5%, 0.15%, 0.25% and 0.35% should be drawn.

Figures 6.13–6.15 show the predicted deviator stresses represented by the diameters of the Mohr circles drawn to just touch the mobilised envelopes. The values of the

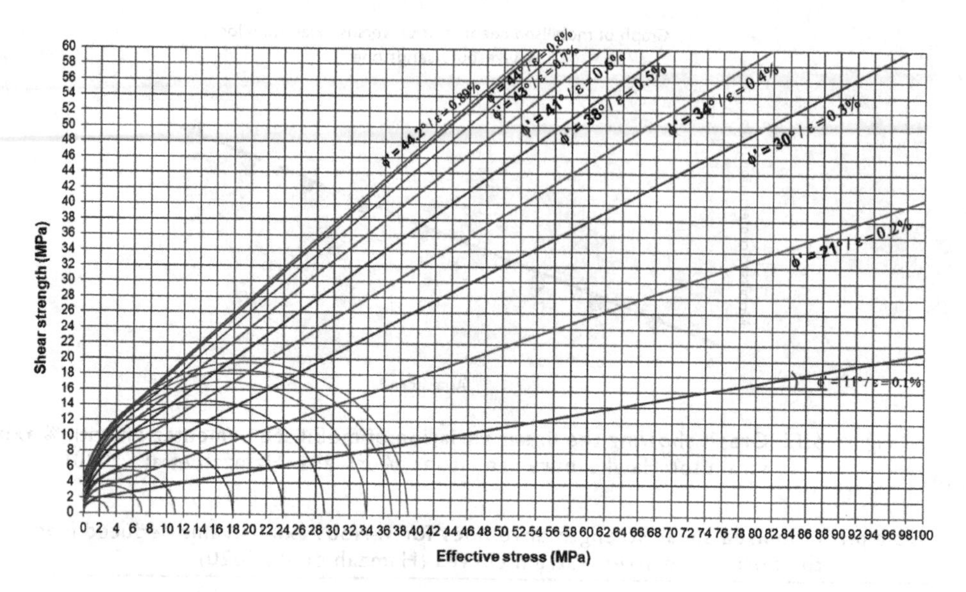

Figure 6.13 Predicted deviator stresses for uniaxial tests on Hawkesbury sandstone using the NSRMYSF at a confining pressure of 0 MPa (Hamzah et al., 2020).

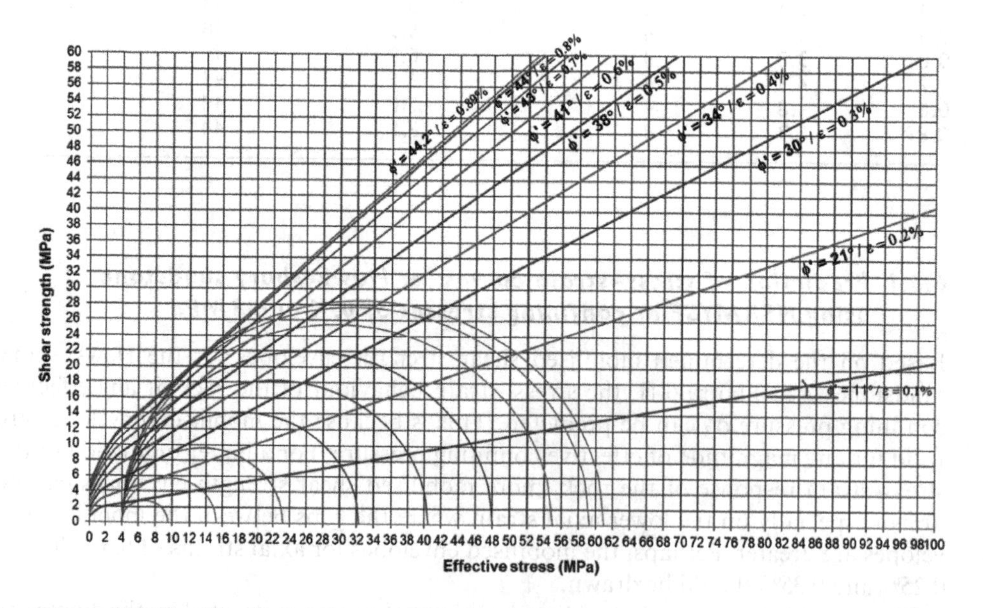

Figure 6.14 Predicted deviator stresses for uniaxial tests on Hawkesbury sandstone using the NSRMYSF at a confining pressure of 4 MPa (Hamzah et al., 2020).

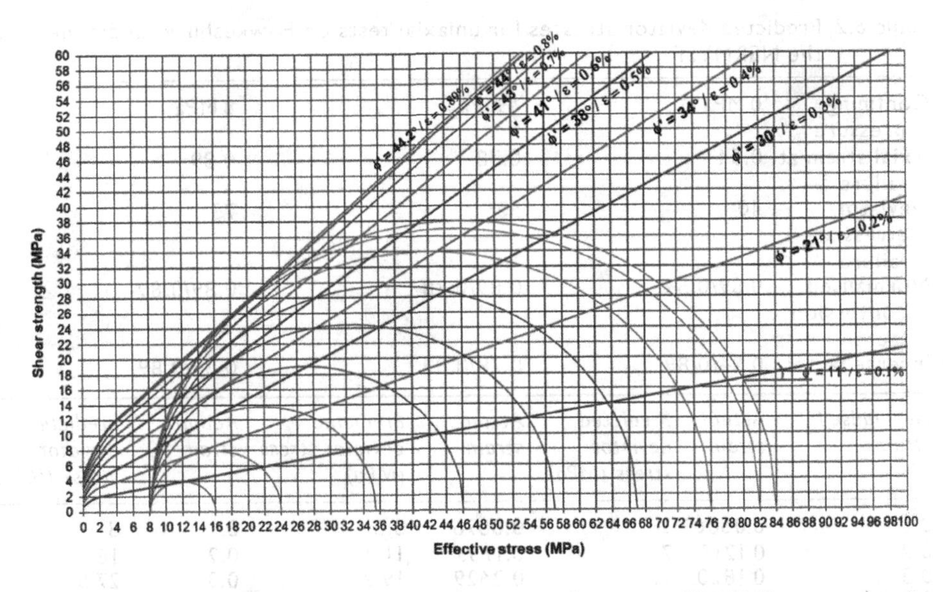

Figure 6.15 Predicted deviator stresses for uniaxial tests on Hawkesbury sandstone using the NSRMYSF at a confining pressure of 8 MPa (Hamzah et al., 2020).

predicted stress–strain are presented in Table 6.2. Then the predicted stress–strain data points are plotted superimposed with the laboratory determined stress–strain curves from the uniaxial tests on the rock specimens as shown in Figure 6.16. Essentially, the predicted data show an excellent fit to the laboratory curves and this has substantiated the applicability of the NSRMYSF to predict stress–strain curves for the rock deformation behaviour. The study also revealed the existence of linear variation between the mobilised cementation and the axial strains as shown in Figure 6.12.

6.6 Development of mobilised shear strength in granite grade II

A rock triaxial machine GCTS Triaxial RTX-3000 in Rock Mechanics laboratory, Faculty of Civil Engineering, Universiti Teknologi MARA as shown in Figure 6.17 has been deployed to obtain the anisotropic stress–strain relationship for weathered granite grade II from Rawang, Selangor sampled at a depth of 20 m and the typical rock specimens are as shown in Figure 6.18 (Md Noor and Jobli, 2018). Three tests have been conducted where the confining pressures applied are 2, 7.5 and 14 MPa and the stress–strain curves obtained are shown in Figure 6.19. The developed mobilised shear strength envelopes within the rock specimens of 50 mm diameter and 100 mm height during the application of the deviator stress are interpreted from the stress–strain curves. These mobilised shear strength envelopes at various axial strains are the intrinsic property and are unique for the rock. Once this property has been established, then it can be used to predict the stress–strain relationship at any confining pressure.

Table 6.2 Predicted deviator stresses for uniaxial tests on Hawkesbury sandstone using the NSRMYSF

Confining pressure	0 MPa		4 MPa		8 MPa	
Axial strain at failure %	0.54		0.78		0.89	
Deviator stress at failure	40		58		73	
Normalised conversion factor	0.89/0.54		0.89/0.78		0.89/0.89	
Inverse factor	0.54/0.89		0.78/0.89		0.89/0.89	
Normalised strain	*Actual strain*	*Predicted deviator stress (MPa)*	*Actual strain*	*Predicted deviator stress (MPa)*	*Actual strain*	*Predicted deviator stress (MPa)*
0.1	0.0606	3	0.0876	5.5	0.1	8
0.2	0.1213	7	0.1753	11.3	0.2	16
0.3	0.1820	11	0.2629	19.2	0.3	27.5
0.4	0.2427	18	0.3506	28.3	0.4	38
0.5	0.3034	24	0.4382	36.5	0.5	49
0.6	0.3640	29	0.5258	44	0.6	59
0.7	0.4247	34	0.6135	51	0.7	68
0.8	0.4854	37	0.7011	55.3	0.8	74
0.89	0.54	39	0.78	57	0.89	76

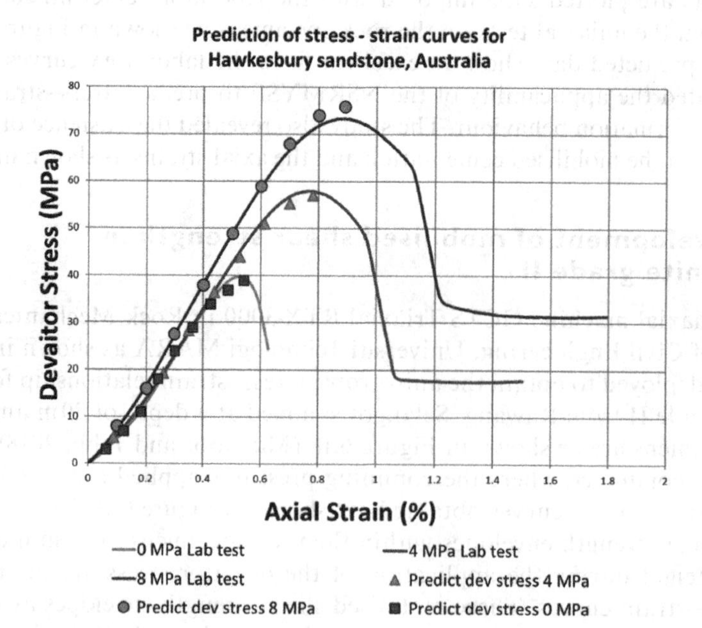

Figure 6.16 Prediction of stress–strain curves for Hawkesbury sandstone using the NSRMYSF (Hamzah et al., 2020).

Figure 6.17 Uniaxial rock testing machine GCTS Triaxial RTX-3000 at Rock Mechanics laboratory, Faculty of Civil Engineering, Universiti Teknologi MARA (Md Noor and Jobli, 2018).

Figure 6.18 Granite grade II from Rawang, Selangor (Md Noor and Jobli, 2018).

The procedure to determine the mobilised shear strength envelopes for soil is described in Section 4.2 and similar steps are applicable for rock. The normalised stress–strain curves for granite grade II at confining pressures of 2, 7.5 and 14 MPa are shown in Figure 6.20. The axial strains of each graph are multiplied by a conversion-normalised factor for each curve. This is to make sure the normalised axial strain at failure is common at an axial strain of 2.36% for every stress–strain curve. The normalised stress–strain curves are nicely spread and ease the reading of the axial strains and the corresponding deviator stresses. From here, the deviator stresses for the three stress–strain curves corresponding to % axial strains of 0.2, 0.3, 0.4, 0.5, 0.6, 0.7, 0.8, 0.9, 1.0, 1.2, 1.4, 1.6, 1.8, 2.0, 2.3 and 2.36 are determined and the Mohr circles are drawn

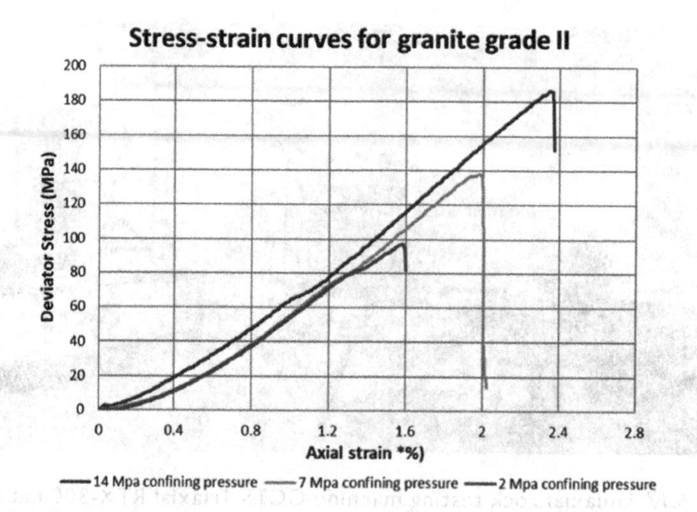

Figure 6.19 Stress–strain curves for granite grade II at confining pressures of 2, 7.5 and 14 MPa (Md Noor and Jobli, 2018).

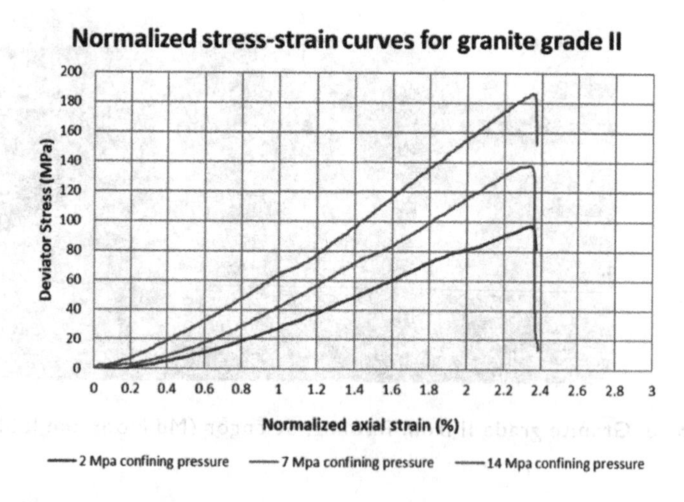

Figure 6.20 Normalised stress–strain curves for granite grade II at confining pressures of 2, 7.5 and 14 MPa with common axial strain at failure of 2.36% (Md Noor and Jobli, 2018).

as in Figure 6.21. Then the mobilised shear strength envelopes for each % axial strain considered are drawn. The parameters applied to draw the mobilised shear strength envelopes and the envelopes at failure are presented in Table 6.3. Figure 6.22 shows the mobilised shear strength envelopes and the failure envelope for granite grade II from Rawang, Selangor as the unique intrinsic property of the rock where the Mohr circles applied to define the mobilised shear strength envelopes are removed. These envelopes

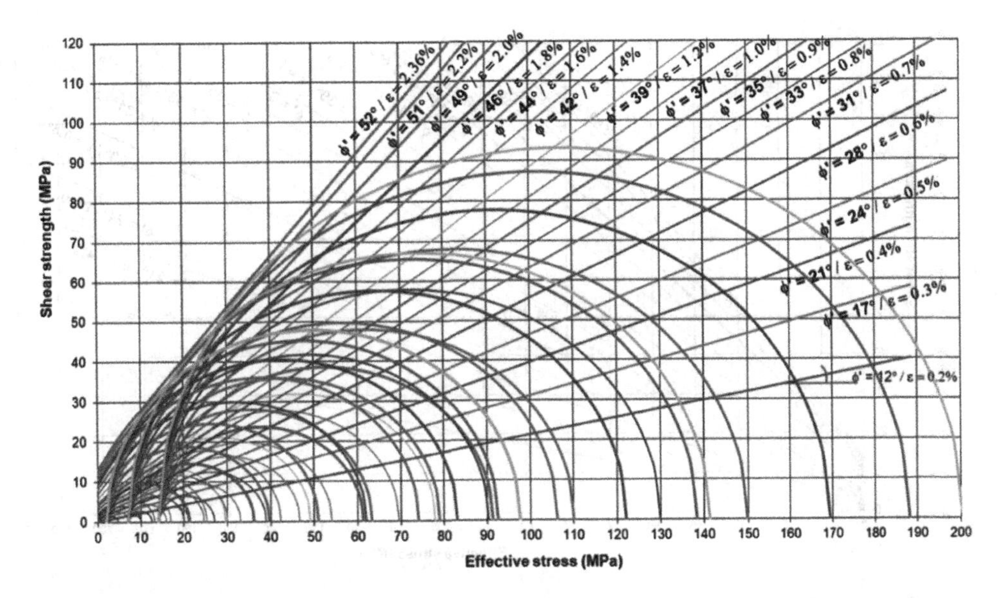

Figure 6.21 Mobilised shear strength envelopes for granite grade II from Rawang, Selangor at axial strains from 0.2% to 2.2% and shear strength envelope at failure at an axial strain of 2.36% (Md Noor and Jobli, 2018).

Table 6.3 Mobilised shear strength envelopes for various axial strains deduced from the normalised stress–strain curves for granite grade II (Md Noor and Jobli, 2018)

Normalised strain %	*Cementation, c' (MPa)*	*Transition effective stress $(\sigma - u_w)_t$ (MPa)*	*Transition shear strength τ_t (kN/m^2)*	*Minimum mobilised friction angle $\phi'_{min mob}$*
0.2	0.2	2	0.4	12
0.3	0.4	2	0.8	17
0.4	0.8	2	1.1	21
0.5	1.4	4	2.3	24
0.6	1.8	4	2.7	28
0.7	2.4	4	3.0	31
0.8	3.3	4	3.2	33
0.9	4.0	4	3.5	35
1.0	4.7	4	3.8	37
1.2	5.8	6	6.4	39
1.4	7.0	6	7.4	42
1.6	8.1	8	10.6	44
1.8	9.6	8	11.4	46
2.0	10.8	8	11.6	49
2.2	12.0	8	12.0	51
2.36	13.0	8	12.4	52

can be applied to determine the stress–strain response at any confining pressures or in other words at any depth in the ground. Figure 6.23 shows the graph of mobilised cementation versus % axial strain of the rock. The mobilised cementation is the intersection of the mobilised shear strength envelope with the vertical shear strength axis.

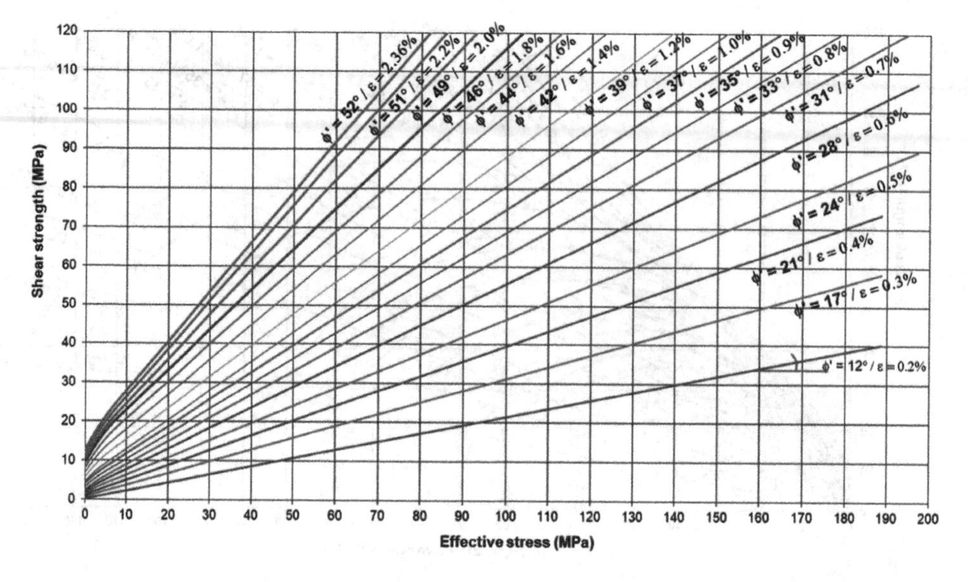

Figure 6.22 Mobilised shear strength envelopes for granite grade II from Rawang, Selangor and shear strength envelope at failure as the unique intrinsic property of the rock (Md Noor and Jobli, 2018).

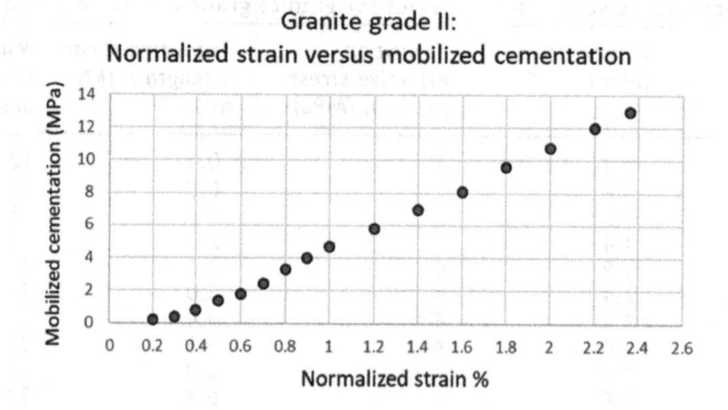

Figure 6.23 Graph of mobilised cementation versus % axial strain for granite grade II from Rawang, Selangor (Md Noor and Jobli, 2018).

This intersection is deduced during the definition of the mobilised shear strength envelopes. Even though the maximum cementation of the rock is 13 MPa, which is the maximum mobilised cementation corresponding to failure point, i.e., maximum deviator stress, its magnitude increases from zero up to this value as the specimen undergoes anisotropic compression. Essentially, the cementation shows an initial slow increase at small strains indicated by the sagging curve and followed by a linear relationship. Largely, the mobilised cementation increases linearly with the magnitude of the axial strains as the specimen is compressed.

6.6.1 Prediction of stress–strain curves for granite grade II

The unique property of the rock as presented in Figure 6.22 can now be applied for the prediction of stress–strain curves at any required effective confining stress. In this case, the predictions are made for effective confining pressures as what have been conducted in the laboratory uniaxial tests which are 2, 7 and 14 MPa so that the predicted curves can be compared with the actual curves. Figures 6.24–6.26 show the predicted Mohr circles corresponding to the axial strains under considerations at effective confining pressures of 2, 7 and 14 MPa, respectively. The diameters of the Mohr circles represent the predicted deviator stresses and the touched mobilised shear strength envelope represent the corresponding axial strain. The predicted deviator stresses are presented in Table 6.4. Figure 6.27 shows the predicted stress–strain curves in the form of data points that are plotted superimposed with the stress–strain curves obtained from laboratory tests. The predicted curves in the form of the data points overlapped the corresponding laboratory curves and thence, this has substantiated the excellent prediction by the NSRMYSF.

6.7 Characterising the volume change behaviour of granite grade III during the anisotropic compression

The applicability of the NSRMYSF to predict the rock stress–strain behaviour has been tested with granite grade III taken from Rawang, Selangor, Malaysia as shown in Figure 6.28. The dimensions of the rock specimens used are 50 mm diameter and 100 mm height. The stress–strain curves obtained from uniaxial tests using a GCTS Triaxial RTX-3000 machine at confining pressures of 5, 10 and 20 MPa are shown in Figure 6.29. The axial strain at failure is 1.33%, 1.36% and 1.46% with the maximum

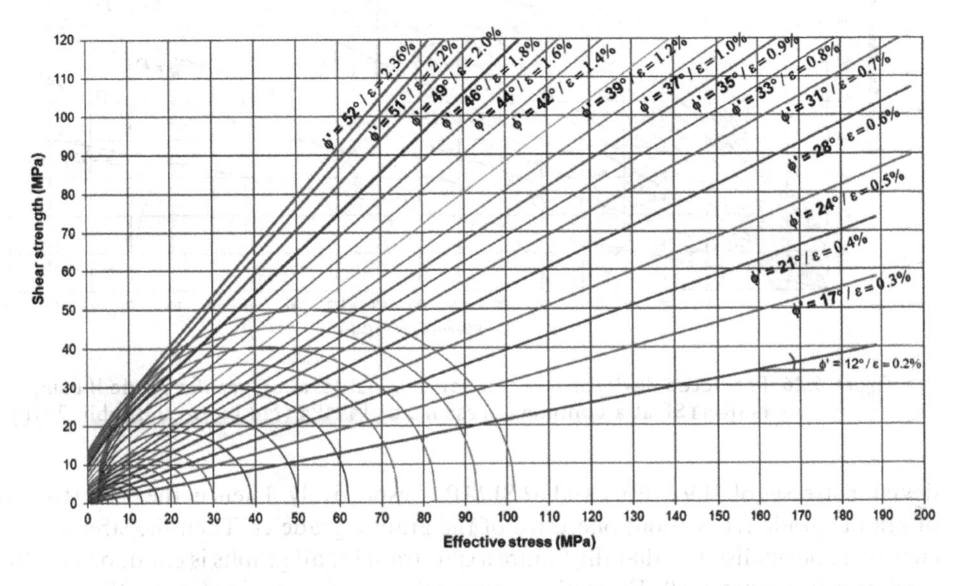

Figure 6.24 Predicted deviator stresses for uniaxial tests on granite grade II using NSRMYSF at a confining pressure of 2 MPa (Md Noor and Jobli, 2018).

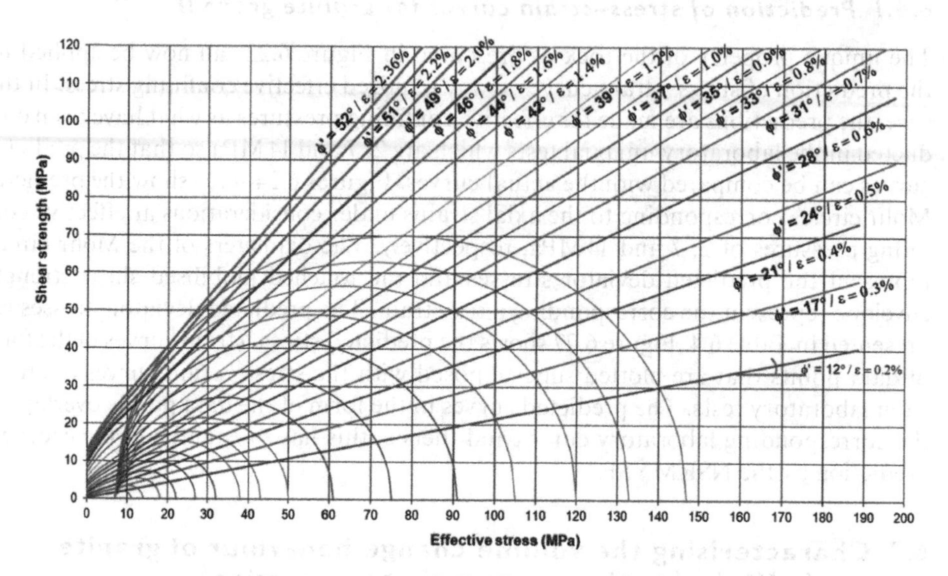

Figure 6.25 Predicted deviator stresses for uniaxial tests on granite grade II using NSRMYSF at a confining pressure of 7 MPa (Md Noor and Jobli, 2018).

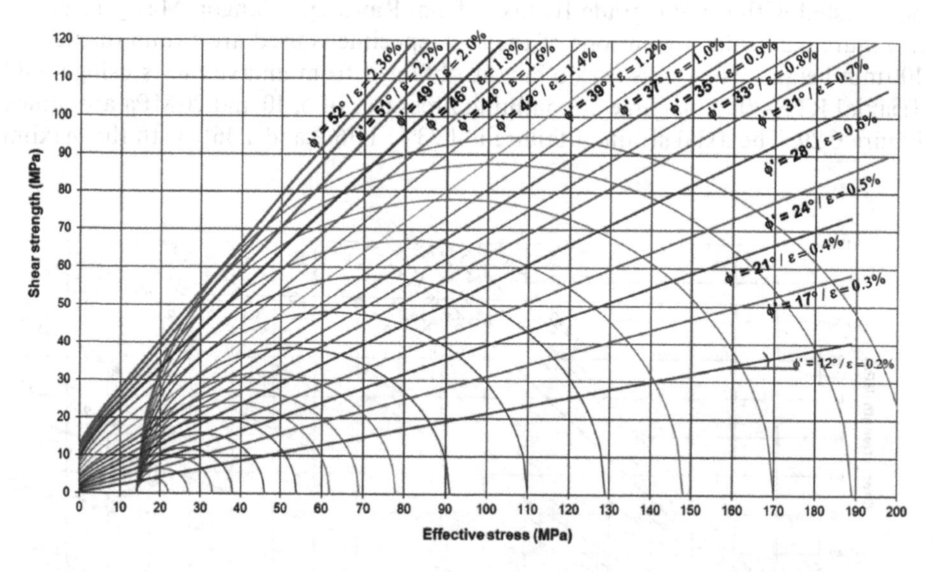

Figure 6.26 Predicted deviator stresses for uniaxial tests on granite grade II using the NSRMYSF at a confining pressure of 14 MPa (Md Noor and Jobli, 2018).

deviator stresses of 31.94, 49.01 and 62.81 MPa, respectively. Thence, the peak strength of granite grade III is about one-third of the granite grade II. Then the stress–strain curves are normalised so that the failure axial strain for all graphs is common at 1.46% as shown in Figure 6.30. From these normalised stress–strain curves, the deviator stresses corresponding to axial strains of 0.2%, 0.3%, 0.4%, 0.5%, 0.6%, 0.7%, 0.8%,

Table 6.4 Predicted deviator stresses for uniaxial tests on granite grade II using the NSRMYSF (Md Noor and Jobli, 2018)

Confining pressure	2 MPa		7 MPa		14 MPa	
Axial strain at failure %	1.59		2.0		2.36	
Deviator stress at failure	95.58		135		185	
Normalised conversion factor	2.36/1.59		2.36/2.0		2.36/2.36	
Inverse factor	1.59/2.36		2.0/2.36		2.36/2.36	
Normalised strain	*Actual strain*	*Predicted deviator stress (MPa)*	*Actual strain*	*Predicted deviator stress (MPa)*	*Actual strain*	*Predicted deviator stress (MPa)*
0.2	0.135	1.5	0.169	4	0.2	8
0.3	0.202	3	0.254	7	0.3	13
0.4	0.269	5	0.339	11	0.4	19
0.5	0.337	8	0.424	15	0.5	24
0.6	0.404	11	0.508	20	0.6	32
0.7	0.472	15	0.593	25	0.7	40
0.8	0.539	19	0.678	31	0.8	48
0.9	0.606	23	0.763	36	0.9	55
1.0	0.674	28	0.847	43	1.0	64
1.2	0.808	37	1.017	54	1.2	77
1.4	0.943	48	1.186	68	1.4	96
1.6	1.078	60	1.356	84	1.6	115
1.8	1.213	72	1.525	98	1.8	134
2.0	1.347	81	1.695	113	2.0	156
2.2	1.482	91	1.864	126	2.2	175
2.36	1.59	100	2.0	140	2.36	189

0.9%, 1.0%, 1.1%, 1.2%, 1.3% and 1.46% are extracted and the three Mohr circles for each axial strain are drawn. From the Mohr circles, the curvilinear mobilised shear strength envelopes and the failure envelope are defined as shown in Figures 6.31 and 6.32. There are four parameters used to define the mobilised envelopes, which are mobilised cementation, transition effective stress, transition shear strength and the minimum mobilised friction angle. The values of these four parameters for each axial strain are presented in Table 6.5. The normalising factor and the inverse factor to revert the normalised axial strains to the actual axial strains are shown in Table 6.6.

The fully mobilised cementation is 1 MPa that corresponds to the failure envelope. The graph of mobilised cementation versus the axial strain for granite grade III is shown in Figure 6.33. Essentially, the graph shows a linear relationship at an axial strain greater than 0.2%. On the lower strain, the cementation is hardly mobilised, may be due to the bedding error where the specimen is adjusting itself as the deviator stress plunger is compressing and closing the gap. This can be seen at the stress–strain curves where it is slightly sagging when the compression starts. Once the specimen is properly seated, the cementation starts to be mobilised linearly. The mobilised cementation

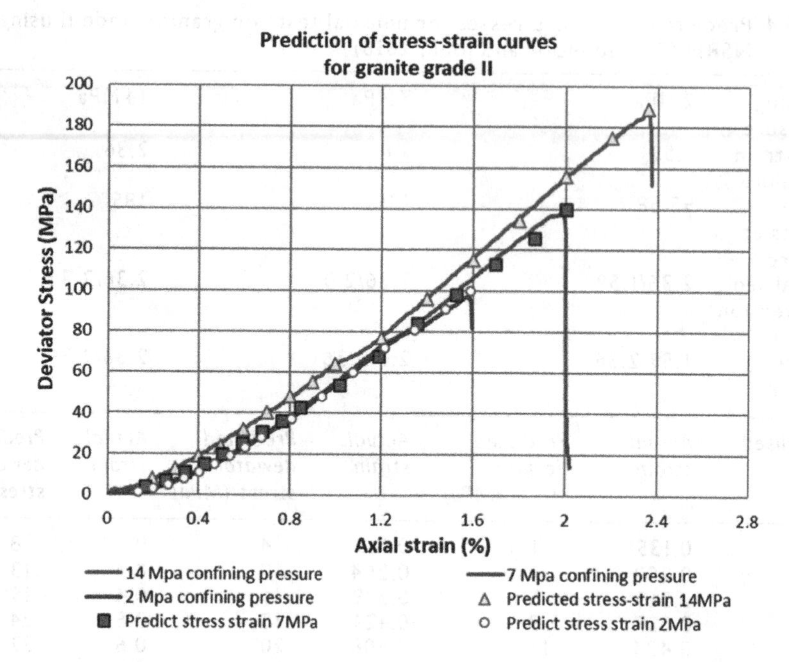

Figure 6.27 The predicted stress–strain curves in the form of data points are plotted superimposed with the stress–strain curves obtained from laboratory tests (Md Noor and Jobli, 2018).

Figure 6.28 Granite grade III.

for every mobilised envelope is extrapolated from higher stress levels to intersect with the vertical axis. The intersection is achieved through a best-fit curve, because the envelope is non-linear at the low-stress levels and it is done manually. In addition, this property of linear variation of mobilised cementation with respect to axial strain considers that the specimen is an intact rock. This behaviour of a rock mass, which includes the complex joints and discontinuities, would be very complex to predict.

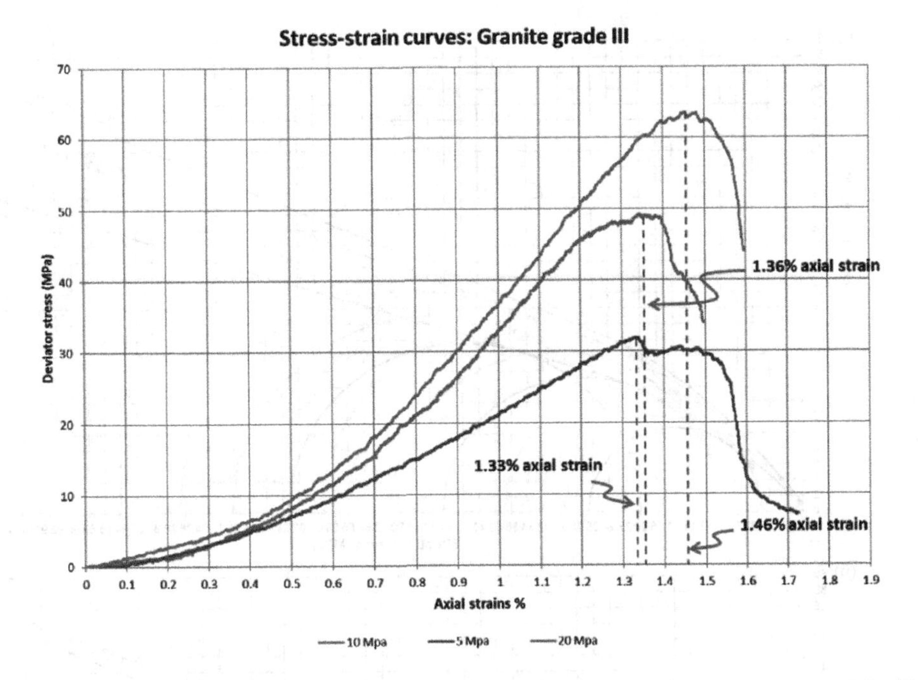

Figure 6.29 Stress–strain curves for granite grade III at confining stresses of 5, 10 and 20 MPa.

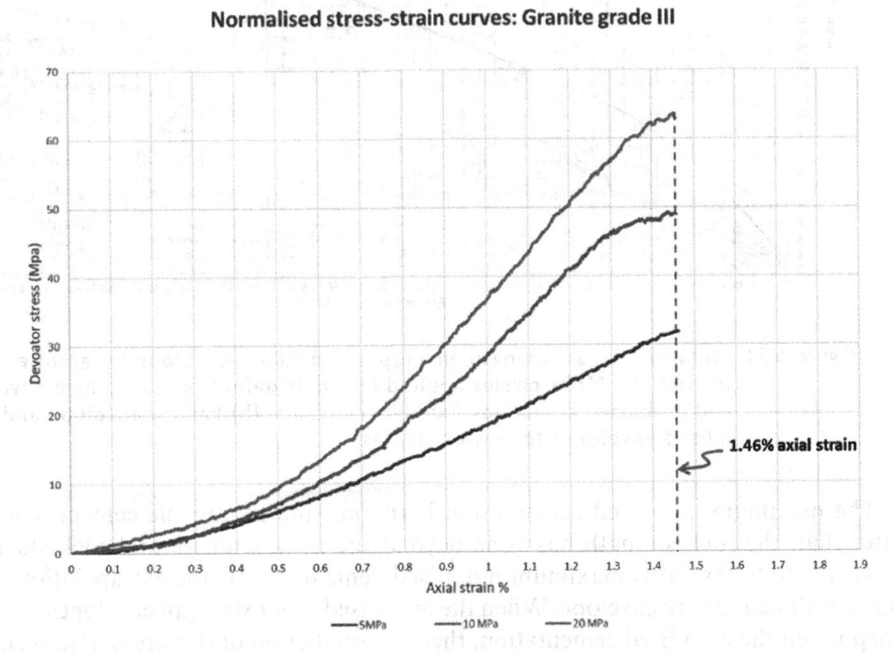

Figure 6.30 Normalised stress–strain curves for granite grade III at confining stresses of 5, 10 and 20 MPa.

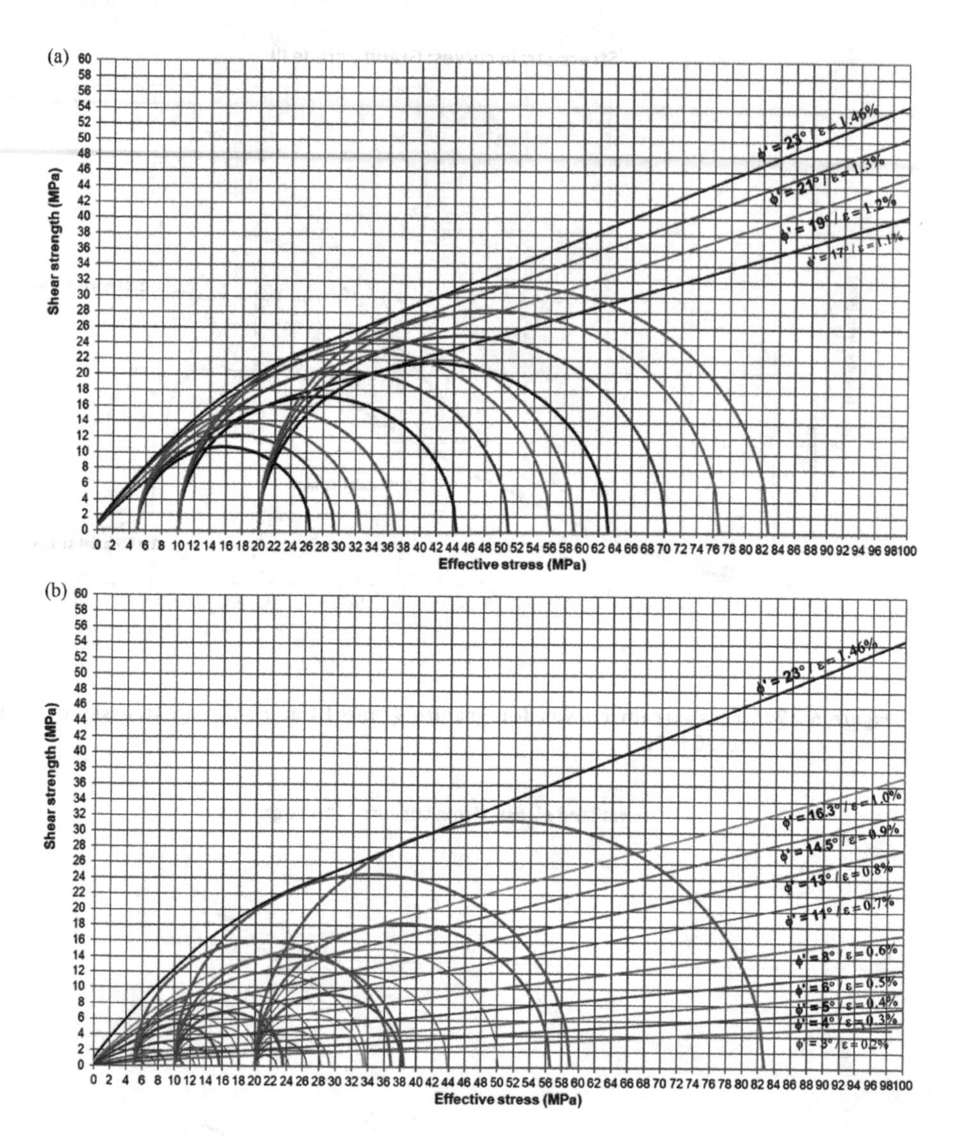

Figure 6.31 Mobilised shear strength envelopes and failure envelope for granite grade III with the Mohr circles applied for their definition. (a) Failure envelope and mobilised envelopes for higher strains. (b) Failure envelope and mobilised envelopes for lower strains.

The maximum mobilised cementation is the maximum possible cement, because after that, the rock strength has gone beyond the peak strength and rock starts to fracture. Note that, this maximum mobilised cementation is the extrapolation from the curvilinear failure envelope. When the mobilised shear strength envelopes have incorporated the mobilised cementation, then the prediction of the stress–strain curves using these envelopes must have also accounted for the existence of the cementation in the rock.

6.7.1 Prediction of stress–strain response for granite grade III

The determined mobilised shear strength envelopes and the failure envelopes for the granite grade III as in Figure 6.32 are applied to predict the rock stress–strain response and the rock peak strength at various confining pressures. Figures 6.34–6.36 show the prediction of the magnitude of deviator stresses for various axial strains considered at confining pressures of 2, 8 and 14 MPa, respectively, and the predicted magnitudes

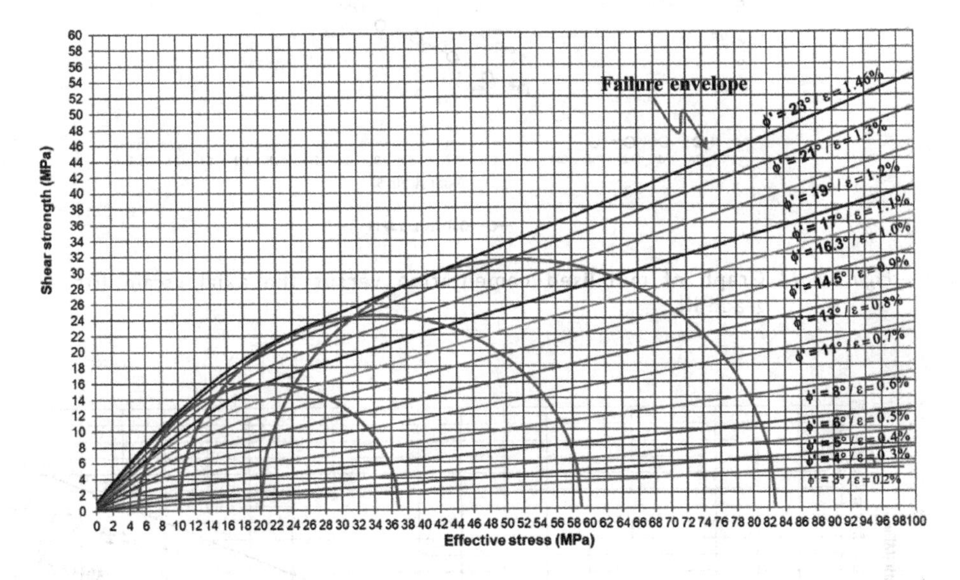

Figure 6.32 Mobilised shear strength envelopes and failure envelope for granite grade III can be applied for the prediction of the rock stress–strain behaviour.

Table 6.5 The parameters for the definition of the mobilised shear strength envelopes for various axial strains deduced from the normalised stress–strain curves for granite grade III

Normalised strain %	*Cementation c' (MPa)*	*Transition effective stress $(\sigma - u_w)_t$ (MPa)*	*Transition shear strength τ_t (N/m²)*	*Minimum mobilised friction angle ϕ'_{minmob}*
0.2	0	10	0.9	3
0.3	0	10	1.4	4
0.4	0.1	10	1.9	5
0.5	0.2	10	3.0	6
0.6	0.25	10	4.2	8
0.7	0.33	15	6.4	11
0.8	0.4	15	7.9	13
0.9	0.45	20	11.4	14.5
1.0	0.54	20	13.2	16.3
1.1	0.6	25	17	17
1.2	0.68	25	19	19
1.3	0.8	25	20.9	21
1.46	1	25	21.8	23

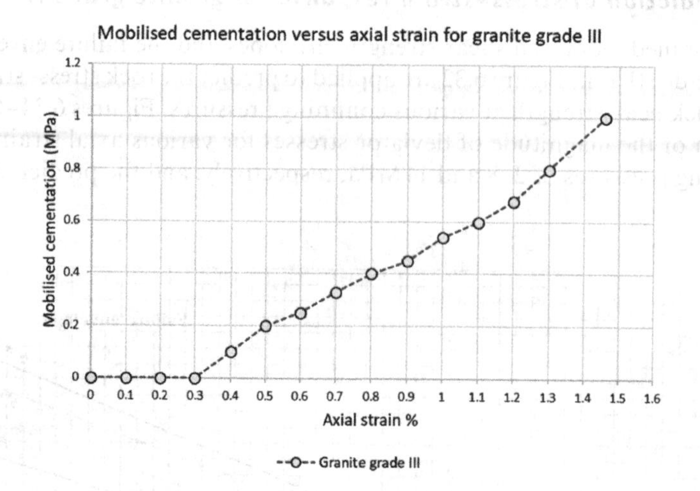

Figure 6.33 Graph of mobilised cementation versus the axial strain for granite grade III.

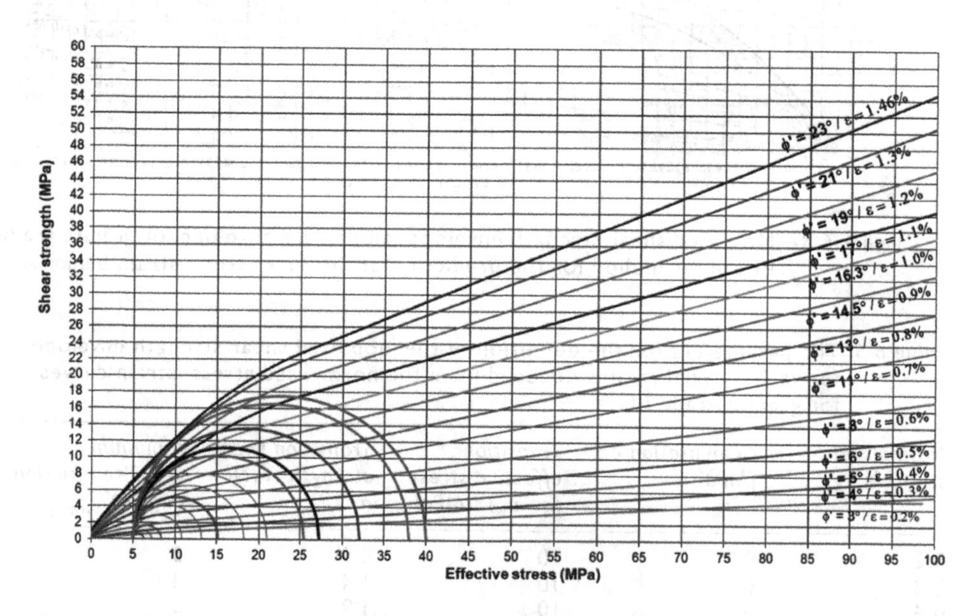

Figure 6.34 Prediction of the magnitude of deviator stresses for various axial strains considered at a confining pressure of 5 MPa.

are presented in Table 6.6. The predicted deviator stress and the corresponding axial strain represented by data points are plotted superimposed with the stress–strain curves obtained from the laboratory uniaxial tests as shown in Figure 6.37. The predicted stress–strain shows an excellent fit with the laboratory curves. This thence has substantiated the applicability of the NSRMYSF to predict accurately the rock stress–strain behaviour and the rock peak strength at any confining pressure.

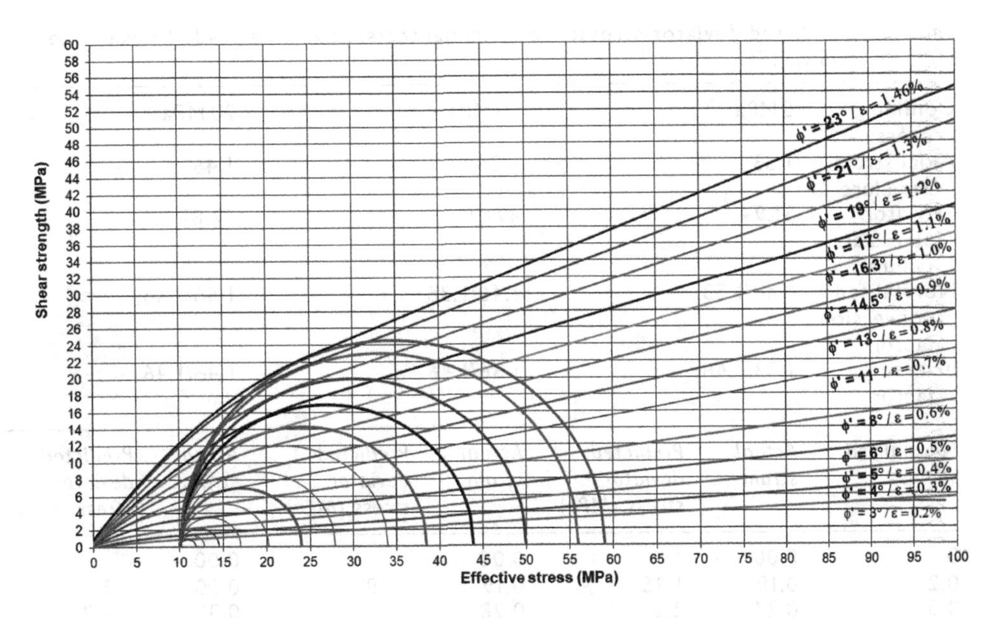

Figure 6.35 Prediction of the magnitude of deviator stresses for various axial strains considered at a confining pressure of 10 MPa.

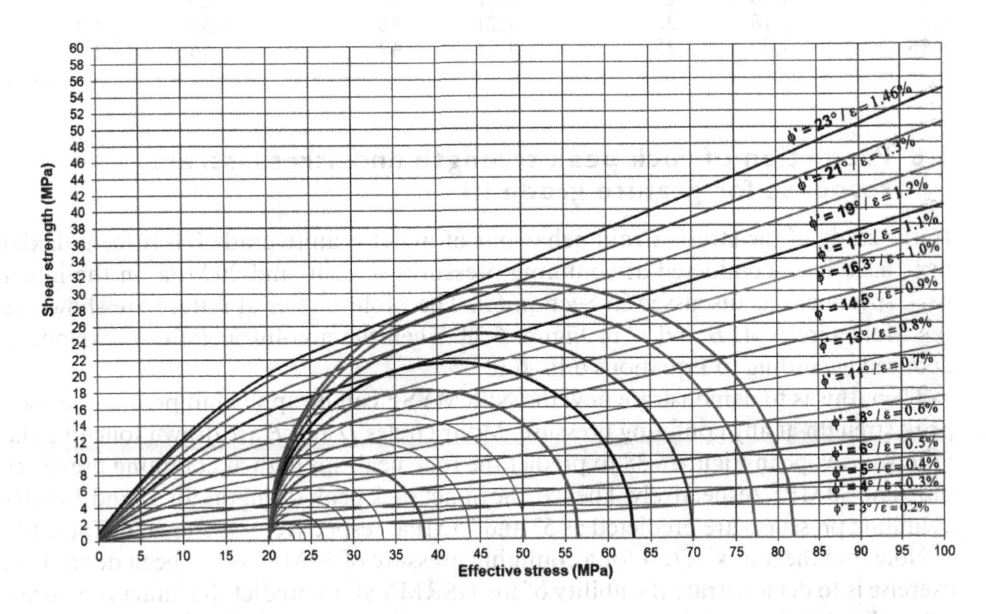

Figure 6.36 Prediction of the magnitude of deviator stresses for various axial strains considered at a confining pressure of 20 MPa.

Table 6.6 Predicted deviator stresses for uniaxial tests on granite grade III using the NSRMYSF

Confining pressure	5 MPa		10 MPa		20 MPa	
Axial strain at failure %	1.33		1.36		1.46	
Deviator stress at failure	31.94		49.01		62.81	
Normalised conversion factor	1.46/1.33		1.46/1.36		1.46/1.46	
Inverse factor	1.33/1.46		1.36/1.46		1.46/1.46	
Normalised strain	*Actual strain*	*Predicted deviator stress (MPa)*	*Actual strain*	*Predicted deviator stress (MPa)*	*Actual strain*	*Predicted deviator stress (MPa)*
0	0.00		0.00		0.00	
0.2	0.18	1.25	0.19	1.8	0.20	3
0.3	0.27	2.2	0.28	2.8	0.30	4.3
0.4	0.36	3.3	0.37	4.5	0.40	6.4
0.5	0.46	5.8	0.47	7	0.50	9.5
0.6	0.55	8.2	0.56	10.2	0.60	13.5
0.7	0.64	10.2	0.65	14	0.70	18.4
0.8	0.73	13.2	0.75	17.9	0.80	23.8
0.9	0.82	16	0.84	24	0.90	30.3
1.0	0.91	20.5	0.93	28.5	1.00	36.5
1.1	1.00	22.3	1.02	33.9	1.10	43
1.2	1.09	27	1.12	40	1.20	50
1.3	1.18	33	1.21	46	1.30	57.3
1.46	1.33	35	1.36	49	1.46	62

6.8 Prediction of rock peak strength and stress–strain response for granite grade III

In the study of the stress–strain behaviour of intact granite grade III, three uniaxial tests have been conducted at confining pressures of 5, 10 and 20 MPa on the intact rock specimens as discussed in Section 6.7. The Mohr circles at failure are shown as the Mohr circles *A*, *B* and *C* in Figure 6.38. Then the non-linear failure envelope is deduced according to Equations 6.16–6.18.

Then, this is to demonstrate how the NSRMYSF can be applied to predict the rock peak strength at any confining pressure. Mohr circles *D* and *E* are drawn touching the failure envelope in Figure 6.38 to predict the rock peak strength at confining pressures of 15 and 30 MPa, respectively. Thence, the intact rock peak strengths for 15 and 30 MPa confining pressures are predicted as 57 and 75 MPa, respectively, as shown in Table 6.7.

Note that the uniaxial test for a confining pressure of 30 MPa is not been done. This exercise is to demonstrate the ability of the NSRMYSF to predict the intact rock peak strength with the complete stress–strain curve. The predicted Mohr circles for the 30 MPa confining pressure is as shown in Figure 6.39 applying the earlier determined

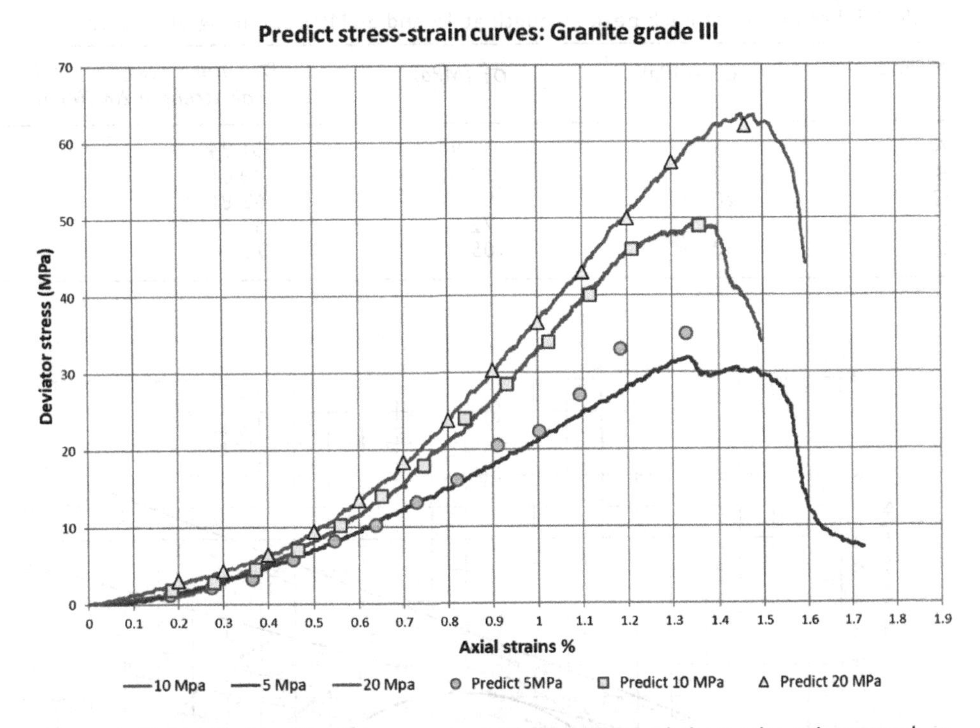

Figure 6.37 Prediction of the stress–strain data points and the rock peak strength at confining pressures of 5, 10 and 20 MPa shows an excellent fit with the laboratory curves.

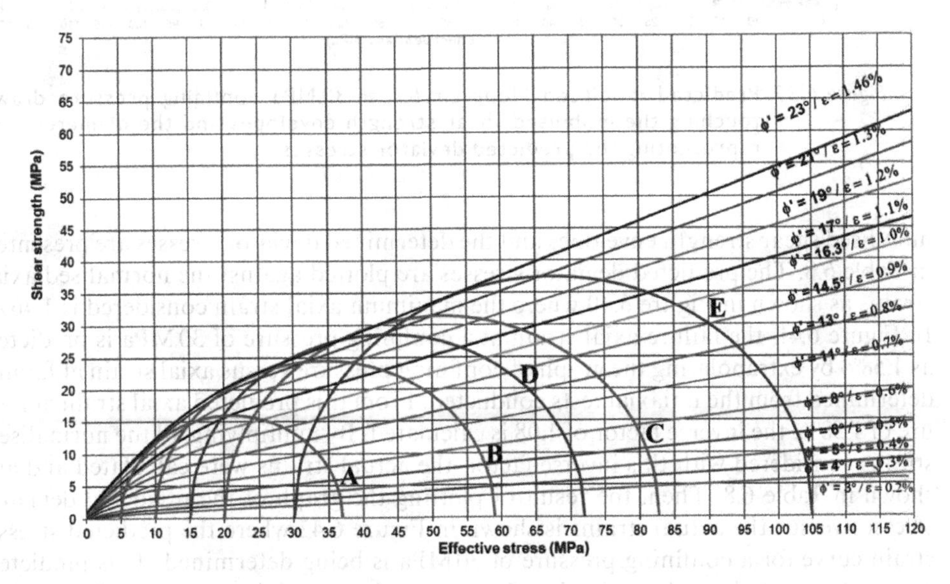

Figure 6.38 Predicted Mohr circles D and E to represent the rock peak strength at 15 and 30 MPa confining pressure.

Table 6.7 Prediction of rock peak strength at 15 and 30 MPa confining pressures

Mohr circle	σ_3' *(MPa)*	σ_1' *(MPa)*	*Deviator stress, i.e., rock peak strength* $\Delta\sigma$ *(MPa)*
A	5	36.94	31.94
B	10	59.01	49.01
C	20	82.81	62.81
D	15	72	57
E	30	105	75

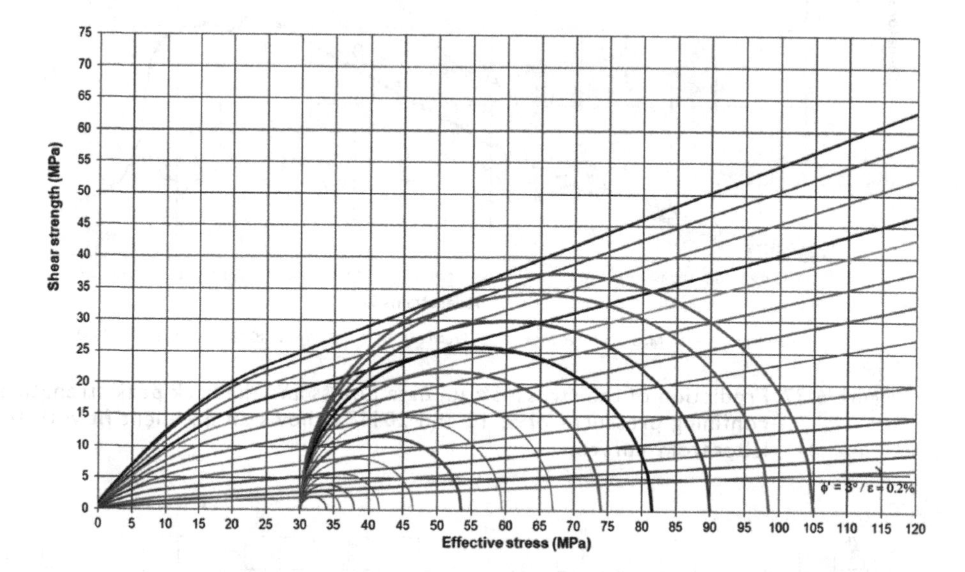

Figure 6.39 Predicted mobilised Mohr circles at 30 MPa confining pressure drawn touching the mobilised shear strength envelopes and the diameters are representing the predicted deviator stresses.

mobilised shear strength envelopes and the determined deviator stresses are presented in Table 6.8. The predicted deviator stresses are plotted against the normalised axial strains as shown in Figure 6.40 where the maximum axial strain considered is 1.46%. In Figure 6.41, the failure axial strain at a confining pressure of 30 MPa is predicted as 1.58% by extrapolating the graph of confining pressure versus axial strain at failure determined from the uniaxial tests conducted. From this predicted axial strain at failure of 1.58%, the inverse factor of 1.08 is calculated. By multiplying all the normalised strains considered with this inverse factor, the actual strains were calculated and are shown in Table 6.8. Then, the result of plotting the graph of the predicted deviator stresses versus the actual strains is shown in Figure 6.42 where the predicted stress–strain curve for a confining pressure of 30 MPa is being determined. This predicted stress–strain graph can be verified by conducting the uniaxial tests at a confining pressure of 30 MPa after this prediction.

Table 6.8 Predicted deviator stresses for a confining pressure of 30 MPa for granite grade III

Normalised strain inverse factor	*1.58/1.46 = 1.08*	
Normalised strain%	*Actual strain%*	*Predicted deviator stress (MPa)*
0.00	0	0
0.20	0.22	4
0.30	0.32	6
0.40	0.43	8
0.50	0.54	11.5
0.60	0.65	16.5
0.70	0.76	23.5
0.80	0.87	29.5
0.90	0.97	37
1.00	1.08	44
1.10	1.19	51.5
1.20	1.30	60
1.30	1.41	68.5
1.46	1.58	75

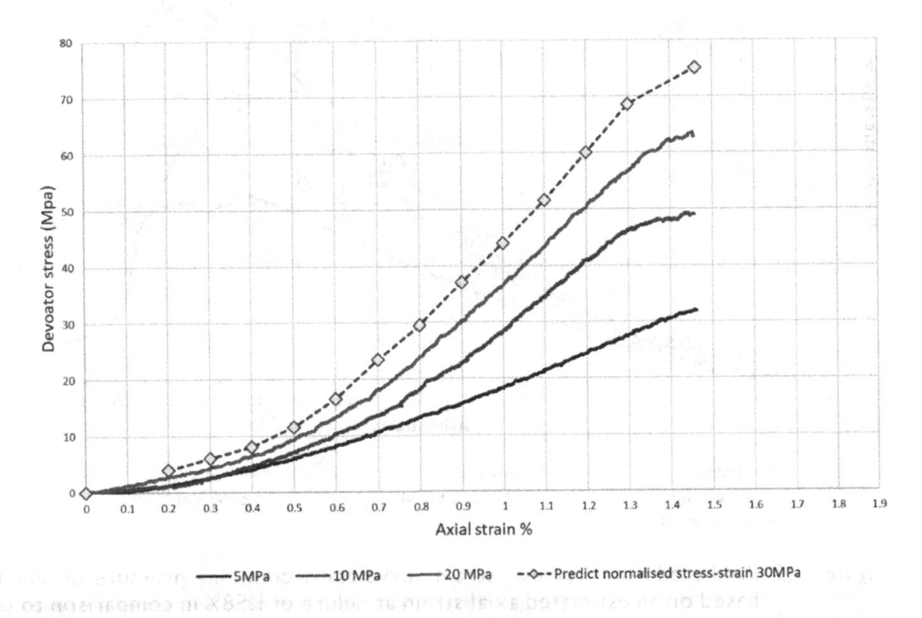

Figure 6.40 Predicted normalised stress–strain curve for a confining pressure of 30 MPa.

In soils, the unique relationship $\phi'_{\min_{\text{mob}}} - \%\varepsilon_a$ represents the rotation of the mobilised shear strength envelope about the origin because there is no cementation in soils. The rotation indicates the increase in the mobilised shear strength in the soil mass as the soil is anisotropically compressed. Therefore, the increase in the $\Delta\phi'_{\min_{\text{mob}}}$

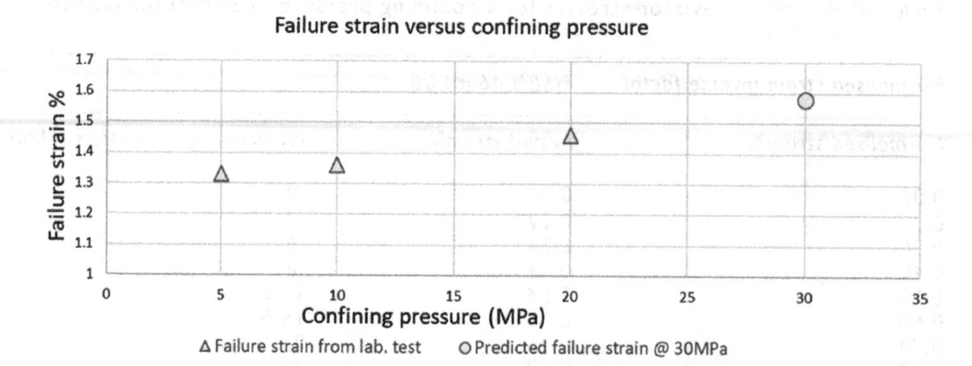

Figure 6.41 Graph of confining pressure versus failure axial strain for granite grade III and estimation of failure axial strain of 1.58% for 30 MPa confining pressure.

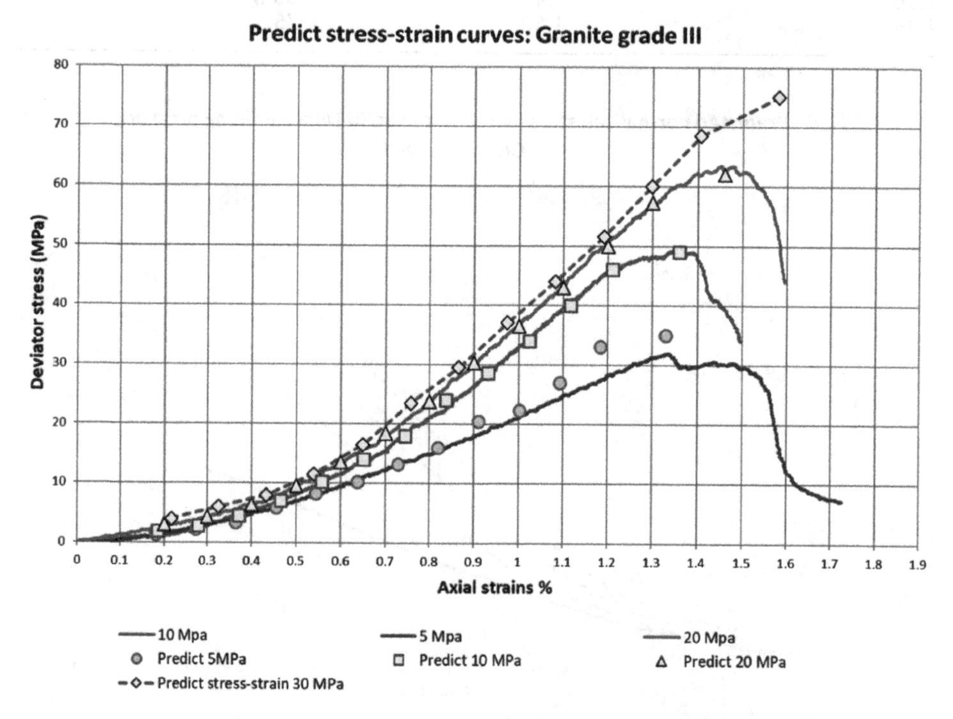

Figure 6.42 Predicted actual stress–strain curve for a confining pressure of 30 MPa based on an estimated axial strain at failure of 1.58% in comparison to the predicted and actual stress–strain curves at 5, 10 and 20 MPa.

is directly indicating the relative increase in the mobilised shear strength. However, in rocks when the rock underwent anisotropic compression, the mobilised envelope rotates while the mobilised cementation is also increasing. Therefore, the increase in the mobilised shear strength within the rock mass cannot be represented by the increase in the $\Delta\phi'_{min_{mob}}$. Thence, in rocks, there is no unique relationship in the form of $\Delta\phi'_{min_{mob}} - \%\varepsilon_a$. Nevertheless, the mobilised shear strength envelopes still represent the intrinsic property of the rock and are unique for the tested rock.

6.9 Summary

Rock is frequently treated as a linearly elastic, homogeneous and isotropic material. The true rock response to stress is in an elastic–plastic manner as discussed in Section 6.3. This section also proves that rock responds in an anisotropic manner when subjected to uniaxial compression in Hoek's cell. Thence, significant errors can be encountered in stress and deformation analysis when assuming anisotropic rock to be isotropic and the non-linear elastic–plastic response to be linearly elastic. Section 6.4 describes various rock deformation models and the most prominent model that has been widely applied by the rock mechanics community is the Hoek and Brown failure criterion. Hoek and Brown also clearly demonstrated the difficulties involved in characterising the rock mass peak strength from conducting laboratory tests on intact rock specimens. Hoek and Brown criterion has incorporated the non-linear material constant m and s to make approximation according to the Mohr–Coulomb linear criterion c' and ϕ'.

Then the application of the anisotropic and elastic–plastic rock deformation model known as NSRMYSF has been tested on stress–strain behaviour of Hawkesbury sandstone from Australia and granite grade II and III from Rawang, Malaysia. Essentially, this rock deformation framework can make an excellent prediction of the rock non-linear stress–strain behaviour when compared with the stress–strain curves of those rocks obtained by conducting laboratory rock uniaxial compression tests. The NSRMYSF has deployed the development of the non-linear mobilised shear strength envelopes with mobilised cementation intercept with the shear strength vertical axis when the rock is subjected to anisotropic compression. The exercise also substantiated that the mobilised cementation increases linearly with the % axial strain. The non-linear mobilised shear strength envelopes are the rock intrinsic property that has not been utilised for the characterisation of rock deformation behaviour and this property is very much related to the rock anisotropic stress–strain behaviour at any subjected confining pressure. Section 6.8 has demonstrated that the NSRMYSF can make use of the mobilised shear strength envelopes for the prediction of the rock stress–strain behaviour at any confining pressure and the failure axial strain needs to be estimated beforehand.

References

Barla, G. (1974). "Rock anisotropy: theory and laboratory testing." In: Müller, L. (eds), *Rock mechanics*, Springer, Vienna, 132–169.

Barla, G., and Wane, M. T. (1970) "Stress-relief method in anisotropic rocks by means of gauges applied to the end of borehole." *International Journal Rock Mechanics Mineral Science*, 7, 171–182.

Becker, R. M., and Hooker, V. E., (1967). "Some anisotropic considerations in rock stress determinations." U.S.B.M. Report of Investigation 6965.

Berry, D. S., and Fairhurst, C. (1966) "Influence of rock anisotropy and time dependent deformation on the stress-relief and high-modulus inclusion techniques of in situ determination." *Testing Technique of Rock Mechanics*, STP 402, ASTM, 190–206.

Chen, Z. H., Tham, L. G., Yeung, M. R., and Xie, H. (2006). "Confinement effects for damage and failure of brittle rocks." *International Journal Rock Mechanics and Mining Sciences*, 43, 1262–1269.

Eberhardt, E., Stead, D., Stimpson, B. (1999), "Quantifying progressive pre-peak brittle fracture damage in rock during uniaxial compression." *International Journal of Rock Mechanics and Mining Sciences*, 36(1999) 361–380.

Eberhardt, E. (2012). "The Hoek-Brown failure criterion." *Rock Mechanics Rock Engineering*, 45, 981–988. doi:10.1007/s00603-012-0276–4.

Hamzah, N., Mat Yusof, N., and Md Noor, M. J.. (2020). "Anisotropic deformation model for progressive damage mechanism of Hawkesbury sandstone incorporating the inherent mobilized shear strength." *5th Symposium on Damage Mechanism in Materials and Structures (SDMMS 2020)*, Penang Island, Malaysia.

Hoek, E. (1968). "Brittle failure of rock." In: Stagg, K. G., and Zienkiewicz, O. C. (eds), Rock *mechanics in engineering practice*, Wiley, London, England, 99–124.

Hoek, E., and Brown, E. T. (1980a). "Empirical strength criterion for rock masses." *Journal of the Geotechnical Engineering Division*, 106(GT9), 1013–1035.

Hoek, E., and Brown, E. T. (1980b). *Underground excavations in rock*, Institution of Mining & Metallurgy, London, England.

Hoek, E., and Brown, E. T. (1990). "Technical note: Estimation of Mohr – Coulomb friction and cohesion values from Hoek-Brown failure criterion." *International Journal of Rock Mechanics Mineral Science and Geomechanics*, 27(3), 227–229.

Hoek, E., Carranza-Torres, C. T., and Corkum, B. (2002). "Hoek–Brown failure criterion—2002 edition. In: Hammah, R., Bawden, W., Curran, J., and Telesnicki, M. (eds), *Proceedings of the Fifth North American Rock Mechanics Symposium (NARMS-TAC)*, University of Toronto Press, Toronto, 267–273.

Hoek, E., Wood, D., and Shah, S. (1992). "A modified Hoek Brown criterion for jointed rock masses." In: Hudson, J. A. (eds), *Rock Characterization, ISRM Symposium: Eurock '92*, British Geotechnical Association, London, England, 209–214.

Md Noor, M. J., and Anderson, W. F. (2006). "A comprehensive shear strength model for saturated and unsaturated soils." *Proceedings of the 4th International Conference on Unsaturated Soils*, Carefree, Arizona, ASCE Geotechnical Special Publication No. 147, Vol. 2, 1992–2003.

Md Noor, M. J., and Jobli, A. F. (2018). "A state-of-the-art anisotropic rock deformation model incorporating the development of mobilised shear strength." 4th *International Conference on Civil and Environmental Engineering for Sustainability*, Langkawi, Malaysia, IOP Conference Series: Earth and Environmental Science. doi:10.1088/1755–1315/140/1/012074.

Xiao, J. Q., Ding, D. X., Jiang, F. L., and Xu, G. (2010). "Fatigue damage variable and evolution of rock subjected to cyclic loading." *International Journal Rock Mechanics and Mining Sciences*, 47, 461–468.

Index

9780367639563